Management of Technology Innovation and Value Creation

Selected Papers from the 16th International Conference on Management of Technology

Management of Technology

Series Editor: Tarek Khalil *(University of Miami, USA)*

Published

Vol. 1 Challenges in the Management of New Technologies
edited by Marianne Hörlesberger, Mohamed El-Nawawi & Tarek Khalil

Vol. 2 Management of Technology Innovation and Value Creation
edited by Mostafa Hashem Sherif & Tarek M. Khalil

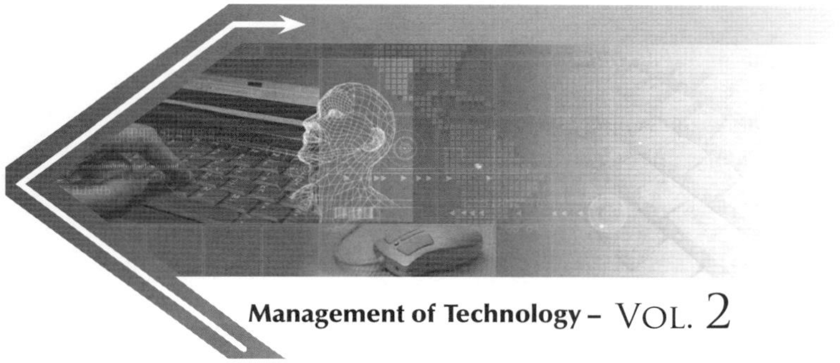

Management of Technology Innovation and Value Creation

Selected Papers from the 16th International Conference on Management of Technology

edited by

Mostafa Hashem Sherif

AT&T, USA

Tarek M. Khalil

University of Miami, USA

World Scientific

NEW JERSEY • LONDON • SINGAPORE • BEIJING • SHANGHAI • HONG KONG • TAIPEI • CHENNAI

Published by

World Scientific Publishing Co. Pte. Ltd.
5 Toh Tuck Link, Singapore 596224
USA office: 27 Warren Street, Suite 401-402, Hackensack, NJ 07601
UK office: 57 Shelton Street, Covent Garden, London WC2H 9HE

British Library Cataloguing-in-Publication Data
A catalogue record for this book is available from the British Library.

Management of Technology — Vol. 2
MANAGEMENT OF TECHNOLOGY INNOVATION AND VALUE CREATION
Selected Papers from the 16th International Conference on Management of Technology

ISBN-13 978-981-279-053-8
ISBN-10 981-279-053-5

Printed in Singapore by World Scientific Printers

Contents

Introduction

This book is part of the series on Management of Technology (MOT) that presents trends and advances in MOT research and practice. This volume consists of selected papers from the 16th Annual Conference of the International Association for Management of Technology (IAMOT) held in Miami Beach, Florida, USA, from May 13–17, 2007. The main topic of the conference was "Management of Technology for the Service Economy." Following two rounds of reviews 156 papers out of the 413 proposals received for the conference were retained and published in the CD of the conference proceedings.

It has been an IAMOT tradition to share some of these papers with a wider audience and 2007 was no exception. However in the interest of disseminating the greatest amount of information about recent work in MOT research and practice, IAMOT Executive Council decided to publish the selected papers into two volumes: A special issue on Management of Technology in the Service Sector to be published in IAMOT's official journal *Technovation*, and papers in diversified areas of MOT research to be included in this volume. Thus, 25 papers were selected for publication in this book. The final selection was guided by the authors' readiness to further improve on their contributions based on the comments provided in the conference and afterwards.

The book consists of six major sections. The first is on the general context of the knowledge economy and includes two chapters. Hidalgo and Albors attempt to predict the way management of technology will change due to the growing contribution of knowledge in the design and development of products and services. Their paper presents the results of a balanced survey highlighting the respondents' views on the business relevance of various techniques currently in use. They also give several suggestions for companies to improve their preparedness and take advantage of the opportunities ahead. Kuittinen et al. contribute to the discussion on inter-firm cooperation by bringing together two factors: decision under uncertainty and industry dynamics. Using three industries with different degrees of maturity, the authors show how the type and degree of uncertainty affect the cooperative arrangement and its governance.

The next section contains 5 papers dealing with the complex relations among R&D intensity, innovation, productivity and economic performance. Kuittinen et al. show that these relations are sector-related, i.e., they are not the same across industries. In particular, R&D investment decisions should consider the difference in the nature of R&D activities and the variations of the time lag between the investment and the returns. Ouchi and Watanabe study how Canon has been consistently profitable compared to other members of its peer groups. Using a System Dynamics model, they show the dynamic relations among technological diversification, learning by doing and new functionality development. Tou's paper supports the same argument for Japan's electric machinery firms. In other words, the benefits from R&D investment in terms of quality improvement and market evaluation depend on the corporate institution and the degree that its governance structure induces information sharing and institutional learning. Yanagisawa presents ways to calculate the benefits from studies that Japan Atomic Energy Research Institute (JAERI) had conducted over 45 years concerning various types of nuclear reactors. The last paper of this section proposes a method consistent with the accounting principles of the International Accounting Standards Board (IASB) to value intangible technological assets throughout the whole life cycle.

The third sections deals with the vital subject of green technologies and sustainable development. Matsumoto et al. use multi-agent modeling to analyze the diffusion of clean energy vehicles in Japan under various scenarios of oil prices. Nunes and Bennett draw on three major fields of research (environmental management, operations management and automotive production) to show how environmental concerns could be integrated in a company's decision process through the modification of the traditional SWOT analysis. Pätäri et al. argues that biofuel can help the Finnish forest sector, which is facing tough competition from emerging suppliers. In the same vein, Chéry and Marcandella present a methodology to evaluate the sustainability of a product before the actual design starts. This would allow managers to evaluate product innovations taking into account environmental factors. All these considerations have to be reflected in new practices in the management of technology that capture the dynamic interactions between nature and society, which is the subject that Brent and Pretorius have addressed in their contribution.

The largest section of the book is Section 4, which deals with the commercialization of knowledge for economic development. Ortt et al. study the cases of the photocopier, the videocassette recorder and the microwave oven. Their conclusion is that successful companies adapt their strategy according to the phase of the technology life cycle. Buys presents a method for introducing a new technological system and which was used in South Africa's program for a new generation of nuclear reactors. In the following paper, He and Fallah track the mobility of inventors in two clusters in New Jersey and Texas, respectively. They show that changes in the network properties can predict future regional economic and social conditions. Of course, development depends also on the availability of unique competencies that can be commercialized. Trumbach et al. make an inventory of the intellectual capital in the states of Louisiana, Mississippi and Alabama along the so-called I-10 Corridor as a pre-requisite for economic recovery after Hurricane Katrina of 2005. When intellectual assets are lacking, they have to be developed one step at a time. In her paper, El Fakir examines how South Korea was able to create the appropriate learning spaces at each phase of its catching-up process from assimilation to adaptation and then improvement of imported technologies. She suggests that another institutional transformation is needed to give South Korea the ability to contribute to radically new knowledge. Tang et al. depict how the Chinese pulp and paper manufacturers have learned from global players that invested in China and are now positioned to increase the scale of their production and export aggressively to meet the global demands. The final two chapters of this section give two different, but complementary, perspectives on technological innovations in Brazil. Rodrigues et al., explain how Petrobras developed its R&D capabilities in four phases so that it is able to build on its competence in deep-sea exploration to establish an international collaborative network for innovation. Zawislak et al. track the overall R&D expenditures in Brazil over three decades and arrive at the conclusion that the "Brazilian Way" of doing business is not conducive to radical innovations because it puts more emphasis on operational management practices to extract the maximum amount of revenues from what has been mastered in the past.

The papers of Section 5 underline the importance of organizational capabilities. Shah et al. identify three key elements for a successful venture: a thorough necessity analysis, clear objectives and the right environment. Their conclusions are based on an analysis of venturing initiatives and in particular the successes within Shell, Nokia and IBM. Grimpe and Sofka explain how a persistent R&D engagement or a highly dynamic environment (but not both) can encourage the development of a "rapid response" capability, in the face of technological volatility. Finally, Morel and Boly define two kinds of technological audits to evaluate the innovation processes in a given firm. The first is a self-assessment that gives a basic view of a company's innovative capabilities. The in-depth evaluation is conducted by a consultant and is based on a multi-objective optimization function.

The sixth and last section of the book is on foresight and forecasting. Weissenberger-Eibl and Speith explain how a firm could anticipate forecast disruptive technologies that have the potential of harming its business by combining technology roadmapping and indicator-based forecasting. Tseng adds fuzzy relationships to the traditional logistics and Gompertz models to estimate future volume sales.

In closing, the editors would like to express their appreciation for each of the individual contributors who graciously accepted to make the many revisions that were requested.

M. Hashem Sherif, AT&T
Tarek M. Khalil, University of Miami
October 2007

Section I

Paradigms for the Knowledge Economy

Chapter 1

New Innovation Management Paradigms in the Knowledge-Driven Economy

Antonio Hidalgo and José Albors***

**Universidad Politécnica de Madrid, Madrid, Spain*
Email: ahidalgo@etsii.upm.es
***Universidad Politécnica de Valencia, Valencia, Spain*

The growing importance of knowledge as a production factor and as a determinant of innovation can be explained by the continuous accumulation of technical knowledge over time. Innovation Management Techniques (IMTs) are critical to support the process of innovation in firms and help them in a systematic way to meet new market challenges.

1. Introduction: The Knowledge Economy

The paradigm of the knowledge economy originally appeared as a consequence of new trends in the economy and of new categories of statistical data on economic activity (Machlup, 1962). In the mid-1990s, the concept evolved to refer to two presumed characteristics of the new economy: the increased relevance of abstract knowledge, both quantitatively and qualitatively, and the prevalence of applications of information and communication technologies as economic drivers (David and Foray, 1995). The OECD (1996) defines knowledge-based economies as "*economies, which are directly based on the production, distribution and use of knowledge and information.*" Thus, the knowledge economy is based on an efficient system of knowledge access and distribution, as a *sine qua non* condition for increasing the amount of innovative opportunities (Godin, 2003).

This increasing importance of knowledge is changing the way firms compete as well as the sources of competitive advantage between countries. For the leading countries in the world economy, the balance between knowledge and resources has shifted so much towards the former that knowledge has become one of the most important determinants of the standard of living (World Bank, 1998). Today's most technologically advanced economies are knowledge-based in the sense that knowledge is increasingly considered to be a commodity (Boulding, 1996), that advances in ICTs (Information and Communication Technologies) have reduced the cost of many aspects of knowledge activity (Howells, 2000), and the degree of connectivity between knowledge agents has increased dramatically (Aridor et al., 2000).

The paper has three basic objectives:

1. To provide a comprehensive review of the scope, characteristics, trends and business relevance of the main innovation management methodologies developed by significant actors in this field (those seeking to provide advice to firms and focused on knowledge as the most important benefit to a firm) across the European Union, USA and Japan.
2. To clarify a conceptual framework in this area and to facilitate a consensus among the relevant actors developing and using these methodologies.
3. To analyze the perceptions of various key players–the promoters and users of such methodologies.

The methodology followed in this research is based both on a literature research and a survey carried out on a balanced sample (geographically and activity wise) of firms, academic centers, business schools, consulting firms and business support organizations. The research was financed by the European Commission and was carried out among respondents from the 15 Member States of the European Union, Japan and the United States. In total, 433 completed questionnaires were returned. The information collected from the survey was completed via phone interviews with the most representative stakeholders, which went into more detail on certain issues of relevance for the study and clarified some outstanding questions.

2. Knowledge and Innovation Management

The conception of innovation has evolved significantly over the last forty years. During the 1950s, innovation was considered a discrete development resulting from studies carried out by isolated researchers. Nowadays, innovation is no longer conceived as a specific result of individual actions, but more as a problem-solving process (Dosi, 1982), an interactive process involving relationships between firms with different actors (Kline and Rosenberg, 1986), a diversified learning process (Cohen and Levinthal, 1990), a process involving the exchange of codified and tacit knowledge (Patel and Pavitt, 1994), and an interactive process of learning and exchange where interdependence between actors generates an innovative system or an innovation cluster (Edquist, 1997). Other authors (Garcia and Calantone, 2002; McDermott and O'Connor, 2002) have outlined other aspects of innovation more related to the final consumer of the innovation and to the innovation process itself.

The evolution from a technological network perspective of innovation management to a social network perspective has been led by the challenge to transform information into knowledge (e.g. information contextually connected to the development or improvement of products or processes). Knowledge-based innovation requires the convergence of many different kinds of knowledge retained by a variety of actors (Kipping and Engwall, 2001; Smits and Moor, 2004).

The increasing importance of knowledge as an economic driver has major implications for innovation management, which is, in turn, a key determinant of national and regional competitiveness in the global, knowledge-driven economy. The contribution of knowledge to innovation is achieved in part by reducing transaction costs between firms and other actors, most notably in the areas of research and information, buying and decision-making, policy and enforcement (Maskell, 1999).

Innovation and knowledge generation have been analysed from a specific systemic approach considering the market role, the knowledge architecture and the innovation alternatives (process, product, radical, incremental) outlining a parallel comparison between both processes. The systemic approach to innovation recognizes that innovation and knowledge generation take place as a result of a variety of activities, many of them outside the formal research process. Knowledge is thus generated not just in universities and research centers, but also in a very wide variety of locations within the economy, and notably as a product (learning-by-doing) or of consumption (learning-by-using). In the current economic context, growth must mainly originate from increasing the productivity of knowledge work, and increasing this productivity is the most important contribution management can make (David and Foray, 1995; Kay, 1999).

In comparison to the traditional mechanical versus organic approach to management (Sine et al., 2006), these characteristics involve a fundamental change in the strategic perception of the organization, which accordingly has to consider the following management challenges: to manage human capabilities in a strategic manner (Lengnick-Hall, 2002), to generate networks with internal and external partners (Pittaway et al., 2004), to create adaptive and interactive organizational structures and to balance individual and corporate motivation (Gioia et al., 2000).

Finally, the challenges of the new knowledge-driven economy can be classified into the following groups:

New characteristics of the market. The market is constantly changing, it is becoming more global and new competitors are emerging. In addition technology complexity is increasing, product life-cycles are shortening, and knowledge is consolidating as a crucial input.

New types of innovation. Innovation takes many forms. It emerges where the market offers incentives to introduce new products and production methods, and where people are willing to take risks and experiment with new ideas (Tidd et al., 2005).

New needs of stakeholders. Customers, owners and stock markets increasingly equate an organization's worth with its ability to get winning products to market on time, every time (Magleby and Todd, 2005).

New approach to innovation management. The capacity of a firm to implement innovation management revolves around its success in dealing with these two main challenges, top-line growth and bottom-line efficiency (Aggeri and Segrestin, 2007).

New technology innovation assessment skills. The rapid development of new technologies prompts firms to assess and implement the most appropriate technology according to their need to keep their competitiveness (Libutti, 2000).

Need for new innovation management tools. The development of knowledge-based innovation management requires the capacity to implement technical and relational tools. Technical tools refer to the acquisition and utilization of new information and communication technologies – they do not create competitive advantage because they are readily available to others. The creation of competitive advantage rests in relational tools – the way of doing business, both in the internal and external environments of firms (Lengrand and Chartrie, 1999; Hidalgo, 2004; Thomke, 2006).

3. Innovation Management Techniques

Innovation does not always mean employing the very latest cutting-edge technology. On the contrary, it is less a question of technology and more a way of thinking and finding creative solutions within the company. In this context, innovation management techniques (IMTs) can be seen as a range of tools, techniques and methodologies that help companies to adapt to circumstances and meet market challenges in a systematic way (Cordero, 1991; Hidalgo, 2004).

In innovation management, there are a wide range of IMTs available on the market. This study focused on IMTs that complied with the following parameters:

1. They were sufficiently developed and standardized, and had fairly systematic methods of application. In other words, the implementation procedures and the benefits for the IMT were generally known and recognized in the market.
2. They are aimed at improving the competitiveness of firms by focusing on knowledge as the most important benefit.

3. They were freely accessible and not subject to any copyright or licensing agreement.

The application of a group of selection criteria resulted in ten groups of IMTs called "IMT typologies". The table 1 summarizes the 10 IMT typologies and their associated methodologies/tools.

There is no single correlation between a firm's specific business problem and the methodology that solves it. As a result, it cannot be claimed that there is a closed set of developed and proven IMTs for solving all challenges faced by business as a whole. Furthermore, IMTs do not usually act in a deterministic, unique manner and the diversity of firms and business circumstances means that there is not a single ideal model for innovation management, though there are some principles of good practice.

For these reasons, an innovation management technique cannot be considered in isolation. The usefulness of one IMT for a particular business challenge is normally measured in combination with other IMTs, this combination being adapted to varying degrees for each specific case. The benefit gained by the company depends on a combination of IMTs and the firm itself, and the mix of these two elements is what determines an effective outcome.

Table 1: IMT typologies and associated methodologies.

IMT typologies	Methodologies and tools
Knowledge management tools	- Knowledge Audits - Knowledge Mapping - Document Management - IPR Management
Market intelligence techniques	- Technology Watch - Patents Analysis - Business Intelligence - Customer Relationship Management - Geo-marketing
Cooperative and networking tools	- Groupware - Team-building - Supply Chain Management - Industrial Clustering
Human resources management techniques	- Teleworking - Corporate Intranets - On-line Recruitment - e-Learning - Competence Management
Interface management approaches	- Marketing Interface Management - Concurrent Engineering
Creativity development techniques	- Brainstorming - Lateral Thinking - TRIZ - Scamper Method - Mind Mapping
Process improvement techniques	- Benchmarking - Workflow - Business Process Re-engineering - Just in Time
Innovation project management techniques	- Project Management - Project Appraisal - Project Portfolio Management
Design management tools	- CAD Systems - Rapid Prototyping - Usability Approaches - Value Analysis
Business creation tools	- Business Simulation - Business Plan - Spin-off (from research to market)

4. Key Perceptions of the Leading Actors

4.1. *Role of each actor*

For the purpose of the study, "major actors" were defined as those bodies that play an important role in the development and/or promotion of methodologies to support innovation management in the knowledge-driven economy. These actors were classified into four groups: Business schools, Consultancies, Academic Centers and Research and Technology Organizations (RTOs), and Business Support Organizations (BSOs).

The study produced the following overall views on the roles of the major actors:

Academic Centers, including Research and Technology Organizations (RTOs), are *promoters* of IMTs and, in some specific cases, *developers* of them. In that case, they only adapt specific tools for SMEs. Their capacity to develop IMTs is concentrated sometimes in the development of strategies to raise the level of R&D activity among local or regional governments and some evaluation of R&D public programs.

Business schools are *developers* and *promoters* of IMTs. From the development perspective, it is the academic specialists with a high research orientation and high specialization that integrate business schools, because many of them develop part of their research activity directly in academic centers and combine academic and research work with consulting activities. As promoters, business schools use a great deal of tools. The most interesting mechanisms used to disseminate methodologies are the organization of seminars and workshops.

Consultancy firms consider themselves *more as developers than promoters* of IMTs and, for that reason, some of them in Europe were founded to support the regional economy or to diversify national economic activities. Some individual consultancy firms stressed the importance of motivation. These firms considered it one of their main objectives to motivate people to run their business, and to motivate SMEs to diversify activities.

Business Support Organizations are *promoters and users* of IMTs: they make available some tools to the SME members of their organization. They also act as a link between SMEs and innovation consultants and try to encourage the use of IMTs among third-party organizations. BSOs also consider themselves as *developers* of IMTs, but only when adapting IMTs in cooperation with consultants.

The opinion of managers within the companies was that consultancies are the main actors promoting the use of IMTs (27%), jointly with business schools (20%), and business support organizations (20%). With respect to helping firms use IMTs, consultancies are seen as the major agents (41%), while business schools (16%) and BSOs (15%) have less importance. The companies themselves consider their role to be more as *users* than *developers* of such methodologies.

All the major actors agree that only a few IMTs are widely recognized, and most are unidentifiable and inaccessible by firms. Over 37% of the actors declared that most firms are not aware of the existence of IMTs, while 34% stated that few IMTs are sufficiently defined to be successfully applied within firms. Consultancy firms and business schools generally believe that most firms are not aware of the existence of IMTs. Academic centers and industry generally see IMTs as systematically applied only in firms that want to be market leaders. Business support organizations mostly believe that very few IMTs are sufficiently well defined to be successfully applied within firms. All actors are convinced that new challenges coming from the knowledge-driven economy require new IMTs.

4.2. *Difficulties and challenges in facing the knowledge-driven economy*

The main difficulties seemed to revolve around the fact that introducing an IMT within an organization means an extra effort that requires time, motivation and money. The challenge is to motivate management support, to think of the future and foster creativity, to install a culture of innovation, to formulate an innovation strategy, to implement the innovation process and to overcome the pressure for meeting quarterly results in preventing experimentation.

IMTs are sometimes considered to have a more academic than practical role, because they are subject to a lack of awareness and motivation, and consequently a widespread ignorance about how IMTs can help companies to survive in the new knowledge-driven economy. On the other hand, many actors stressed the lack of an innovative culture in firms, as well as the uncertainty in predicting the conditions for competitive performance in new markets. Another difficulty is that innovation management cannot be handled as a product or as production management. The reason is that many firms do not have the capacity to identify innovations and introduce them into the normal production process. Further difficulties include: bureaucratic complexity, low awareness of innovation technology amongst managers, lack of suitable metrics, and unwillingness to share knowledge.

From the challenges point of view, actors highlighted four specific areas as presenting the greatest obstacles: financial investment needed, difficulty of accepting failure, excessive bureaucracy and uncertainty, and the need to support training schemes and to overcome intercultural complications, particularly when knowledge sharing is necessary.

5. Business Relevance of IMTs

In the knowledge economy, products and companies live or die by information–the most successful companies are those that use their intangible assets better and faster. Corporate reporting is still founded on a financial and management accounting model. This model was developed for the industrial economy and is not able to deal with today's knowledge economy, where most corporate value creation is based on knowledge assets rather than on physical resources and financial capital.

As a means of quantifying the business relevance of the different IMTs, the survey questionnaire detailed a list of benefits for the IMTs that respondent were invited to evaluate. The list of benefits is as follows: increasing flexibility and efficiency, managing knowledge effectively, increasing productivity and reducing time to market, facilitating teamwork, enabling online gathering of marketing information, improving relationships with suppliers, integrating differing sources of customer information, making client relationships more effective, eliminating redundant processes, reducing costs by implementing IT-based solutions, reducing bureaucratic tasks (those that did not add value), using e-learning, exploring e-commerce, increasing the market range of goods and services, and improving relationships with employees.

The Business schools point of view is that the main advantages that IMTs give firms are increased flexibility and efficiency, an understanding about how to use e-learning, facilitated teamwork and improved gathering of on-line marketing information (Fig. 1). Business schools consider creativity development, business plan development, e-learning techniques and customer relationship management (CRM) as the IMTs most used within their organizations.

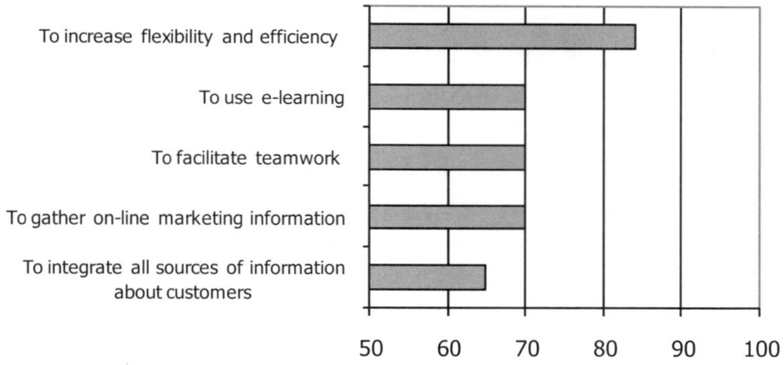

Figure 1: Business relevance for Business Schools.

From the perspective of the Academic centers, IMT benefits tend to be in the areas of managing knowledge effectively, reducing costs by using IT-based solutions, increased productivity and shorter time-to-market, increased flexibility and efficiency, better gathering of on-line market information, and improved teamwork (Figure 2). Project management, corporate intranet, spin-off and e-learning are the IMTs most successfully applied by the academic centers and RTOs.

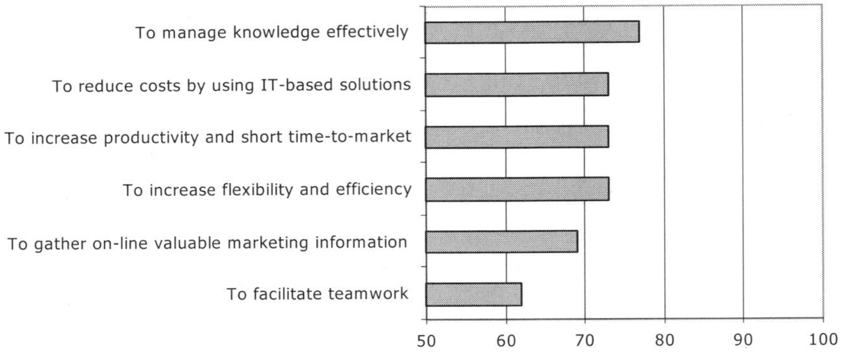

Figure 2: Business relevance of Academic Centres.

Consultancy firms tend to the view that the most important benefits are managing knowledge effectively, increased flexibility and efficiency, facilitating teamwork, reduced bureaucratic tasks, increased productivity and improved relationships with suppliers (Fig. 3). Consultancies consider business plan development and project management as the IMTs most used within their organizations.

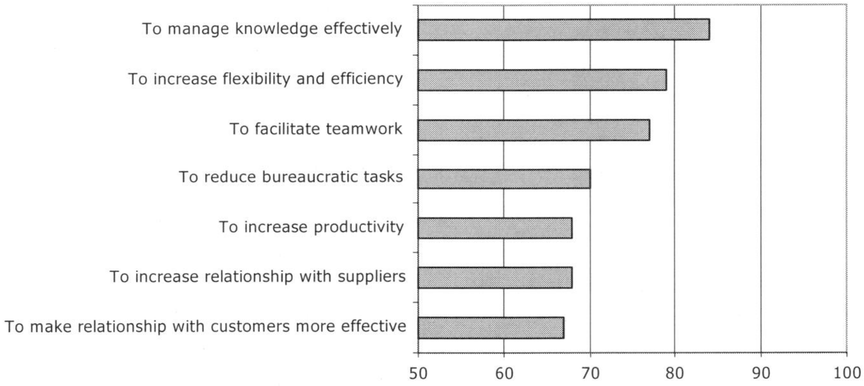

Figure 3: Business relevance for Consultancies.

From the perspective of BSOs, IMTs serve mainly to increase flexibility and efficiency, increase productivity and reduce time-to-market, gather on-line marketing information, manage knowledge effectively, and increase the effectiveness of relationships with suppliers (Figure 4). BSOs are more oriented towards project management, corporate intranets, business plan development and outsourcing.

Figure 4: Business relevance for Business Support Organisations (BSOs).

Within the firms that actually implement IMTs, the perspective of the managers involved is that IMTs can help their firms to foster competitive advantages in the following ways: increasing flexibility and efficiency (86%), managing knowledge effectively (76%), improving productivity and time-to-market (73%), improving relationships with suppliers (72%), gathering on-line marketing information (69%), facilitating teamwork (67%), integrating different sources of customer information (66%), reducing costs by using IT-based solutions (65%), and eliminating redundant processes (64%).

Innovation is seen as a key business opportunity for many industrial partners, but not for all of them. For some managers, IMTs do not seem to be central to their business concerns. To them, the importance of IMTs would be part of their culture or overall approach to innovation; their appreciation of IMTs seems to be very superficial. They all agree to recognise that IMTs are not well known, not readily identifiable and are inaccessible.

On the other hand, the lack of a clear and homogeneous view of innovation makes it difficult to relate it to the knowledge economy; the relationship between the two concepts is far from obvious and its relevance is not easy to demonstrate. In fact, managers are themselves asking for new inputs to better understand the extent and the scope of this question. Encouraging staff to disperse their acquired knowledge within the firm is a big challenge, and possibly one that can be encouraged within the knowledge-driven economy by application of technology-based tools to support this process.

6. Conclusions

The growing importance of knowledge as a production factor and as a determinant of innovation can be explained by the continuous accumulation of technical knowledge over time, and by the use of communications technologies that make that knowledge available very rapidly on a worldwide scale.

IMTs are critical to increasing the competitiveness. Participants in the study found that the main IMTs used were project management (82%), followed by business plan development (67%), corporate intranets (66%) and benchmarking (60%). Less used IMTs included Delphi method and lateral thinking. Some 43% of the actors in the study stated that they have successfully used IMTs in their own organization. Another 32% said that they do not use IMTs.

This study shows that proper application of IMTs facilitates a company's ability to introduce appropriate new technologies in products or processes, as well as the necessary changes to the organisation. However, most companies do not have an innovation culture that encourages the introduction of change within the organisation, more often there is a strong resistance from staff and sometimes from management. Companies can use consulting firms to get advice in this area, but generally have no tradition of asking consultancies for their help, a practice that has resulted in a limited range of operational models.

Finally, the following suggestions are intended to help promote an innovation culture, to assist companies to increase their competitiveness through innovation, and to help take advantage of the opportunities of the knowledge-driven economy:

1. Set up an overall scheme together with national and regional governments to promote innovation management. The objective is to

improve the know-how of actors promoting innovation management methodologies and tools within firms, in particular to SMEs. Also to promote the development of global networking among the various actors to encourage the exchange of knowledge and experience.

2. Support for well-designed awareness initiatives to enhance citizens' confidence in innovation as a means to foster competitiveness in companies and well being in our societies.
3. Support the development of common certification systems in innovation management. Certain preparatory work would be necessary to define practices and standards in this area.

References

Aggeri, F. and Segrestin, B. (2007). Innovation and project development: an impossible equation? Lessons from an innovative automobile project development. *R&D Management*, 37(1), 37–47.

Aridor, Y., Carmel, D., Lempel, R., Soffer A. and Maarek, Y. S. (2000). Knowledge Agents on the Web. In: Cooperative Information Agents IV—The Future of Information Agents in Cyberspace. Proceedings of the 4th International Workshop, (Klusch, M. and Kerschberg, L., eds.), CIA, Boston, MA, pp. 15–26, Springer Berlin/Heidelberg.

Boulding, K. E., (1996). The economics of knowledge and the knowledge of economics. *The American Economic Review*, 56, pp. 1–13.

Cohen, W. M. and Levinthal, D. A. (1990). Absorptive Capacity: a New Perspective on Learning and Innovation. *Administrative Science Quarterly*, 35, pp. 128–152.

Cordero, R. (1991). Managing for Speed To Avoid Product Obsolescence: A Survey of Techniques. *Journal of Product Innovation Management,* 8(4), pp. 283–294.

David, P. and Foray, D. (1995). Assessing and Expanding the Science and Technology Knowledge Base. *STI Review*, 16, pp. 55-68.

Dosi, G. (1982). Technological Paradigms and Technological Trajectories, *Research Policy,* 11 (3), pp. 147–162.

Edquist, D. (1997). Systems of Innovation. Technologies, Institutions and Organisations. Pinter, London.

Garcia, R. and Calantone, R. (2002). A critical look at technological innovation typology and innovativeness terminology: A literature review. *Journal of Product Innovation Management*, 19(2), pp. 25-37.

Gioia, D. A., Schultz, M. and Corley, K. G. (2000). Organizational Identity, Image, and Adaptive Instability. *The Academy of Management Review*, 25(1), pp. 63–81.

Godin, B. (2003). The Knowledge-Based Economy: Conceptual Framework or Buzzword? Project on the History and Sociology of S&T Statistics. *Journal of Technology Transfer*, 31, pp. 17–30.

Hidalgo, A. (2004). Innovation management and the Knowledge-Driven Economy. European Commission, Brussels-Luxembourg

Howells, J. (2000). International coordination of technology flows and knowledge activity in innovation. *International Journal of Technology Management*, 19(7-8), pp. 806–819.

Kay, J. (1999). Business Strategy in the Knowledge-Driven Economy. Conference organized by Department of Trade and Industry and the Centre for Economic Policy Research, January 1999, London.

Kipping, M. and Engwall, L. (2001). Management Consulting. Emergence and Dynamics of a Knowledge Industry. Oxford University Press, New York.

Kline, S.J. and Rosenberg, N. (1986). An Overview of Innovation. In: The Positive Sum Strategy. Harnessing Technology for Economic Growth (Landau, R. and Rosenberg, N.. eds.), National Academy Press, Washington D.C., pp. 275–306.

Lengnick-Hall, C.A. (2002). Strategic human resources management: a review of the literature and a proposed typology. In: Human Resource Management: A Critical Perspective (Poole, M., ed.), Business & Economics, London.

Lengrand, L. and Chartrie, I. (1999). Business Networks and the Knowledge-Driven Economy. European Commission, Brussels.

Libutti, L. (2000). Building competitive skills in small and medium-sized enterprises through innovation management. *Journal of Information Science*, 26(6), pp. 83-95.

Machlup, F. (1962). The Production and Distribution of Knowledge in the United States. Princeton University Press, Princeton.

Magleby, S.P. and Todd, R.H. (2005). Creating a Successful Capstone Program by Considering the Needs of Stakeholders. *European Journal of Engineering Education*, 30(2), pp. 203 – 214.

Maskell, P. (1999). Social Capital, Innovation and Competitiveness. Oxford University Press, Oxford.

McDermott, C. M. and O'Connor, G. C. (2002). Managing radical innovation: an overview of emergent strategy issues. *Journal of Product Innovation Management*, 19(6), pp. 424–438.

OECD (1996). The Knowledge-based Economy. STI Outlook, Paris.

Patel, P. and Pavitt, K. (1994). National Innovation Systems: why they are important and how they might be measured and compared, *Economics of Innovation and New Technology*, 3, pp. 77–95.

Pittaway, L., Robertson, M., Munir, K., Denyer, D., and Neely, A. (2004). Networking and innovation: a systematic review of the evidence. *International Journal of Management Reviews*, 5-6(3-4), pp. 137–168.

Sine, W.D., Mitsuhashi, H. and Kirsch, D.A. (2006). Revisiting burns and stalker: formal structure and new venture performance in emerging economic sectors. *Academy of Management Journal*, 49(1), 121–132.

Smits, M. and Moor, A. (2004). Effective knowledge management in knowledge-intensive organizations. Proceedings Organizational Knowledge, Learning and Capabilities Congress, Innsbruck.

Thomke, S. H. (2006). Capturing the real value of innovation tools topic: management of technology and innovation. *MIT Sloan Review*, 2, pp. 24–32.

Tidd, J., Bessant, J. and Pavitt, K. (2005). Managing Innovation: Integrating Technological, Market and Organizational Change. John Wiley & Sons, Hoboken.

World Bank (1998). Knowledge for Development. World Development Report. Oxford University Press, New York.

Chapter 2

The Many Facets of Uncertainty and the Structure of Cooperation

Hanna Kuittinen, Ari Jantunen**, Kalevi Kyläheiko***, and Jaana Sandström*****

Lappeenranta University of Technology, School of Business
P.O. Box 20, Lappeenranta, FIN-53851, Finland
**E-mail: hanna.kuittinen@lut.fi*
***E-mail: ari.jantunen@lut.fi*
****E-mail: kalevi.kylaheiko@lut.fi*
*****E-mail: jaana.sandstrom@lut.fi*

This paper considers uncertainty as a significant determinant of the structure of cooperation. In the early phase of industrial development, there are many possibilities as to how the industry can evolve in terms of technological opportunities. The rise of technological standards and dominant designs narrow the range of technologies in use, hence decreasing radical uncertainty. Our hypothesis is that this evolvement has an effect on the choice of alliance governance mode. Following our uncertainty logic, we assume that the more mature an industry is, the less is the role of radical uncertainty and the more dominant is the role of parametric uncertainty, and vice versa. Our statistical analysis draws from three industries that are in different phases of development: (i) embryonic (biotechnology), (ii) developing (telecommunication equipment manufacturing), and (iii) mature (pulp and paper industry).

1. Introduction

In recent years, there has been an enormous growth of alliances especially in the high-tech sectors such as biotechnology and information and communication technology. As the number of hybrid forms of organization has increased, the determination of efficient organizational boundaries has gained a lot of theoretical and practical attention. Traditionally transaction cost economics (TCE) has dominated the discussion of the organizational boundary selection. Between the two extremes of governance structures, namely the market and the hierarchy, there are intermediate "hybrids" or networks, such as joint ventures, strategic alliances and other cooperation modes. In this paper we pick up two specific hybrid forms from the continuum of the market and hierarchy: (i) joint ventures and (ii) strategic alliances. Joint ventures occur when two or more firms pool a portion of their resources into a common legal organization (Kogut, 1988). Because a joint venture is a legal organization itself, its structure is closer to a hierarchy (i.e., a vertically integrated corporation) in the continuum. A strategic alliance does not involve any equity ownership, and hence it can be considered as a looser form of cooperation and closer to a market-like cooperation structure than a joint venture. This paper contributes to this discussion by paying special attention to the many faces of uncertainty as important determinants of the choice between an equity joint venture and strategic alliance.

The main question of transaction cost economics is whether a certain transaction is performed more efficiently inside the firm boundaries or outside (Coase, 1937; Williamson, 1975). The underlying assumption is that because of the vital role of competition, the open market mode would always be a more efficient form compared to a hierarchy, unless there were costs to use the market called transaction costs. The pre-contractual transaction costs arise when negotiating and monitoring a contract and when using the market as a coordination mechanism.

When transaction costs are high, the transaction may more preferably be performed inside the firm boundaries. The standard TCE explanations have neglected the benefits related to cooperation. However, some recent advances and extensions of TCE have considered the governance structure choices by also taking into account the benefits of cooperation.

Our paper explores the role of uncertainty as a significant determinant and motivation for the rise and structure of cooperation between the firms. David and Han (2004) found in their systematic assessment of empirical work on TCE that the relationship between uncertainty and governance mode is nothing but clear. This has led to more refined definitions of uncertainty, and some recent studies on governance mode selection (e.g., Santoro and McGill, 2005; Carson et al, 2006) have noted that different types of uncertainties have different effects on organizational boundary choices. In this paper we divide uncertainty into (i) parametric uncertainty and (ii) radical (structural) uncertainty (see Langlois, 1984) and propose that the effects of these two types of uncertainties depend on the maturity of the industry.

In our view, uncertainty is strongly related to the evolution of an industry: as the industry matures and the dominant design becomes viable, the radical uncertainty decreases through the eliminative selection of competing technologies. In the early phase of an industry's development, the technological opportunities are numerous. However, the rise of technological standards and dominant designs narrow the range of technologies in use, hence decreasing the radical part of technological uncertainties.

We suggest that radical and parametric uncertainties have an effect on the kind of cooperation structures. We examine the use and structure of cooperation in three industries that are in different phases of their development: (i) embryonic (biotechnology), (ii) developing (information and communication technology, especially telecommunication equipment manufacturers), and (iii) mature (pulp and paper). We assume that the more mature an industry is, the lesser is the role of radical uncertainty (U_r) and the more dominant is the role of parametric uncertainty (U_p), and vice versa. The maturity of an industry has an impact on possibilities to attain transaction benefits (TB) as well. As the industry evolves from embryonic to mature the role of transaction benefits and radical uncertainty diminishes whereas the importance of static transaction cost (TC) determinants and parametric uncertainty

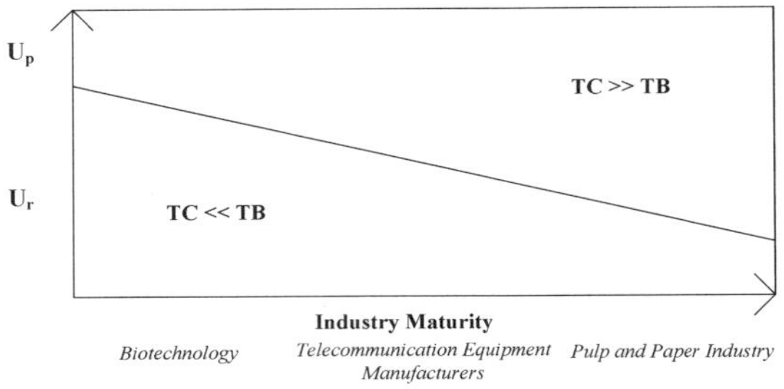

Figure 1: Parametric and radical uncertainty, industry maturity, and relative role of transaction costs and benefits.

increases. Figure 1 illustrates in a simplified manner this relation between the many facets of uncertainty and industry maturity.

2. Hypotheses

We start our hypotheses building by means of some basic transaction cost economics arguments. We first take asset specificity and (parametric) uncertainty and predict how the occurrence of these two affects the alliance form. Then we enlarge the framework with the concept of radical uncertainty, which is in our view closely related to industry evolution as well as to possibilities to attain transaction benefits. We then see how this dynamic element changes the predictions made with the first set of hypotheses.

2.1. *Asset specificity as a transaction cost determinant*

The concept *asset specificity* deals with the problem of transferability of assets to alternative uses (Williamson, 1985). When assets are specific to a certain transaction they might have very little value in alternative uses. This situation creates opportunities for post-contractual opportunistic behavior and costly bargaining as well, thus increasing transaction costs. When an alliance involves investments that are alliance-specific, one partner may threaten the other with disengagement from the alliance which would lead to potential losses. Thus the type of cooperation where the risk of opportunistic behavior is high due to significant asset specificity is likely to lead to a governance mode that replicates the hierarchy i.e., a joint venture. On the other hand, when asset specificity is expected to be lower, a hybrid form closer to an open market option will be utilized. Therefore, our first hypothesis is:

Hypothesis 1a: High asset specificity increases the likelihood of a more hierarchical governance mode in alliances.

2.2. *Many facets of uncertainty as determinants of transaction costs and benefits*

In the static transaction cost view, uncertainty is seen as originating from changes in the operating environment and the rate and unpredictability of the changes create uncertainty about future conditions (Williamson, 1985). In the transaction cost approach, this type of market and technology related uncertainty is called *parametric uncertainty.* Parametric uncertainty is based on the agent's subjective belief regarding the probabilities of events and the consequences of their actions (Hirshleifer and Riley, 1979). In sum, an agent has *certain* knowledge as to the structure of the decision problem, but *uncertain* knowledge as to the (probability) parameters of the problem (Kyläheiko 1995). Relevant events can be traced back to the states of the world that are supposed to be *independent* from the actions of market players and institutional actors.

Transactions that are based on the contracting and the use of markets always involve opportunism-based risks in uncertain conditions. Since contracts are always incomplete by nature but also inflexible under significant uncertainty, they are problematic. Parametric uncertainty (Carson et al., 2006; Folta, 1998), in other words, the speed of change and unpredictability of changes in operating environments result in difficulties when trying to evaluate future conditions. Because future conditions cannot be known ex ante, adjustments are needed ex post. The more parametric the uncertainty, the more control and decision-making power are needed in the governance of operations. This suggests the following hypothesis:

Hypothesis 1b: High parametric uncertainty increases the likelihood of the emergence of a more hierarchical governance mode in alliances.

On the other hand, besides parametric uncertainty there also exists other type of uncertainty. This type can generally be called *structural* by nature (Kyläheiko 1995, 38–40). This structural uncertainty is based on imperfect knowledge as to the structure the future can take. The sources of uncertainty can be found either from the lack of necessary information concerning future outcomes or from the lack of sufficient computational and other cognitive capabilities of a decision maker (Dosi and Egidi, 1991). In conditions where structural uncertainty is very high, one can speak about *radical uncertainty*. We propose that the higher the radical uncertainty, the more preferable it is to use loose cooperation and/or market operations, since the more important it is to open up as many strategic options as possible for different kinds of technologies. This can best be realized by loose cooperation that exploits hard-power incentives typical for efficient markets. In our view, radical uncertainty is typically related to the nature of technology.

Proposition 1: High radical uncertainty increases the likelihood of the emergence of a more market-based governance mode in alliances.

This proposition brings us to the core of the so-called transaction benefits issue and presupposes a more thorough analysis of the maturity of industries.

2.3. *Industry evolution and radical technological uncertainty as determinants of transaction benefits*

Recently, *transaction benefits* have also been taken into account when explaining the rise of hybrid network solutions. To put it briefly, transaction benefits depend on the possibilities (i) to use a common knowledge pool through cooperation in knowledge sharing, (ii) to share the risks of large fixed sunk costs among the partners, and (iii) to utilize market-based variation of ideas through high-powered incentives of independent entrepreneurs (Blomqvist et al., 2002).

The extended TCE argument suggests that the governance structure should be chosen in such a way that it minimizes the difference between the firm's transaction and management costs (i.e., the governance costs) at the same time when the value of transaction benefits is maximized (see Blomqvist et al., 2002).

This notion of transaction benefits comes very close to Langlois' (1992) conception of dynamic transaction costs: "the cost of not having the capabilities you need when you need them." Especially when the circumstances of operating environments involve not only parametric uncertainty but also radical uncertainty, the central determinants of the choice on the governance mode may be the factors that help to cope with radical uncertainty and take advantage of cooperation benefits. In the conditions of radical uncertainty, loose cooperation forms can best help to scan technological changes and to identify major technological breakthroughs in their early phases. Loose alliances help to get access to relevant information sources in the condition where new knowledge is scattered and tacit by nature. Compared to joint ventures, looser alliances have benefits when radical uncertainty is high. When radical uncertainty decreases, a strong dependence on partners may be a dead weight. Decrease of radical uncertainty winnows some viable solutions from the range of many alternatives. This leads to the conclusion that keeping all strategic options open by means of loose cooperation is advisable when radical uncertainty about future development is high.

In our view, the role of radical uncertainty is related to the predictability and maturity of industries. When new technologies and technology-based industries come into existence, their development is often rapid and discontinuities are common. That is, radical uncertainty is high in the beginning. When industries mature, they attain the phase in which technological development stabilizes and radical uncertainty decreases. This development also has consequences as for the organization of cooperation. When an industry is in the embryonic phase, radical uncertainty is high resulting in cooperation forms that are based on loose alliances. As the industry matures and the dominant design grows up, the role of normal static transaction cost factors arises and the choice of cooperation form is determined mainly by ordinary transaction cost determinants, such as asset specificity and parametric uncertainty.

To conclude, we see that the relationship between asset specificity and (parametric) uncertainty and the governance mode crucially depends on the maturity of an industry, and hypothesize:

Hypothesis 2a: The positive relationship between asset specificity and the likelihood of a more hierarchical governance mode in alliances increases along with the maturity of an industry.

Hypothesis 2b: The positive relationship between parametric uncertainty and the likelihood of amore hierarchical governance mode in alliances increases along with the maturity of an industry.

3. Data and Methods

3.1. *Data*

The Securities Data Company (SDC) database on Alliances and Joint Ventures was our database for collecting alliances in the industries chosen. Our sample consists of all the alliances from the *biotechnology*, *telecommunication equipment* and *pulp and paper industries* (PPI) in a ten-year period in 1996–2005.

The criteria for our data search were set so that at least one partner of the alliance was from one of these industries. For the PPI and telecommunication equipment industry, standard industry classification (SIC) codes were directly used as search determinants, with SIC codes beginning with 26 for the PPI, and 366, 367 for the telecommunication equipment industry. For the biotechnology industry we used the high technology code of the SDC database. These selection criteria led us to 3,515 alliances in biotechnology, 2,683 in the telecommunication equipment industry and 429 in the pulp and paper industry.

3.2. *Measures*

The *alliance governance* mode was defined on a continuum of market and hierarchies, joint ventures being closer to the hierarchy whereas strategic alliances are closer to the market option. Using the SDC database, each alliance observation was coded as 1 equaling a joint venture and 0 for a contractual strategic alliance.

To capture the level of *asset specificity* we used the SDC database information about the main activity of the alliance. If it was marketing, the asset specificity measure was coded as 1. When the alliance main activity was manufacturing, it was coded as 2, and for research and development the asset specificity measure was 3. For all the alliances whose main activity was none of the already mentioned, asset specificity was coded as 0.

Parametric uncertainty was intended to capture the role of exogenous parametric technological uncertainty, specific to each industry. Folta (1998) has used the volatility of an industry index as a measure of uncertainty, and we follow his lead by using the volatilities of each industry's index returns. We used the Nasdaq Biotechnology Index, Nasdaq Telecommunications Index and S&P Paper Products Index derived from the Thomson Financials database. To separate the general market uncertainty from the unique technological uncertainty of an industry we took into our calculations the market index (S&P 500 Composite).

Our measure of parametric technological uncertainty was calculated from the one-year standard deviation of the log of daily returns for each of the three industry indices and the market index. Consistent with the unique risk in finance models (see Sharpe, 1999, p. 185), the following formula was used:

$$\sigma_{\varepsilon i}^2 = \sigma_i^2 - \beta_{im}\sigma_m^2,$$

where $\sigma_{\varepsilon i}$ is unique risk of the industry index, σ_i is standard deviation of returns on the industry index, β_{im} slope term of the industry index returns to the market index returns and σ_m is standard deviation of the returns on the market index. We used this unique industry uncertainty as a measure of parametric technological uncertainty. It was calculated for each industry and for each year during 1996–2005. Although this measure could have been calculated for shorter sub-periods as well, we thought that the perceived parametric uncertainty does not change in a shorter period of time.

To capture the effect of variables unrelated to uncertainty and asset specificity, we included two controlling dummy variables in our models. The alliance partner nationality difference was taken into account with the measure *International*, coded as 1 when alliance partners were from different countries and given value 0 when they were from the same country. The second control variable *Same Industry* was given value 1 when the firms in the alliance were from the same industry (at the two digit SIC level) and when their industries differed this measure was coded as 0.

4. Results

Table 1 shows descriptive statistics and correlations between independent variables. The sample consisted of altogether 6,266 alliances of which 1,517 were joint ventures. The analyses were made for each industry separately. In addition we wanted to estimate the change that might have occurred during the observation period and thus we conducted two models for each industry. The first model was delimited to a five-year period 1996–2000 and the second one for the remaining five-year period 2001–2005.

Table 1: Descriptive statistics.

	Pulp and paper			Telecom. equip. manufacturing			Biotechnology		
	n	mean	std.	n	mean	std.	n	mean	std.
Joint venture	428	0.630	0.483	2683	0.210	0.410	3515	0.190	0.394
Same industry	428	0.210	0.406	2683	0.220	0.415	3515	0.430	0.495
International	428	0.620	0.486	2683	0.640	0.479	3515	0.560	0.497
Asset specificity	428	1.230	0.922	2683	0.670	0.975	3515	2.180	1.045
Uncertainty	428	0.015	0.006	2683	0.013	0.005	3515	0.015	0.015

We performed logistic regression in order to preserve the effect of our control and independent variables on the choice of the alliance governance mode. In logistic regression a binary variable was assigned to both types of alliances, with joint ventures taking 1 and strategic alliances taking value 0. Thus a positive coefficient estimate indicates that the variable in question predicts the governance mode towards joint venture and a negative coefficient predicts a governance mode towards strategic alliance.

Table 2: Results of logistic regression on the choice between a joint venture and strategic alliance in the pulp and paper industry (standard errors in parenthesis).

Pulp and paper	Years 1996-2000				Years 2001-2005			
	Model 1a		Model 1b		Model 2a		Model 2b	
	B	**s.e.**	**B**	**s.e.**	**B**	**s.e.**	**B**	**s.e.**
Constant	0.302	0.204	0.615**	0.232	-0.677*	0.296	-0.331	0.330
International	0.629*	0.259	0.502	0.268	1.003**	0.372	0.975**	0.408
Same industry	0.610	0.336	0.381	0.346	0.408	0.478	0.614	0.514
Asset specificity			0.662**	0.185			0.704**	0.252
Uncertainty			-0.024	0.138			0.833**	0.262
Chi-square	9.601*		23.189**		9.242*		11.555*	
-2 Log-likelihood	353.464		339.875		177.900		88.530	
Nagelkerke R^2	0.045		0.107		0.088		0.194	
n	293		293		135		135	

* sig.<0.01, **sig.<0.05

It can be seen from Model 1b in Table 2 that asset specificity is significantly and positively related to the choice of alliance governance in the pulp and paper industry giving support to Hypothesis 1. Our measure of parametric uncertainty has a positive and significant impact on the propensity to choose joint venture only in the latter time period (see Model 2b). Uncertainty, however, was not a significant predictor in the earlier time period.

Our results about the choice of the governance mode in the telecommunication equipment manufacturing industry (Table 3) show that asset specificity is not significantly related to the governance mode in 1996-2000. In the time period in 2001–2005 it had a positive and significant effect on the governance mode. This supports our Hypothesis

2a. What comes to our uncertainty measure, a similar pattern can be seen. During the years 1996–2000 increasing parametric uncertainty led to a more market-based governance mode, which can be seen as a significant and negative coefficient of the variable uncertainty from Model 3b in Table 3. Even more interestingly, in 2001–2005 uncertainty was already showing a positive and significant effect on the alliance governance mode (see Model 4b), which is in line with Hypothesis 2b.

Table 3: Results of logistic regression on the choice between a joint venture and strategic alliance in the telecommunication equipment manufacturing industry.

Telecom. equip. manuf.	Years 1996-2000 Model 3a		Model 3b		Years 2001-2005 Model 4a		Model 4b	
	B	**s.e.**	**B**	**s.e.**	**B**	**s.e.**	**B**	**s.e.**
Constant	-1.698**	0.114	-1.789**	0.131	-2.373**	0.196	-1.739**	0.209
International	0.938**	0.131	1.034**	0.133	0.766**	0.214	0.697**	0.222
Same industry	-0.016	0.136	-0.127	0.141	0.364	0.200	0.099	0.213
Asset specificity			0.112	0.073			0.834**	0.105
Uncertainty			-0.476**	0.074			0.249*	0.116
Chi-square	56.485***		107.621***		18.318***		84.813***	
-2 Log-likelihood	1770.528		1719.391		895.897		829.402	
Nagelkerke R^2	0.051		0.095		0.030		0.133	
n	1610		1610		1073.000		1073.000	

* sig.<0.01, **sig.<0.05

Models 5b and 6b show that asset specificity has a significant and negative effect on the alliance governance mode giving support to our radical uncertainty hypothesis. Uncertainty was also showing a negative impact, not significantly however, in 1996–2000. During the years 2001–2005 our parametric uncertainty measure had a positive and significant coefficient implying that traditional transaction cost determinants are getting some support.

Table 4: Results of logistic regression on the choice between a joint venture and strategic alliance in the biotechnology industry.

Biotechnology	Years 1996-2000				Years 2001-2005			
	Model 5a		Model 5b		Model 6a		Model 6b	
	B	**s.e.**	**B**	**s.e.**	**B**	**s.e.**	**B**	**s.e.**
Constant	-1.161**	0.109	-1,070**	0,114	-1.776**	0.107	-1,575**	0,115
International	-0.345**	0.130	0.434**	0.129	0.42**	0.125	0.391**	0.127
Same industry	0.443**	0.128	-0.362**	0.006	-0.527**	0.126	-0.467**	0.129
Asset specificity			-0.228**	0.069			-0.359**	0.073
Uncertainty			-0.006	0.049			0.372**	0.073
Chi-square	19.043***		30.156***		27.628***		75.694***	
-2 Log-likelihood	1504.813		1493.699		1823.288		1775.223	
Nagelkerke R^2	0.021		0.033		0.022		0.060	
n	1332		1322		2183		2183	

* sig.<0.01, **sig.<0.05

5. Discussion and Conclusions

Our aim in this paper was to investigate more closely the role of asset specificity and industry specific uncertainty as determinants when a firm chooses the governance mode of cooperation. Our attention was especially focused on how these effects evolve according to the maturity of the industry. Our empirical findings indeed suggest that when an industry is in the developing phase, such as biotechnology currently or telecommunication equipment manufacturing a decade ago, static transaction cost determinants fail to explain the cooperation governance mode.

However, as the industry matures the normal static Williamsonian transaction cost determinants can explain the choice of the governance mode much better. In the mature pulp and paper industry asset specificity determines the choice between a strategic alliance and joint venture – the higher the asset specificity, the higher the likelihood of a tighter governance mode i.e. a joint venture. In the telecommunication equipment manufacturing industry the effect of the asset specificity was positive as well. However, the effect is stronger in the later time period;

thus indicating in our view the decrease of radical uncertainty. In the embryonic biotechnology industry the high asset specificity led to a looser governance structure in cooperation. This result implies that under radical uncertainty the benefits created by loose cooperation exceed the importance of static transaction cost determinants, such as asset specificity. Our empirical findings indicate that in the earlier time period parametric uncertainty is positively but not significantly related to the hierarchical governance mode only in the pulp and paper industry. However, in the later time period the positive relationship between parametric uncertainty and the use of a joint venture is stronger in all the industries.

This paper contributes to the discussion of organizing inter-firm cooperation by bringing the industry dynamics into the focal point of analysis. Incorporating the industry maturity perspective into the governance mode choice analysis makes it possible to understand why traditional transaction cost determinants are not able to explain the governance choice in certain circumstances, such as the ones dominated by radical uncertainty. When considering industry evolvement phase and cooperation, standardization and standard organizations are playing a central role in reducing radical uncertainty (see e.g., Sherif, 2006). When common rules of certain industry are set, compatibility of different technologies is ensured and it makes in turn cooperation with different industry partners more feasible. Our results also implicate that further research on the crucial relationships between transaction costs and transaction benefits and governance modes should also take into account the industry development phase as a moderating factor.

From the managerial perspective our results underline the importance of the stage of industry evolution. If the industry evolution phase is neglected when considering how to govern alliances, some transactional benefits arising from radical uncertainty may be failed to achieve. Similarly, the costs related to cooperation might be considerably higher if the governance choice is made in ignorance of the industry evolution phase.

References

Blomqvist, K-M., Kyläheiko, K. and Virolainen, V-M. (2002). Filling a gap in traditional transaction cost economics: towards transaction benefits based analysis using Finnish telecommunications as an illustration, *International Journal of Production Economics*, 70, pp. 1-14.

Carson, S. J., Madhok, A., and Wu, T. (2006). Uncertainty, opportunism, and governance: the effects of volatility and ambiguity on formal and relational contracting, *Academy of Management Journal*, 49, 5, pp. 1058-1077.

Coase, R. H. (1937). The nature of the firm, *Economica*, 4, pp. 386-405.

David, R. J. and Han, S.-K. (2004). A systematic assessment of the empirical support for transaction cost economics, *Strategic Management Journal*, 25, pp. 39-58.

Dosi, G. and Egidi, M. (1991). Substantive and procedural uncertainty: an exploration of economic behaviours in changing environments, *Journal of Evolutionary Economics*, 1, pp. 145-168.

Folta, T. B. (1998). Governance and uncertainty: the trade-off between administrative control and commitment, *Strategic Management Journal*, 19, pp. 1007-1028.

Hagedoorn, J. (2002). Inter-firm R&D partnerships – an overview of patterns and trends since 1960, *Research Policy,* 29, pp. 567-586.

Hirshleifer, J. and Riley, J.G. (1979). The analytics of uncertainty and information – an expository survey, *Journal of Economic Studies*, 14, pp. 65-86.

Kogut, B. (1988). Joint ventures: theoretical and empirical perspectives, *Strategic Management Journal*, 9, pp. 319-323.

Kyläheiko, K. (1995). Coping with technology: a study on economic methodology and strategic management of technology, Lappeenranta University of Technology, Research Papers 48/1995, Academic dissertation in economics.

Langlois, R. N. (1984). Internal organization in a dynamic context. In: Communication and information economics: new perspectives (Jussawalla, M. and Ebenfield, H., eds.), North-Holland, Amsterdam, pp. 23-49.

Langlois, R. N. (1992). Transaction cost economics in real time, *Industrial and Corporate Change*, 1, pp. 99-125.

Santoro, M. D. and McGill, J. P. (2005). The effect of uncertainty and asset co-specialization on governance in biotechnology alliances, *Strategic Management Journal*, 26, pp. 1261-1269.

Sharpe, W. F., Alexander, G. J. and Bailey, J. V. (1999). Investments, Prentice Hall, Upper Saddle River.

Sherif, M. H. (2006). Managing projects in telecommunication services, John Wiley & Sons-IEEE Press, Hoboken.

Williamson, O .E. (1975). Markets and hierarchies: analysis and antitrust implications. A study in the economics of internal organization, Free Press, New York.

Williamson, O. E. (1985). The economic institutions of capitalism, Free Press, New York.

Section II

R&D, Innovation and Market Returns

Chapter 3

R&D Intensity and Firm Performance–Sectoral Differences

Hanna Kuittinen, Kaisu Puumalainen**, and Ari Jantunen****

Lappeenranta University of Technology, School of Business
P.O. Box 20, Lappeenranta, FIN-53851, Finland
**E-mail: hanna.kuittinen@lut.fi*
***E-mail: kaisu.puumalainen@lut.fi*
****E-mail: ari.jantunen@lut.fi*

This paper investigates the relation of research and development intensity to firm performance. We believe that this relation is strongly sector-related, as the logic of innovation differs radically between different industries and apply the Pavitt taxonomy to classify different industries into more homogenous sectors. We use a large data set of public companies worldwide over the 1996–2005 period. An interesting question is the delay between an R&D investment and its results. We assume that this lag is dependent on the technological environment (technology regime, uncertainty, intensity) and hence differs between industry categories. There is also reason to believe that lag structures are different depending on the performance measure used. The effects on growth are expected to have a shorter time lag when compared to profitability measures. Our results imply that sectoral differences do exist in firm-level R&D intensity as such, with science-based industries having the highest R&D intensity. Sectoral differences are also evidential when estimating the time lag effect between R&D expenditures and performance of the firm.

1. Introduction

The relation between research and development investments and innovation output is a widely studied phenomenon in the realm of innovation studies. Studies concerning R&D productivity on the firm level widely concentrate on either a single industry or on multiple industries, utilizing industry control variables in their analyses. Thus, studies of this type investigate R&D productivity in a given industry or examine the differences between industries. However, even though this type of research strategy enables us to analyze *how* industries differ from each other in terms of their R&D input and output ratio, the interesting question *why* industries differ in this respect remains unexplored. In other words, if there are industry-specific factors that have an effect on the R&D productivity of firms operating in a given industry, intriguing is what those industry characteristics or mechanisms are that mediate the relationship between R&D inputs and outputs. In this study, we try to look behind industry codes and explain R&D productivity differences with differences between industry sectors in the nature and sources of innovation. In this endeavor, we take the Pavitt categorization of industrial sectors (Pavitt 1984) as our starting point.

The goals of this paper are twofold. *First*, we estimate the time lag between R&D investment and the performance of a firm in different industry taxonomies. We expect that the length of the time lag varies between the taxonomies since the logic of innovation activities is different. Performance is measured as (i) growth (sales growth 2000–2005), (ii) profitability (return on invested capital) and (iii) value added (market to book value). It is also expected that the time span of R&D intensity and its effect on performance varies depending on the performance measure used. Our *second* goal is to estimate the effect of R&D intensity on performance and especially differences between industrial sectors in this respect.

2. R&D Productivity and Sectoral Differences in the Nature of Innovation

The relationship between R&D investments and firm performance has been of interest to researchers at least since Griliches (1981). In general, several empirical studies have found a positive relationship between R&D inputs and firm performance (see, e.g. Crépon et al. 1998; Parisi et al. 2006). However, the studies and also results are different in many respects. Studies cover different countries and industries, and differences in macroeconomic conditions or national-level innovation systems may have an impact on R&D productivity (Delmas 2002; Hall & Mairesse 2006). Some of the studies measure the output side of innovation activities with patents (see, e.g. Hall et al. 1986). Further, in previous studies the performance variable has been measured by turnover (Mairesse & Hall, 1996; Kafouros, 2005), innovative sales (Crépon et al. 1998,), profitability, (Jefferson et al. 2006, Fryxell 1990), market value (Hall & Oriani 2006, Bae & Kim 2003, Jaffe 1986), or various combinations of these or related measures. Many studies have examined R&D productivity as elasticity – the ratio of the growth rate of total factor productivity to the growth rate of R&D capital, and derived their models from the Cobb-Douglas production function (e.g. Hall & Mairesse 1995; Tsai 2005), while others have used more simple ways to measure productivity. For instance Billings et al. (2004) measured R&D productivity as current sales per dollar of R&D expense lagged by three years.

The Pavitt (1984) taxonomy of industrial sectors categorizes industries according to determinants and patterns of technological change, including the nature and sources of innovation, technology user types and means of appropriability as central factors used in industry classification. The Pavitt (1984) taxonomy categorizes firms into four groups: 1) supplier dominated, 2) scale-intensive, 3) specialized suppliers and 4) science-based. This classification has been utilized in previous innovation research by Laursen and Meliciani (1999), Marsili and Verspagen (2001) and Souitaris (2002) among others.

Similarly with Pavitt (1984) and others (e.g. Dosi 1982) we see that the essentially different conditions of technological development in different industry sectors have an effect not only on the average R&D intensity of industries but also on the type of innovation activities of firms operating in these sectors. In other words, R&D costs consist of different types of input costs in different sectors, and outcomes of innovation activities are also dependent on industry characteristics. In our view, this should also have consequences on R&D productivity, reflected in differences between industry sectors in terms of returns to innovation, market-value or other performance indicators. However, to our knowledge, there exist no previous studies that approach R&D productivity from the perspective which examines input-output effects by comparing industry sectors (i.e. Pavitt taxonomy classes) to each other.

In our view, one very critical aspect of R&D productivity studies is the time lag R&D expenditure has on productivity change. Investments in research and development are highly uncertain and time-consuming endeavors and this further complicates the measurement of the performance effects. This delay can create problems and lead to biased estimates of R&D's importance (Balcombe et. al, 2005). This time-lag structure of R&D inputs and outputs has previously been studied using patents as the output measure (Kondo, 1999; Hall et. al., 1986). The results vary from a time lag of 1 to 2 years, and Kondo (1999) found that the time lag was even longer in processing industries, such as chemical products. Others (Hall & Oriani, 2006; Harhoff, 1998) have used R&D capital stocks computed from past and present R&D expenditures over a certain time. The shortcoming of this measurement is that it does not take into account the possibilities of different lag structures.

3. Methodology

3.1. *Data*

Our sample consisted of global publicly traded companies from 20 different manufacturing industries. The data was drawn from the World Scope Database for the observation period 1996–2005. The financial data include the usual items, such as sales, profits, assets, standard industrial classification (SIC) and also R&D expenditure. One major constraint of the sample size was the availability of R&D expenditures. We allowed all the companies having at least one measure of R&D stay in our sample. To capture the effect of size in the R&D measure, we used R&D intensity (R&D expenditures divided by net sales) in our models as an input measure. We used three measures of performance: (i) market to book value, (ii) return on invested capital and (iii) sales growth. Market to book value was calculated as fiscal year end market value divided by fiscal year end common equity.

Return on invested capital was calculated using the following formula:

ROIC = Net Operating Profit after Taxes / Invested Capital,

where Invested Capital includes last year's total capital, short term debt and current portion of long term dept.

Our third measure, sales growth was calculated between the years 2000–2005 and it is simply net sales 2005 divided by the net sales of the year 2000.

We first removed all incomprehensible values, such as negative R&D expenditures, or negative market to book values from the data as well as considerably large outliers. After this procedure, plots of R&D intensity for any given year against any of the performance measures indicated that there were still some outliers leading to removal of 5% from each R&D intensity measure. Similarly we restricted the values of return on investment, market to book value and sales growth.

We classified the industries into groups under the Pavitt taxonomy as described in Table 1. Previous studies (e.g Laursen and Meliciani, 1999) have used a similar grouping of industries adaptation of the original Pavitt (1984) classification.

Table 1: Industry groups in the sample.

Industry	SIC code	Pavitt Taxonomy
Food, drink and tobacco	20, 21	Supplier dominated
Textiles, footwear and leather	22, 23, 31	Supplier dominated
Pulp and paper	26	Scale intensive
Industrial chemicals	280-282, 284-287, 289	Science-based
Pharmaceuticals	283	Science-based
Petrolenium	291	Supplier dominated
Rubber and plastics	30	Scale intensive
Stone, clay and glass	32	Scale intensive
Ferrous metal	331,332	Scale intensive
Non-ferrous metal	333-336	Supplier dominated
Fabricated metal	34	Scale intensive
Non-electrical machinery	351-354, 356, 358, 359	Specialized supplier
Office machines and computers	357	Science-based
Electrical machinery	361-365, 369	Specialized supplier
Communication equipment and semiconductors	366,367	Science-based
Shipbuilding	373	Scale intensive
Other transport	374, 375, 379	Scale intensive
Motor vehicles	371	Scale intensive
Aerospace	372, 376	Scale intensive
Instruments	38	Specialized supplier

3.2. *Results*

Table 2 shows the descriptive statistics of the key variables. The market to book value ratios somewhat varies across the sectors; on an average the lowest ratios are in supplier dominated sectors (.78), whereas the average ratios amount to 1.1 in specialized supplier sectors. The profitability averages range between 6.5% for the supplier dominated category and 7.3% in specialized suppliers. The supplier dominated sector seems to have the lowest average performance also in terms of sales growth: the average five-year sales growth is 165%, compared to the high of 192% in the science-based industries.

Table 2: Descriptive statistics.

	Scale intensive			Supplier dominated			Specialized supplier			Science based		
	n	Mean	Std.	n	Mean	Std.	n	Mean	Std.	n	Mean	Std.
rd05	1227	1.520	2.090	797	2.124	2.124	901	4.254	4.425	2190	4.797	5.228
rd04	1209	1.507	1.945	795	1.781	1.781	921	4.318	4.539	2139	4.917	5.292
rd03	1164	1.613	2.247	752	2.026	2.026	1015	4.553	4.715	2067	5.320	5.861
rd02	968	1.759	2.636	644	1.899	1.899	1205	5.128	5.633	1768	5.901	6.551
rd01	863	1.621	2.220	573	0.957	1.899	1008	5.111	5.556	1549	6.033	6.033
rd00	790	1.568	2.225	522	0.85	1.648	891	4.624	4.61	1247	5.573	5.573
rd99	694	1.500	2.055	445	1.009	2.348	800	4.333	4.322	1089	5.596	5.596
rd98	636	1.511	2.093	415	0.887	1.87	743	4.322	4.414	977	5.672	5.672
rd97	595	1.433	1.835	372	0.748	1.277	638	3.672	3.482	812	4.456	4.456
rd96	551	1.345	1.631	357	0.698	1.138	570	3.496	3.121	724	3.957	3.957
MTB	1777	0.811	1.039	1453	0.779	1.022	1414	1.105	1.234	2374	0.906	1.154
ROIC	2061	6.608	5.944	1707	6.463	5.969	1529	7.272	5.897	2420	6.673	6.477
SGR	1722	1.834	1.677	1447	1.651	1.545	1489	1.825	1.87	2203	1.922	2.174

In scale intensive sectors, R&D intensities have remained quite stable, ranging from the average of 1.35 in 1996 to the averages of 1.76 in 2002, and 1.52 percent of sales in 2005. However, the variation between companies is large, as the standard deviations are about 1.5 times as high as the means. A somewhat similar effect has occurred in the supplier dominated sectors, where average R&D intensity rose from 0.7% in 1996 to 1.0% in 2002. In the specialized supplier industries, R&D intensities have been notably higher, ranging from 3.5% to 5.1% during the period of interest. Variation within the Pavitt classes is relatively low in this category as the coefficients of variation are around 1. Finally, the science-based industries are characterized by slightly higher average R&D intensities compared to the specialized suppliers. A one-way analysis of variance with R&D intensities as dependent variable and the Pavitt class as grouping variable resulted in F test values ranging from 174.18 to 297.78 with 96–2198 and 3 degrees of freedom, and $p<.01$. The post hoc tests verified that the science-based and specialized supplier sectors had significantly higher average R&D intensities than scale intensive or supplier dominated sectors.

In order to define the most appropriate lag length for the R&D productivity analysis for each Pavitt class and each dimension of performance, we first examined the correlational structure of the variables. The columns I in Tables 3 shows the correlations between R&D intensity and market to book value of year 2005. In this regard the specialized supplier and science-based sectors show a very similar pattern. The cross-sectional correlations are all positive, about the same magnitude, and statistically significant. The highest correlations are found between current market to book value and 1998 R&D expenses, implying a seven-year lag for these Pavitt classes. In the supplier dominated sectors, there was no significant correlation between previous or current R&D expenses and the current market to book value. Scale intensive industries showed weak positive correlations at six to nine-year lags, but interestingly no significant correlation with shorter lags.

Table 3: Correlations of R&D inputs and performance.

	Scale intensive			Supplier dom.			Specialized sup.			Science based		
	I	II	III	I	II	III	I	II	III	I	II	III
rd05	.02	.03	.03	.07	-.06	-.08	0.20**	-.01	0.12**	.28**	-.04	.03
rd04	-.01	.04	.00	.05	-.04	-.09	0.22**	.00	.07	.28**	-.04	.12**
rd03	-.02	.05	.11*	.05	-.03	-.08	0.24**	.02	0.10*	.27**	-.03	.07
rd02	-.05	.10*	.16**	.06	-.05	-.07	0.22**	-.01	0.12*	.29**	-.07	.04
rd01	-.01	.11*	.05	.07	-.08	-.05	0.27**	.00	0.17**	.28**	-.09*	-.01
rd00	.01	.08	-.03	.05	-.08	-.01	0.26**	-.02	.01	.30**	-.08	-.04
rd99	.15**	.02	.01	.08	-.13	.07	0.30**	.09	.08	.31**	-.06	-.05
rd98	.13*	.03	-.08	.05	-.10	-.02	0.32**	.06	.03	.34**	-.06	-.03
rd97	.14*	.11	-.11	.06	-.09	.00	0.30**	.08	.04	.30**	-.02	-.05
rd96	.13	.05	-.06	.04	-.07	-.07	0.28**	.09	.02	.31**	.04	-.07

I = Market to book ratio 2005, II = ROIC 2005, III = Sales growth 2000-2005
**p<.01, *p<.05

The columns II in Table 3 contain correlations between previous R&D intensity and current profitability as measured by return on invested capital. In general, the correlations are rather weak in all sectors, but there are some very interesting differences across the Pavitt classes in terms of lag length and the sign of the correlations. Firstly, the significant correlations are positive in the scale intensive and specialized supplier sectors, but negative in the supplier dominated and science-based industries. The lags vary from 3–5 years in the scale intensive and science-based sectors to 6-9 years in other two industries. Correlations between previous R&D intensity and subsequent sales growth are weak in all Pavitt classes and mostly not statistically significant. In the scale intensive sectors, companies with higher R&D intensity in 1997 tended to have less sales growth during 2000–2005. On the other hand, the specialized suppliers who had invested more in R&D in 1999, had stronger sales growth in 2000–2005 than their less R&D intensive counterparts. In all Pavitt classes, except for the supplier dominated sectors, there are some positive correlations between concurrent R&D and sales growth. This seems natural in the light of findings of for example Fryxell(1990), who found that companies tend to invest a constant proportion of sales in R.

The effects of R&D intensity on performance were explored using a linear regression analysis with a cross-sectional design, i.e. the dependent performance variables were measured at only one point in time, and only one year of R&D intensity was used as the independent variable in order to avoid multicollinearity problems. The appropriate year for the independent variable in each Pavitt class and dependent variable combination was chosen on the basis of the correlational analysis presented above. The regression models were composed in two blocks: Model 1 regresses the dependent performance indicator on the control variable sales, and Model 2 adds the hypothesized R&D intensity to the model. The results are shown in Tables 4-6. Table 4 shows that the effect of R&D intensity on the market to book ratio occurs best at the lag of six or seven years. In the specialized suppliers and science-based sectors R&D accounts for 11–12% of the variation in market value, but only 1–2% in the scale intensive and supplier dominated sectors.

Table 4: Regression results, dependent variable = market to book value.

Scale intensive					Supplier dominated				
	Model 1		Model 2			Model 1		Model 2	
	B	s.e	B	s.e		B	s.e	B	s.e
Constant	.696**	.060	.575**	.069	Constant	.666**	.074	.615**	.080
Sales05	.000**	.000	.000**	.000	sales05	.000*	.000	.000**	.000
rd99			.078**	.023	Rd99			.042	.026
R^2	.022		.046		R^2	.023		.032	
R^2 Change	.022**		.024**		R^2 Change	.023*		.009	
N	456		456		N	274		274	
Specialized suppliers					Science-based				
	Model 1		Model 2			Model 1		Model 2	
	B	s.e	B	s.e		B	s.e	B	s.e
Constant	1.067**	.065	.638**	.082	Constant	.885**	.053	.463**	.068
Sales05	.000	.000	.000*	.000	Sales05	.000	.000	.000	.000
rd98			.093**	.012	Rd98			.071**	.008
R^2	.003		.112		R^2	.001		.119	
R^2 Change	.003		.109***		R^2Change	.001		.118***	
n	495		495		n	641		641	

**p<.01, *p<.05

According to Table 5, the effects on profitability are strongest with a 4-year lag in the scale intensive and science-based sectors, but at 6-year lag among the supplier dominated and specialized suppliers industries. R&D alone accounts for less than 1.5% of the variation in ROIC among the firms in all Pavitt classes, but the sign and magnitude of the regression slopes differ. The *positive* effect in the scale intensive sectors is much stronger than that in the specialized suppliers industry. On the other hand, the *negative* effect in the supplier-dominated sector is stronger than in the science-based industries.

Table 5: Regression results, dependent variable = ROIC05.

Scale intensive	Model 1		Model 2		Supplier dominated	Model 1		Model 2	
	B	s.e	B	s.e		B	s.e	B	s.e
Constant	6.036**	.294	5.525**	.341	Constant	6.000**	.407	6.346**	.441
Sales05	.002**	.001	.002**	.001	Sales05	.002*	.001	.001	.001
rd01			.308**	.106	Rd99			-.288*	.143
R^2	.017		.030		R^2	.013		.025	
R^2 Change	.017**		.013**		R^2Change	.013*		.013*	
n	625		625		N	313		313	

Specialized suppliers	Model 1		Model 2		Science-based	Model 1		Model 2	
	B	s.e	B	s.e		B	s.e	B	s.e
Constant	6.955**	.280	6.347**	.379	Constant	6.422**	.227	6.936**	.290
Sales05	.001*	.001	.002*	.001	sales05	.001*	.001	.001*	.001
rd99			.132*	.056	Rd01			-.083**	.029
R^2	.008		.017		R^2	.004		.011	
R^2 Change	.008*		.009*		R^2Change	.004*		.007**	
n	598		598		n	1104		1104	

**p<.01, *p<.05

Table 6 shows the regression results for sales growth. The effect of R&D intensity is not significant in the supplier dominated industries, but there are weak positive effects in other sectors. The largest contribution to R square occurs in the specialized suppliers category, but it is only about 3 percent of the between company variation in sales growth.

The differences in R&D effect time lags varied generally a couple of years depending on the Pavitt class. The specialized supplier category seemed to have the longest lags in all performance indicators. However, we cannot conclude that the effects would generally occur faster in any of the four classes. For example in the science-based industries effects on growth and profitability occur relatively faster, but effects on market value occur later than in other industries.

Table 6: Regression results, dependent variable = sales growth.

Scale intensive					Supplier dominated				
	Model 1		Model 2			Model 1		Model 2	
	B	s.e	B	s.e		B	s.e	B	s.e
Constant	1.684**	.085	1.496**	.096	Constant	1.509**	.085	1.576**	.093
Sales05	.000*	.000	.001**	.000	Sales05	.000**	.000	.000**	.000
rd02			.102**	.025	rd04			-.070	.040
R^2	.015		.040		R^2	.018		.024	
R^2 Change	.015**		.026**		R^2Change	.018**		.006	
n	601		601		n	474		474	

Specialized suppliers					Science-based				
	Model 1		Model 2			Model 1		Model 2	
	B	s.e	B	s.e		B	s.e	B	s.e
Constant	1.794**	.080	1.446**	.106	Constant	1.791**	.074	1.572**	.093
Sales05	.000	.000	.000	.000	Sales05	.001**	.000	.001**	.000
rd01			.061**	.012	rd04			.046**	.012
R^2	.001		.033		R^2	.010		.023	
R^2 Change	.001		.032**		R^2Chang	.010**		.013**	
n	744		744		n	1155		1155	

**p<.01, *p<.05

Another objective of our analysis was to examine, whether the sign and/or magnitude of the R&D effects on performance differ across Pavitt categories. In this regard, the results for market to book value and sales growth were consistent: the effects of R&D intensity in the supplier dominated industries were not significant, while they were positive, significant, and about the same magnitude in all other Pavitt categories. However, the results for profitability showed a clearly distinct pattern: we observed significant negative effects in two categories (a stronger one in the supplier dominated and a weaker one in the science-based sectors), and significant positive ones in two categories (a stronger one in the scale intensive and a weaker one in the specialized supplier industries).

4. Conclusions

In this paper we have investigated the relation between R&D intensity and firm performance by using the Pavitt taxonomy as a grouping variable. In addition we explored the time lags between R&D investment and performance effect in different Pavitt groups. Performance was measured as growth, profitability and market to book value. Our empirical findings indicate that Pavitt taxonomy groups differ significantly, namely, in terms of first R&D intensity and second time lag effects. The science-based and specialized supplier groups have the highest R&D intensities. The impacts of R&D intensity on performance vary depending on the industry category and performance measure used. In the scale intensive group, R&D intensity was significantly related to sales growth, profitability and market to book ratio. In the supplier dominated group, R&D intensity did not have a significant impact on sales growth or market to book ratio but had a negative effect on profitability. The specialized suppliers group was characterized by long effect lags and positive R&D intensity effects on all performance metrics. In the science-based industries, R&D intensity was positively related to market to book value and sales growth, but negatively related to profitability.

Our results indicate that the use of the Pavitt taxonomy in categorizing firms is a useful way to study the productivity of R&D. The four Pavitt categories turned out to be internally relatively homogeneous in terms of R&D intensity, effect magnitudes, and lag structures, whereas we were able to uncover major differences across the categories. The differences reflect the nature of the R&D activities; e.g. in the science-based industries the role of basic research is emphasized, and the research costs often form a major part of the firm's total costs. Thus the mildly negative effect on profitability within the time span of ten years seems logical. Another example is the specialized supplier sector, which typically focuses on product innovations for the use of the scale intensive and supplier dominated industries. Therefore, the effects are manifested after a longer time span. On the other hand, the expectations of future potential (reflected in market to book value) are most strongly related to R&D in the science-based and specialized supplier sectors. Our results imply that R&D investment decision makers should consider the fact that the delay between investment and return varies across industries due to differences in the nature of R&D activity, but also due to the differences in output measures applied.

References

Bae, S.C. and Kim, D. (2003). The effect of R&D investments on market value of firms : Evidence from the U.S., Germany, and Japan. *The Multinational Business Review*, 11(3), pp. 51–75.

Billings, B.A., Musazi, B. G. N. and Moore, J. W. (2004). The effects of funding source and management ownership on the productivity of R&D. *R&D Management*, 34(3), pp. 281–294.

Crépon, B., Duguet, E. and Mairesse, J. (1998). Research, innovation and productivity : and econometric analysis at the firm level. *Economics of Innovation and New Technology*, 7, pp. 115–158.

Delmas, M.A. (2002). Innovating against European rigidities – Institutional environment and dynamic capabilities. *Journal of High Technology Management Research*, 13, pp. 19–43.

Dosi, G. (1982). Technological paradigms and technological trajectories: A suggested interpretation of the directions of technological change. *Research Policy*, 11, pp. 147–162.

Fryxell, G.E. (1990). Multiple outcomes from product R&D: Profitability under different strategic orientations. *Journal of Management*, 16(3), pp. 633–645.

Griliches, Z. (1981). Market value, R&D and patents. *Economic letters*, 7, pp. 183–187.

Hall, B.H.; Griliches, Z; Hausman, J. (1986). Patents and R and D: Is there a lag? *International Economic Review*, 27(2), pp. 265–283.

Hall, B. H. and Mairesse, J. (1995). Exploring the relationship between R&D and productivity in French manufacturing firms. *Journal of Econometrics*, 65, pp. 263–293.

Hall, B. H. and Mairesse, J. (2006). Empirical studies of innovation in the knowledge-driven economy. *Economics of Innovation and New Technology*, 15(4/5), pp. 289–299.

Hall, B. H. and Oriani, R. (2006). Does the market value R&D investment by European firms? Evidence from a panel of manufacturing fFirms in France, Germany, and Italy. *International Journal of Industrial Organization*, 24, pp. 971–993.

Harhoff, D. (1998). R&D and productivity in German manufacturing firms. *Economics of Innovation and New Technology*, 6, pp. 29–49.

Jaffe, A.B. (1986). Technological opportunity and spillovers of R&D: Evidence from firms' patents, profits, and market value. *The American Economic Review*, 76(5), pp. 984–1001.

Jefferson, G. H., Huamao, B., Xiaojing, G. and Xiaoyun, Y. (2006). R&D performance in Chinese Industry. *Economics of Innovation and New Technology*, 15, 4/5, pp. 345–366.

Kafouros, M.I. (2005). R&D and productivity growth: Evidence from the UK. *Economics of Innovation and New Technology*, 14, 6, pp. 479–497.

Kondo, M. (1999). R&D dynamics of creating patents in the Japanese industry. *Research Policy*, 28, pp. 587–600.)

Laursen, K. and Meliciani, V. (1999). The importance of technology based inter–sectoral linkages for market share dynamics. Conference paper, DRUID (what is this?), June 1999.

Marsili, O. and Verspagen, B. (2001). Technological regimes and innovation: Looking for regularities in Dutch manufacturing. Working Paper, ECIS, Eindhoven University of Technology.

Pavitt, K. (1984). Sectoral patterns of technical change: Towards a taxonomy and a theory. *Research Policy*, 13, pp. 343–373.

Souitaris, V. (2002). Technological trajectories as moderators of firm-level determinants of innovation. *Research Policy*, 31, pp. 877–898.

Tsai, K.-H. (2005). R&D productivity and firm size: A nonlinear examination, *Technovation*, 25, pp. 795–803.

Chapter 4

An Analysis of High Profitability Mechanism by Means of Dynamism between Technological Diversification, Learning and Functionality Development

Noritomo Ouchi and Chihiro Watanabe

Department of Ind. Engineering & Management, Tokyo Institute of Technology, Tokyo, Japan

Contrary to the low profitability of Japanese electric machinery firms, Canon has consistently kept high profitability. Canon has incrementally developed innovations in related technological areas. This diversification strategy seems to be related to Canon's high profitability. In particular, learning and functionality development, which are important in making a profit under the current severe price competition, can be attributed to technological diversification. This study will demonstrate the dynamism between technological diversification, learning and functionality development.

1. Introduction

Although many Japanese electric machinery firms are not able to generate enough profit under the severe price competition, Canon has consistently kept high profitability.

Many researchers have pointed out that the source of Canon's high profitability is its business model with a high-profit structure of office supplies in the business machines market. That is, once firms sell copying machines and printers, they can consistently earn profits from the maintenance service and consumption of office supplies. However, the profitability of the business machine segment of Canon is much higher than that of its rivals in the business machine market (e.g., Ricoh, Fuji Xerox, Epson etc.). This suggests that the high profitability of Canon cannot be explained only with the business model of office supplies.

Canon has actively diversified technologies in related technology areas, such as copying machines and laser beam printer (LBP) (Suzuki and Kodama, 2004; Watanabe et al., 2004). As a result, only Canon has a high market share in both the copying machines and printers and that seems to be related to Canon's high profitability.

Under current severe price competition, learning and functionality development are important in making a profit. Following Arrow (1962)'s pioneering postulate on "learning-by-doing," many empirical analyses have demonstrated that learning is an important factor in cost reduction (e.g., Hatch and Mowery, 1998). Creating new products which have new value and functionality (i.e., functionality development) is important to increase sales price (i.e., slow down the speed of decreasing sales price). Watanabe et al. (2003) pointed out that the spillover effect is significant for functionality development. By using the concept of technological position proposed by Jaffe (1986), many researchers have analyzed the spillover effect (e.g., Odagiri and Kinukawa, 1997). Learning and functionality development can be attributed to technological diversification. However, no study has analyzed the dynamic link among technological diversification, learning, and functionality development.

This study will demonstrate this dynamism. First, we analyze the relationships between copying machines and LBP from the point of view of the learning curve. Second, we consider the learning process and the functionality development based on concrete examples of Canon's copying machines and printers. Finally, we simulate this dynamism by means of System Dynamics approaches.

2. Learning Curve of Copying Machines

In this section, we analyze the relationships between copying machines and LBP. We measure the learning effect of full color copying machines, and compare learning effects of copying machines before and after LBP being introduced by using domestic production data in Japan.[1] The date can be acquired from "*Machinery statistics*" released by METI (Ministry of Economy, Trade and Industry).

While Japan's production volumes of black and white copying machines have decreased, their production prices demonstrated increase trend. This can be attributed simply to the change in the places of their production. Japanese firms came to produce low-priced products in foreign countries such as China and they produced only high-priced production in Japan. As for black and white copying machines, it is difficult to measure the learning effect because the quantities of production and prices are strongly affected by production allocation. On the other hand, regarding full color copying machines, they had been produced mainly in Japan until recently because of their production complexity and high price.[2] Therefore, we analyze full color copying machines from 1997 to 2004, when the price had decreased continually, because the effect of production allocation is assumed to be small.

The learning curve is expressed as follows:

$$P = P_0 V^{-\lambda} \tag{1}$$

$$\ln P = A - \lambda \ln V \ \ (A = \ln P_0) \tag{2}$$

where P: production price, P_0: initial price, V: cumulative production.

Based on the equation (2), by means of incorporating the dummy variables corresponding to the development stage, leaning coefficients in each respective stage are estimated.

$$\ln P = A_0 + \sum_{i=1}^{n} A_i D_i - \sum_{i=1}^{n} \lambda_i D_i \ln V \tag{3}$$

where $D_1 = 1$ (period 1) and 0 (others), $D_2 = 1$ (period 2) and 0 (others), $D_3 = 1$ (period 3) and 0 (others).

---

[1] It is best to analyze at the firm level. However, acquisition of data at the firm level is impossible because firms have not released figures on the numbers of production and cost. Thus, in this section, we focus on the characteristics of copying machines and printers

[2] The statistics of JMBIA shows that the proportion of full color copying machines' trilateral trade was 25% in 2004.

Period 1 (Jan. 97 - Jul. 98) is the period before color LBP. Period 2 (Aug. 98 - Jun. 02) is the period after color LBP was introduced. Period 3 (Jun. 02 - Dec. 04) is the period of digitalizing and networking.

The result of the regression analysis is summarized in Table 1.

Table 1: The result of the regression analysis.

	A_i	λ_i
Period 1	3.049	0.251
	(0.99)	(1.00)
Period 2	9.435	0.750
	(2.01)	(13.64)
Period 3	6.872	0.558
	(1.11)	(5.25)
adj. R^2	0.903	

[a] Figures in parentheses indicate the t-value.

The learning coefficient of the period after color LBP (λ_2 = 0.750) is higher than one of the period before color LBP was introduced (λ_1 = 0.251). That is, the learning effect of copying machines increased after the introduction of color LBP. Copying machines and LBP have the same technologies, such as drum, laser transmitter unit, intermediate transfer belt and fixing roller. Sharing these technologies makes it possible to increase the learning effect. This result suggests that Canon can reduce cost faster than its rivals because only Canon has a high market shares in both copying machines and LBP.

3. Creating New Functionality due to Effective Utilization of Inter-spillover and Intra-spillover

We take the System Large Scale Integration (LSI) technology[3] as an example of functionality development by effective utilization of inter-spillover and intra-spillover.[4]

---

[3] A system LSI is a large-scale IC that integrates all system functions onto a single chip. This concept is also known as a System-on-a-Chip (SOC).

[4] In this section, the details of technology and history are based on Canon (2006) and Nihon Keizai Shimbun, Inc. (2004).

In 1993, NeXT Computer Inc.[5], dropped their hardware business. Canon acquired their hardware business sector and established FirePower Systems, Inc., a wholly-owned subsidiary of Canon. FirePower Systems developed a chipset packed MPU (Micro Processing Unit) which was called "PowerPC". However, this business was not successful and FirePower Systems was purchased by Motorola in 1997. Although this business was a failure, Canon learned circuit design technology from NeXT Computer. By using this circuit technology, Canon developed System LSI. First, Canon installed System LSI in copying machines and these machines had a characteristic of "multi-function". After the System LSI was improved, Canon installed System LSI in LBP and other products. By effective utilization of spillover and sharing of learning between copying machines and LBP, Canon had created new functionality and responded flexibly to digitization and networking (Figure 1).

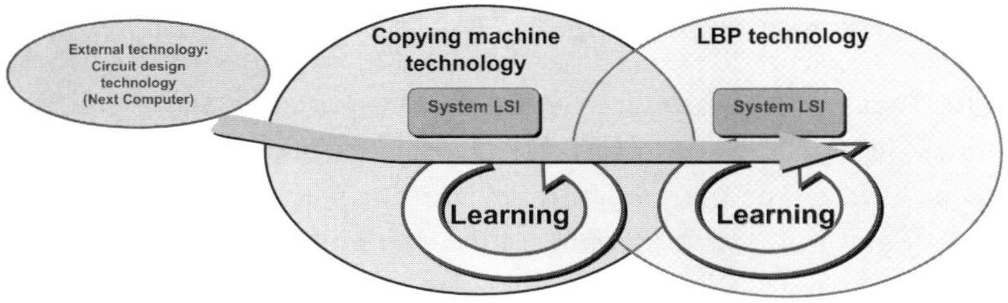

Figure 1: Utilization of inter-spillover and intra-spillover.

4. System Dynamics Model

4.1. *Overview*

The purpose of this section is to provide a detail description for a System Dynamics model used to analyze the dynamism between technological diversification, learning and functionality development. The model focuses on four areas: 1) Research and Development, 2) Production, 3) Finance, and 4) Organizational Experience and Knowledge. The

[5] NeXT was founded in 1985 by Steve Jobs after his resignation from Apple Computer.

relationships between the main model variables are illustrated in Figure 2. Some of the model formulations are draw upon established System Dynamics models (Forrester, 1961; Sterman, 2000; Rydzak, 2006). The System Dynamics model was built using Vensim modelling software version 5.5.[6]

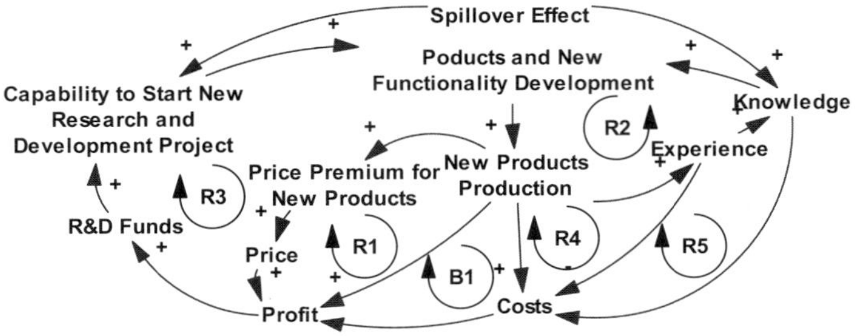

Figure 2: The model structure – causal loop diagram.

4.2. *Research and development*

The first sector in the model represents the research and development activities in organization. New R&D projects accumulate in stock as *Projects in RD*. The completion rate is a third order delay of the *R&D Projects Start Rat*e, with the delay time determined by the research and development time - *RD Time*. Newly developed products accumulate in stock *Ready Products Design*. New products are discarded at a rate determined by the average lifetime of any new product design. A *Possible Projects RD Start Rate* is dependent on available R&D Funds compared to minimal funds level (*RD Project Start Level*) to start the new project. Research and Development time (*RD Time*) is influenced by knowledge (*Effect of Knowledge on RD Time*). The structure of the research and development model area is illustrated in Figure 3.

[6] See Appendix A for equations' details.

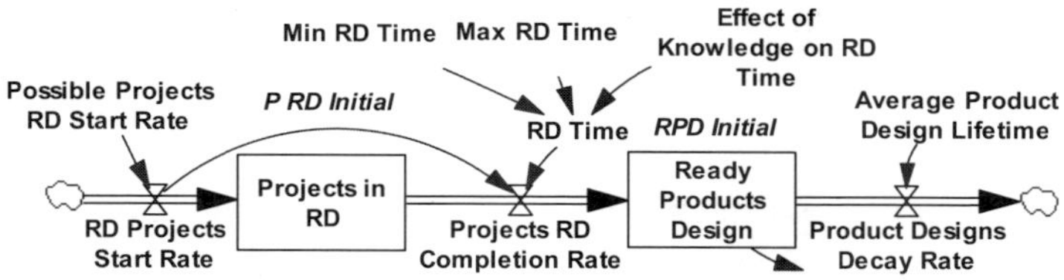

Figure 3: Research and development activities in an organization.

4.3. *Production*

The production sector focuses on average production of each new product design. Production itself is considered a third order delay of *Desired Production Start Rate*, with the delay time determined by the manufacturing time. The amount of manufactured products is calculated in *Cumulative Production* stock. *Desired Production Start Rate* is determined by *Ready Products Design*. The structure of the production sector is illustrated in Figure 4.

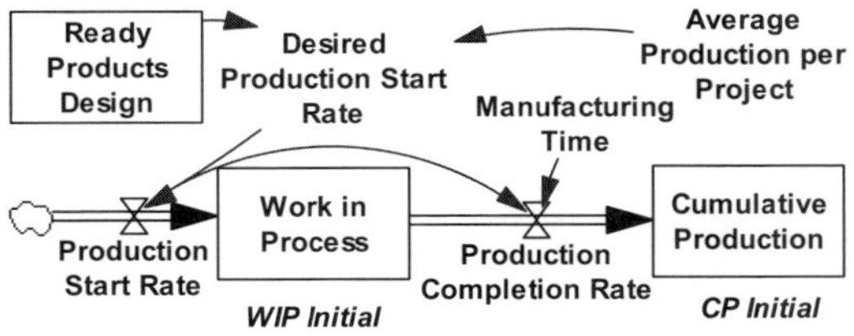

Figure 4: Production in organization.

4.4. *Finance*

The finance sector consists of three sub-sectors: 1) Cost; 2) Price and profit; and 3) R&D funds.

4.4.1. *Cost*

The cost in the model is calculated as the initial cost effected by the learning curve from organizational experience and knowledge. To represent the learning curve a power law was used[7]:

$$C_n = C_i E^{-\alpha} K^{-\beta} \tag{4}$$

where C_n: cost of the n-th unit, C_i: cost of the first unit, E: organizational experience ratio, K: organizational knowledge ratio, α : elasticity of cost to organizational experience, β : elasticity of cost to organizational knowledge. Total cost is calculated in respect to *Cumulative Production*. The cost model structure is illustrated in Figure 5.

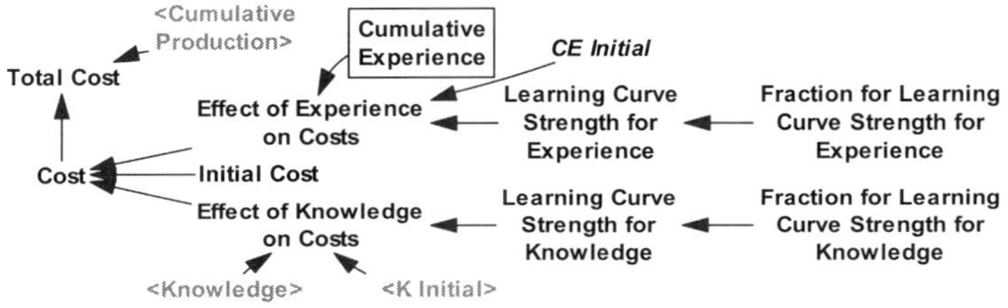

Figure 5: Model structure for cost calculation.

4.4.2. *Price and profit*

The model calculates the *Profit* within an organization as a difference between *Revenue* and *Total Cost*. The *Revenue* is calculated as a *Cumulative Production* multiplied by *Price*. The *Price* increases as a new products start to be produced. Over time the *Price* decreases according to a *Price Decrease Fraction*. The price and profit model structure is presented in Figure 6.

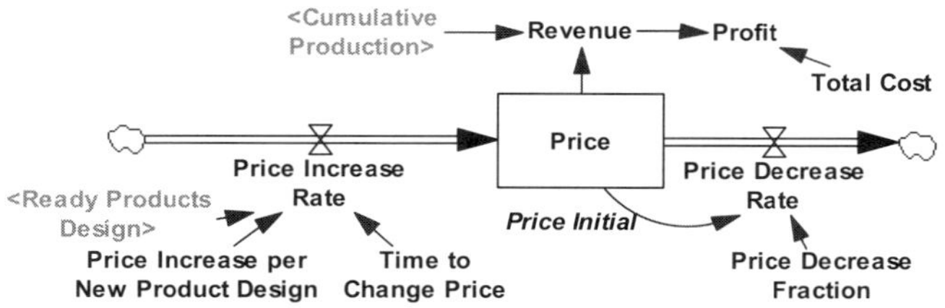

Figure 6: Model structure for profit calculation.

[7] This equation is based on a two-factor learning curve (Klaassen et al., 2005).

4.4.3. *R&D funds*

It is assumed that the organization dedicates a certain percentage of profit for R&D activities (*Percent of Profit for RD*). Every time the organization will save a certain amount of money—indicated by *RD Project Start Level*—the new project is started. At that time the *RD Available Funds* decrease down to 0 and the organization starts to gather funds for the next R&D projects. It is also assumed that the *RD Project Start Level* is influenced by the *Spillover Effect* and can change between its minimal and maximal value. To activate the model and the new R&D activities it is considered that the organization needs a prior investment in R&D (e.g. bank loan indicated in the model as *Additional RD Funds*). The model structure of the whole sub-sector is presented in Figure 7.

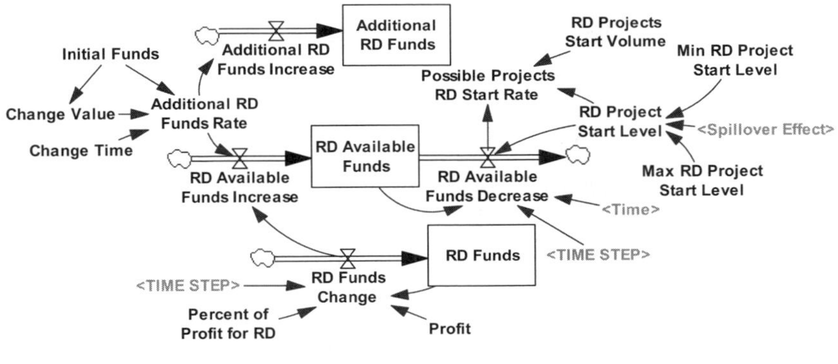

Figure 7: Funds for R&D activities model structure.

4.5. *Organizational experience and knowledge*

4.5.1. *Organizational experience*

An assumption has been made that the organizational experience follows production of new products. The organizational experience decay over time is according to the *Fractional Experience Decay*. A fraction of experience gathered during operations can be formalized through reporting and investigation processes. The model structure of this sub-sector is presented in Figure 8.

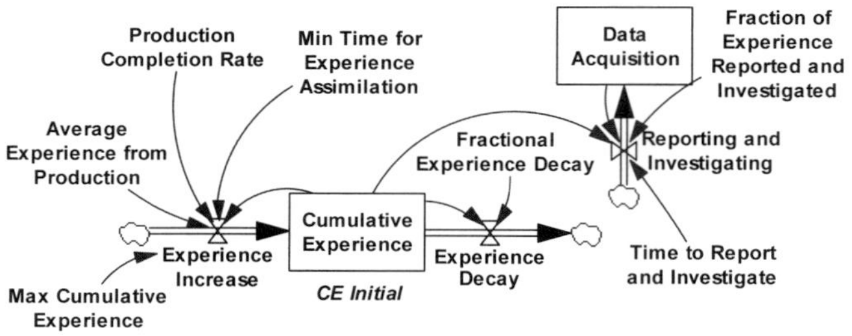

Figure 8: Organizational experience and experience formalization.

4.5.2. *Organizational knowledge*

Data acquired through the reporting and investigation processes can build organizational knowledge. The knowledge increase can be enhanced by the Spillover Effect.

The Spillover Effect is modeled by using the following function.

$$D^{-S}/(D^{-S}+I^{-S}) \qquad (5)$$

where D: Technology Distance, S: Spillover Effect Function Steepness, I: Spillover Effect Function Inflection Point.

Equation (5) indicates that the spillover effect shifts depending on the degree of steepness (S) and inflection point (I). Figure 9 demonstrates three scenarios depending on (S, I) as (2, 0.3), (5, 0.4) and (10. 0.5). The model structure of the organizational knowledge sub-sector is presented in Figure 10.

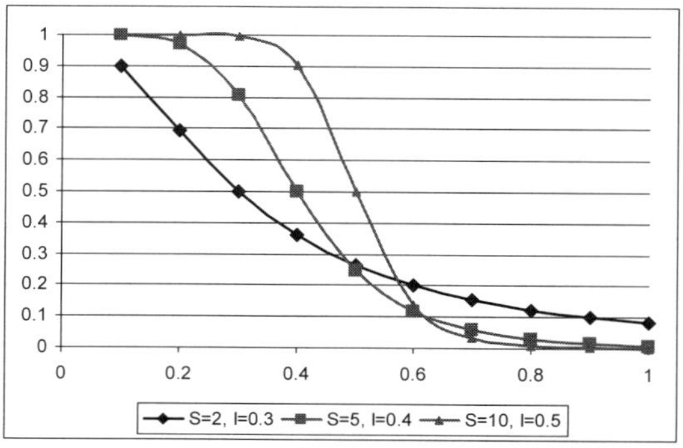

Figure 9: Examples of the Spillover Effect described by the equation (5).

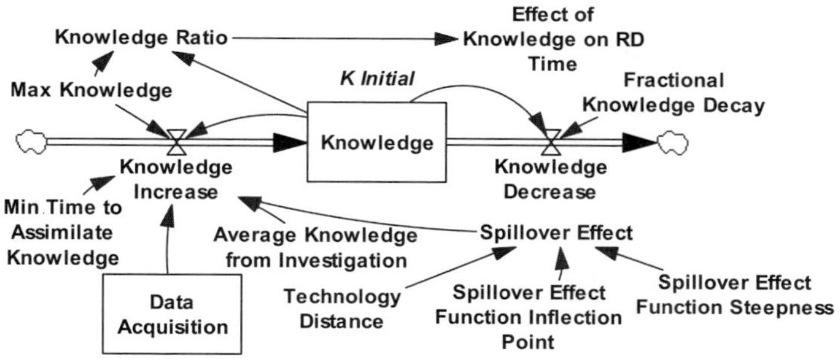

Figure 10: Organizational knowledge and the impact of spillover effect.

4.6. *Simulation*

Integrating each sector, we obtain a System Dynamics model for simulation (Figure 11).

Two scenarios have been simulated:

Scenario 1: Technology Distance = 1.0 (In the case of firms diversified in non-related technologies).

Scenario 2: Technology Distance = 0.3 (In the case of firms diversified in related technologies).

(Technology Distance is set to 0 when the technologies are closely related and is set to 1 when they are completely different).

Results of the two scenarios are presented in Figure 12. The simulation results show that firms diversified in a related area can make profit and get knowledge more than those diversified in a non-related area. These results suggest also that the significance of spillover effect induced by the closeness of technological distance. By utilizing spillover effect, firms can create a virtuous circle to make a profit.

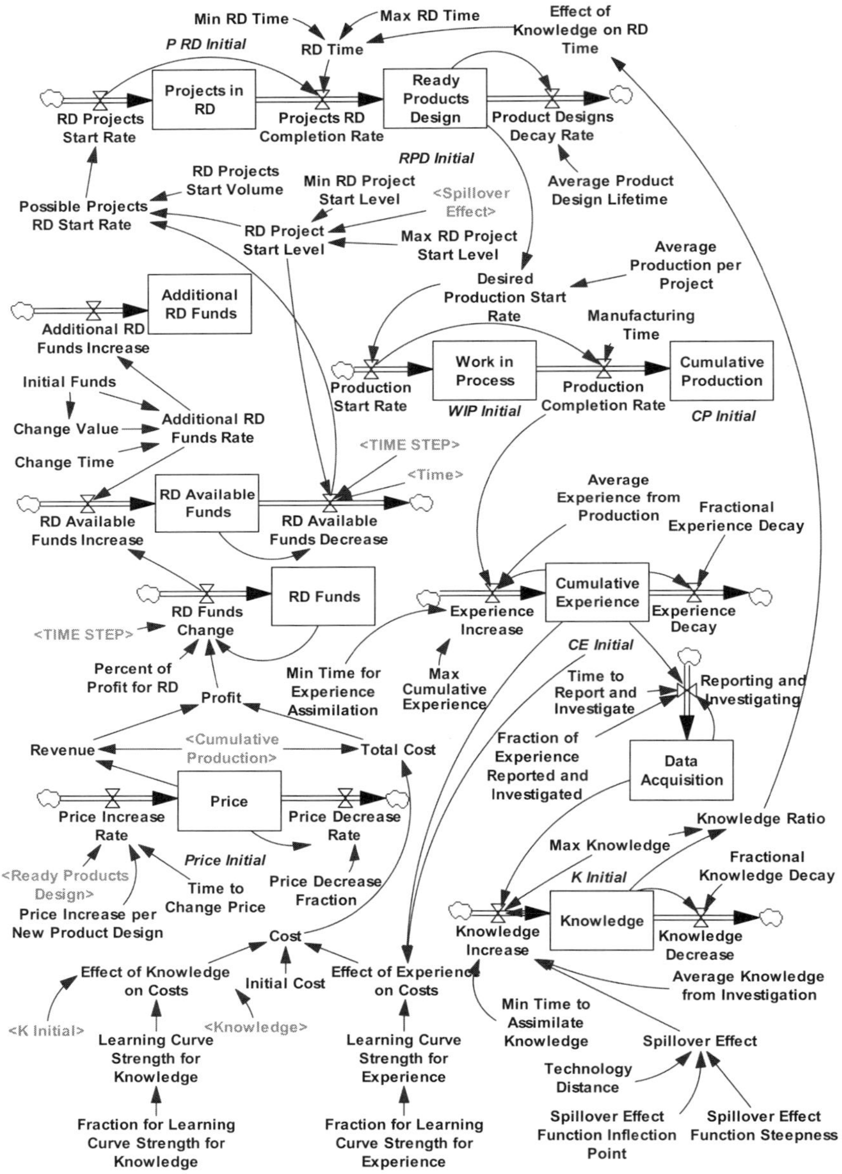

Figure 11: Model for simulation.

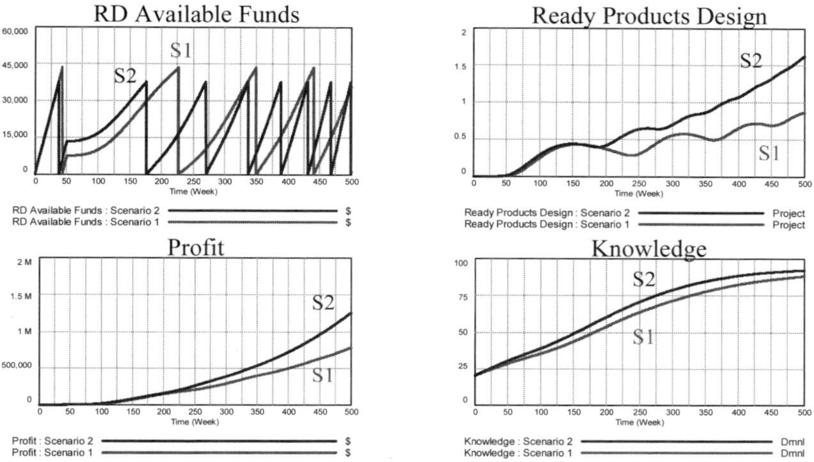

Figure 12: Results for simulation scenario 1 and 2.

5. Conclusions

This study attempted to elucidate the dynamism between technological diversification, learning, and functionality development. We measured the learning effect of color copying machines. It was observed that after color LBP was introduced, the learning coefficient of color copying machines increased. From the case study of copying machines and printers, it was demonstrated that Canon had created new functionalities with the effective utilization of the spillover effect and sharing of learning between copying machines and LBP. We constructed a System Dynamics model and conducted a simulation analysis on the influence of technological distance to the profit which demonstrated a reasonable behavior.

This study suggests that it is important to manage technological diversification. Co-utilization of learning effects and also effective utilization of spillover technologies between copying machines and LBP contributed to accelerate incremental innovations and subsequent sustainable increase in profitability. Thus, firm's R&D strategy considering the leaning and spillover effect is significant for its competitiveness.

In order to generalize these suggestions, analyses of other products and firms should be conducted. In addition, further work on incorporating demand factors into the model is required, since the market and customer sector has not been included in this study.

Acknowledgements

The authors gratefully acknowledge helpful discussions with Dr. Michael Obersteiner and Dr. Felicjan Rydzak on several points in the paper.

Appendix A

The Equations of System Dynamics Model

In our System Dynamics model, we used the following functions which are available in Vensim.
INTEG (rate, initial value): Returns the integral of the rate.
DELAY3 (input, delay time): Returns a 3rd order exponential delay of the input, conserving the input if the delay time changes.
STEP (height, step time): Returns 0 until the step time and then returns height.
IF THEN ELSE (cond, tval, fval): Returns first value (tval) if condition (cond) is true; second value (fval) if condition is false.
TIME STEP: The time interval for the simulation.

The equations of our System Dynamics model are shown below.

[Research and Development]
Average Product Design Lifetime = 80, Unit: Week; ***Max RD Time*** = 100, Unit: Week; ***Min RD Time*** = 60, Unit: Week; ***Product Designs Decay Rate*** = Ready Products Design / Average Product Design Lifetime, Unit: Project / Week; ***P RD Initial***=0, Unit: Project; ***Projects in RD*** = INTEG (RD Projects Start Rate-Projects RD Completion Rate, P RD Initial), Unit: Project; ***Projects RD Completion Rate*** = DELAY3 (RD Projects Start Rate, RD Time), Unit: Project / Week; ***Ready Products***

Design = INTEG (Projects RD Completion Rate-Product Designs Decay Rate, RPD Initial), Unit: Project; ***RD Projects Start Rate*** = Possible Projects RD Start Rate, Unit: Project / Week; ***RD Time*** = Min RD Time + (Max RD Time - Min RD Time) * Effect of Knowledge on RD Time, Unit: Week; ***RPD Initial*** = 0, Unit: Project; ***Possible Projects RD Start Rate*** = RD Available Funds Decrease*RD Projects Start Volume / RD Project Start Level, Unit: Project / Week; ***RD Projects Start Volume*** = 1, Unit: Project; ***Effect of Knowledge on RD Time*** = 1 - Knowledge Ratio, Unit: Dmnl

[Production]
Average Production per Project = 500, Unit: Product / (Project * Week); ***Cumulative Production*** = INTEG (Production Completion Rate, CP Initial), Unit: Product; ***CP Initial*** = 0, Unit: Product; ***Desired Production Start Rate*** = Ready Products Design * Average Production per Project, Unit: Product / Week; ***Manufacturing Time*** = 1, Unit: Week; ***Production Completion Rate*** = DELAY3 (Production Start Rate, Manufacturing Time), Unit: Product/Week; ***Production Start Rate*** = Desired Production Start Rate, Unit: Product / Week; ***WIP Initial*** = 1000, Unit: Product; ***Work in Process*** = INTEG (Production Start Rate - Production Completion Rate, WIP Initial), Unit: Product

[Cost]
Cost = Initial Cost * Effect of Experience on Costs * Effect of Knowledge on Costs, Unit: $ / Product; ***Effect of Experience on Costs*** = (Cumulative Experience / CE Initial) ^ Learning Curve Strength for Experience, Unit: Dmnl; ***Effect of Knowledge on Costs*** = (Knowledge / K Initial) ^ Learning Curve Strength for Knowledge, Unit: Dmnl; ***Fraction for Learning Curve Strength for Experience*** = 0.3, Unit: Dmnl; ***Fraction for Learning Curve Strength for Knowledge*** = 0.3, Unit: Dmnl; ***Initial Cost*** = 5, Unit: $ / Product; ***Learning Curve Strength for Experience*** = LN (1 - Fraction for Learning Curve Strength for Experience) / LN (2), Unit: Dmnl; ***Learning Curve Strength for Knowledge*** = LN (1 - Fraction for Learning Curve Strength for Knowledge) / LN (2), Unit: Dmnl; ***Total Cost*** = Cumulative Production * Cost, Unit: $

[Price and Profit]
Price = INTEG (Price Increase Rate - Price Decrease Rate, Price Initial), Unit: $ / Product; ***Price Decrease Fraction*** = 0, Unit: 1 / Week; ***Price Decrease Rate*** = Price * Price Decrease Fraction, Unit: $ / (Product * Week); ***Price Initial*** = 10, Unit: $ / Product; ***Price Increase per New Product Design*** = 0, Unit: $ / (Product * Project); ***Price Increase Rate*** = Ready Products Design * Price Increase per New Product Design / Time to Change Price, Unit: $ / (Week * Product); ***Profit*** = Revenue - Total Cost, Unit: $; ***Revenue*** = Cumulative Production * Price, Unit: $; ***Time to Change Price*** = 5, Unit: Week; ***Additional RD Funds*** = INTEG (Additional RD Funds Increase, 0), Unit: $

[R&D Funds]
Additional RD Funds Increase = Additional RD Funds Rate, Unit: $ / Week; ***Additional RD Funds Rate*** = Initial Funds – STEP (Change Value, Change Time), Unit: $/Week; ***Change Time*** = 51, Unit: Week; ***Change Value*** = Initial Funds, Unit: $ / Week; ***Initial Funds*** = 1000, Unit: $ / Week; ***Max RD Project Start Level*** = 50000, Unit: $; ***Min RD Project Start Level*** = 30000, Unit: $; ***Percent of Profit for RD*** = 0.2, Unit: Dmnl; ***RD Available Funds*** = INTEG (RD Available Funds Increase - RD Available Funds Decrease, 0), Unit: $; ***RD Available Funds Decrease*** = IF THEN ELSE (RD Project Start Level - RD Available Funds < 0, (RD Available Funds / TIME STEP) * PULSE TRAIN (0, TIME STEP, Time, 500), 0), Unit: $ / Week; ***RD Available Funds Increase*** = RD Funds Change + Additional RD Funds Rate, Unit: $/Week; ***RD Funds*** = INTEG (RD Funds Change, 0), Unit: $; ***RD Funds Change*** = (Profit * Percent of Profit for RD - RD Funds) / TIME STEP, Unit: $/Week; ***RD Project Start Level*** = Min RD Project Start Level + (Max RD Project Start Level - Min RD Project Start Level) * (1 - Spillover Effect), Unit: $

[Organizational Experience]
A***verage Experience from Production*** = 0.001; ***CE Initial*** = 20; ***Cumulative Experience*** = INTEG (Experience Increase - Experience Decay, CE Initial); ***Data Acquisition*** = INTEG (Reporting and Investigating, Fraction of Experience Reported and Investigated * CE

Initial); ***Experience Decay*** = Cumulative Experience * Fractional Experience Decay; ***Experience Increase*** = (Max Cumulative Experience - Cumulative Experience) * Production Completion Rate * Average Experience from Production / Min Time for Experience Assimilation; ***Fraction of Experience Reported and Investigated*** = 1; ***Fractional Experience Decay*** = 0.001; ***Max Cumulative Experience*** = 100; ***Min Time for Experience Assimilation*** = 50; ***Reporting and Investigating*** = (Cumulative Experience * Fraction of Experience Reported and Investigated-Data Acquisition) / Time to Report and Investigate; ***Time to Report and Investigate*** = 10

[Organizational Knowledge]
Average Experience from Production = 0.001, Unit: 1 / (Product / Week); CE Initial = 20, Unit: Dmnl; ***Cumulative Experience*** = INTEG (Experience Increase - Experience Decay, CE Initial), Unit: Dmnl; ***Data Acquisition*** = INTEG (Reporting and Investigating, Fraction of Experience Reported and Investigated * CE Initial), Unit: Dmnl; ***Experience Decay*** = Cumulative Experience*Fractional Experience Decay, Unit: 1 / Week; ***Experience Increase*** = (Max Cumulative Experience - Cumulative Experience) * Production Completion Rate * Average Experience from Production / Min Time for Experience Assimilation, Unit: 1/Week; ***Fraction of Experience Reported and Investigated*** = 1, Unit: Dmnl; ***Fractional Experience Decay*** = 0.001, Unit: 1 / Week; ***Max Cumulative Experience*** = 100, Unit: Dmnl; ***Min Time for Experience Assimilation*** = 50, Unit: Week; ***Reporting and Investigating*** = (Cumulative Experience * Fraction of Experience Reported and Investigated - Data Acquisition) / Time to Report and Investigate, Unit: 1 / Week; Time to Report and Investigate = 10, Unit: Week

References

Arrow, K J. (1962). The economic implications of learning by doing. *Review of Economic Studies*, 29, pp. 155–173.

Canon Inc. (2006). Canon technology highlights, Canon Inc., Tokyo.

Forrester, J.W. (1961). Industrial Dynamics. Productivity Press, Cambridge, MA.

Hatch, N.W. and Mowery, D.C. (1998). Process innovation and learning by doing in semiconductor manufacturing. *Management Science*, 44(11), pp. 1461–1477.

Jaffe A.B. (1986). Technological opportunity and spillover of R&D: Evidence from firm's patents, profits, and market value. *American Economic Review*, 76, pp. 984–1111.

Klaassen, G, Miketa, A., Larsen, K. and Sundqvist, T. (2005). The impact of R&D on innovation for wind energy in Denmark, Germany and the United Kingdom. *Ecological Economics*, 54(2-3), pp. 227–240.

Ministry of Economy, Trade and Industry (METI). (1975-2006). Machinery Statistics, Tokyo, annual issues.

Nihon Keizai Shimbun, Inc. (2004). Canon SHIKI. Nihon Keizai Shimbun, Inc., Tokyo (in Japanese).

Odagiri, H. and Kinukawa, S. (1997). Contributions and channels of interindustry R&D spillovers: An estimation for Japanese high-tech industries. *Economic Systems Research*, 9(1), pp. 127–142.

Rydzak, F. (2006). The impact of weather forecast on oil & gas industry operations - System dynamics model. Report on the project "Global Earth Observation – Benefit Estimation: Now, Next and Emerging." International Institute for Applied Systems Analysis, Laxenburg.

Sterman, J.D. (2000). Business Dynamics: Systems Thinking and Modeling for a Complex World. Irwin/McGraw-Hill, Boston.

Suzuki, J. and Kodama F. (2004). Technological diversity of persistent innovations in Japan -Two case studies of large Japanese firms. *Research Policy*, 33(3), pp. 531–549.

Watanabe, C, Asgari, B. and Nagamatsu, A. (2003). Virtuous cycle between R&D, functionality development and assimilation capacity for competitive strategy in Japan's high-technology industry. *Technovation*, 23(11), pp. 879–900.

Watanabe, C, Matsumoto, K. and Hur, J.Y. (2004). Technological diversification and assimilation of spillover technology: Canon's scenario for sustainable growth. *Technological Forecasting and Social Change*, 71(9), pp. 941–959.

Chapter 5

An Analysis of Dynamism between Market Sensitivity to Technology and Optimal R&D Intensity

Yuji Tou

Department of Ind. Engineering & Management, Tokyo Institute of Technology, Tokyo, Japan

The high-competitiveness firms improve their marginal productivity to technology by appropriate R&D activities for growth, and have raised their market value. On the other hand, in bigger firms, the contribution of the technology knowledge stock to growth decreases by their inappropriate restructuring of R&D activities. Based on the empirical analysis on Japan's electric machinery firms, this chapter attempts to demonstrate the dynamism between firms' R&D, technological development and market reaction. The results of those analyses indicate that the dynamism between improvement of quality of technology through efficient R&D investment and market evaluation can be forming a virtuous circle depending not on quantity but on the corporate institution that promotes structural reform.

1. Background

1.1. *Investment in research and development: The situation in Japan*

It goes without saying that science and technology played an important role in the economic growth of a Japan. However, during the "lost decade" in the 1990s, corporate investment in research and development was reduced as a part of corporate restructuring. In the 1980s and 1990s, although Japanese firms gradually increased R&D investment relative to sales (R&D intensity) in comparison with other OECD countries, this did not lead to increased productivity.

This reduction of the efficiency in Japan's R&D activities affected Japan's international competitive power in the high-tech field. For example, although the Japanese semi-conductor industry was highly competitive in the 1980s, it rapidly withered in the face of a revival of the US industry and the growth of Asian industries.

In order to overcome above mentioned trends, the creation of new technology and added value through R&D activities was particularly necessary. However, as shown in Figure 1, we see an evident stagnation in R&D activities in the electric machinery industry that played a leading role in Japan's high-tech industry.

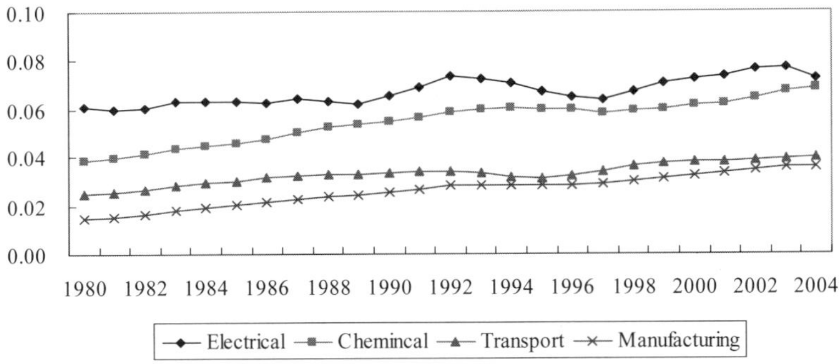

Figure 1: Trend in R&D Intensity in Japan's Major High-tech Manufacturing Industries (1980-2004).

Moreover, although Japan's R&D intensity in the 1990s seemed to show a steady increase in nominal R&D investment, it was in fact diminishing yearly in real terms (Figure 2).

We can see ample evidence of this tendency in the electric machinery industry. This actual decrease of R&D intensity is considered to have been caused by inertia in organizations without a strategy determining a ratio of R&D investment per sales.

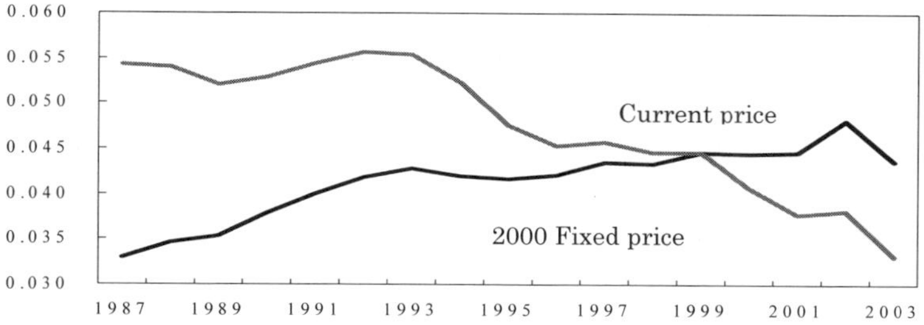

Figure 2: Trend in R&D Intensity in Japan's Electric Machinery Industry (1987-2003).

As shown in Figure 3, the top 10 firms accounted for 80% of R&D activities in the Japanese electric machinery industry.

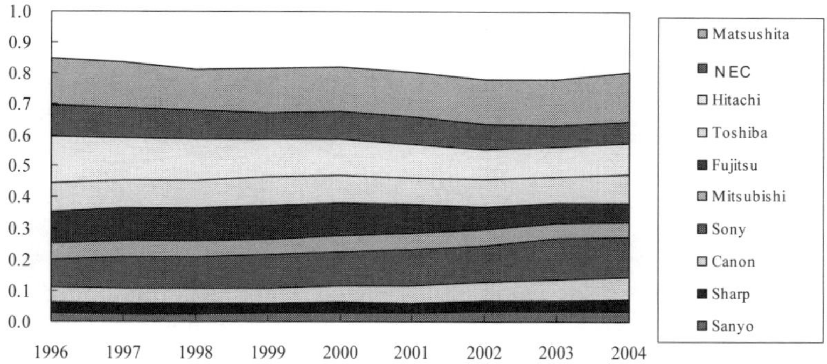

Figure 3: Trend in Share of R&D investment in Japan's Electric Machinery Firms (1996–2004).

As for the transition of the R&D intensity of these 10 firms, in the 2000s, we observe a polarization between those firms with a prominent increase (Matsushita, Canon, Sony, etc.) and those with a decrease (Hitachi, Toshiba, and NEC). This is considered to show a difference in strategy (concentration and selection, diversification of technology, etc.) of each firm toward R&D investment. For example, it is predictable to see a part of the result that NEC and Fujitsu abandoned their images of general electric company (what does this mean?) which handle products in multi-field and implemented concentration and selection. However, such a trend of R&D moving toward inertia of breaking away from firms does not necessarily lead to increased competitive power for all the firms. (not clear)

1.2. *Structural reform of Japanese firms*

The importance of firms' structural reform has been recently suggested (Japanese Electrical Electronic & Information Union 、 2003-2006). However, the capital adequacy ratio (equity capital/total capital) that demonstrates the healthy financial strength of Japanese electric machinery firms, varies depending on the firm. As shown in Figure 4, Canon, which is considered to be a successful firm after 2000, has been rapidly increasing its capital adequacy ratio, while large firms such as Hitachi and Toshiba have been maintaining ratios lower than the industry average since the 1980s.

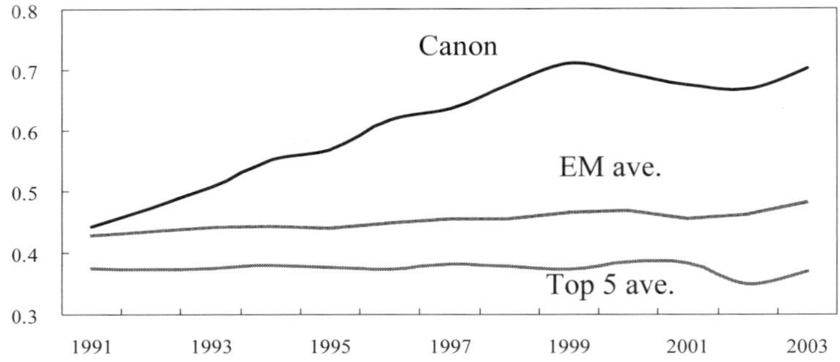

Figure 4: Trends in Capital Adequacy Ratio in Japan's Electrical Machinery Industry (1991–2003).

This situation indicates that Canon has been promoting structural reforms, while large firms have been mired in organizational inertia.

Then, let us compare the correlation between firms' market evaluation and the health of the financial situation. Table 1 shows the historical transition of 15 leading electric machinery firms and the correlation between aggregate market value (Stock prices × issued shares) that honestly indicates the market evaluation of the firm, the capital adequacy ratio and the amount of sales.

Table 1: Correlation between Aggregate Market Value (AMV) and Sales, Capital Adequacy Ratio (CAR) in Japan's Top 15 Electric Machinery Firms (1983–2003).

$$\ln AMV = a + b\ln S + c\ln CAR$$

	a	b	c	$adj.\ R^2$
1980s	-2.10 (-1.35)	0.87 (7.12)	0.41 (1.75)	0.780
1990s	-2.86 (-1.59)	0.94 (7.00)	0.89 (3.12)	0.771
2000s	-6.21 (-2.64)	1.19 (7.09)	1.43 (4.46)	0.793

Table 1 indicates that the market highly values not only the amount of sales but also the health of the financial situation. This demonstrates the fact that the market values the degree of corporate structural reform as well as the fact that it becomes easier to procure investment fund for R&D.

Moreover, Figure 5 shows the correlation between the capital adequacy ratio and operating income to sales (OIS: operating income/amount of sales).

Figure 5 indicates that the improvement of the financial situation began to contribute to an increase in profits in the 2000s, while improvement of financial situation did not affect profits in the 1980s.

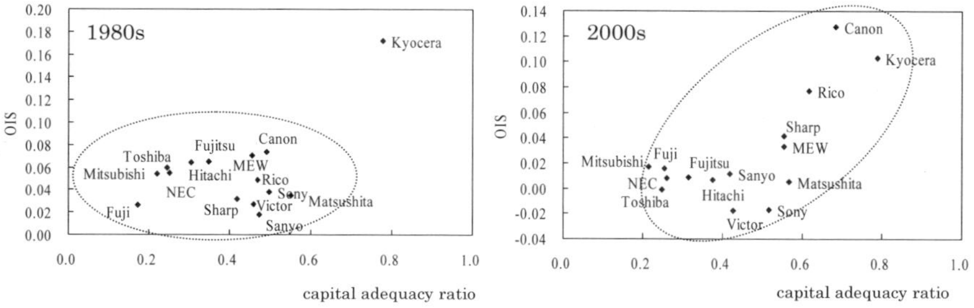

Figure 5: Correlation between Capital Adequacy Ratio and OIS in Japan's Leading 15 Electrical Machinery Firms (1983–2004).

Because improved productivity through technical innovation may lead to an increase in the profit ratio, the phenomenon above may be interpreted as the improvement of quality of technology as a result of efforts to promote investment efficiency of R&D through a structural reform spirit, and generated by the improved financial situation.

1.3. *Improvement of quality of technology and structural reform*

The analysis above indicates that the improvement of the financial situation encouraged technological innovation both in quantity and quality terms. In addition, the improvement in quality of technology depends on the corporate institution promoting the structural reform spirit generating, and generated by the improved financial situation. In particular, since the 1990s, the competitive international environment has been changing dramatically in terms of maturation of economies and the development of globalization (Watanabe and Kondo, 2003). We find that the Japanese high-tech industry has been transforming R&D activities in a qualitative aspect through structural reform.

This paper verifies the following hypothetical perspective by analyzing the relationship between Total Factor Productivity (TFP) (a measure of technological advancement) and firms' structure.

Firms with a high competitive power realized appropriate R&D activities through a structural reform generated by the improved financial situation.

Firms that implemented appropriate R&D activities enjoyed the fruits of technological advancement and strengthened their competitive power.

On the other hand, other firms decreased their competitive power because poor restructuring of R&D due to delay in structural reform decreased their technological power.

2. Measurement of Technological Advancement

2.1. *Measurement of TFP*

1) *Structure of TFP*

Increase of TFP is an increase of productivity that is independent of increase of capital and labor input.

The production function is generally seen in the following way:

$$V = \mathrm{F}\,(L, K, TFP) \tag{1}$$

where V, L, K and TFP are value-added, labor, capital stock and total factor productivity, respectively.

TFP can be decomposed in the following way:

$$TFP = \mathrm{T}\,(T, t) \tag{2}$$

where T: technology knowledge stock and t: time trend.

Technology knowledge stock can be measured in the following way:

$$T_t = R_{t-m} + (1-\rho)T_{t-1} \tag{3}$$

where Rt-m: R&D expenditure in time t-m; m: time lag between R&D and commercialization; and ρ: depreciation rate of technology.

Substituting TFP in Eq. (2) for TFP in Eq. (1), following production function is obtained:

$$V = \mathrm{F}\,(L, K, T) \tag{4}$$

Suppose value-added (V) is a function of capital (K), labor (L), and technology (T), the expansion of V can be measured using the following equation.

$$\begin{aligned} \frac{\Delta V}{V} &= \sum_{X=L,K}\left[\frac{\partial V}{\partial X}\cdot\frac{X}{V}\right]\frac{\Delta X}{X} + \frac{\partial V}{\partial T}\cdot\frac{T}{V}\cdot\frac{\Delta T}{T} \\ &= \sum_{X=L,K}\left[\frac{\partial V}{\partial X}\cdot\frac{X}{V}\right]\frac{\Delta X}{X} + \frac{\partial V}{\partial T}\cdot\frac{R}{V} \end{aligned} \tag{5}$$

where V, L, K and T are value-added, labor, capital stock and technology knowledge stock and R is R&D expenditure which is nearly equal to flow of technology knowledge stock (ΔT).

Eq.(5) shows that TFP comprises the marginal productivity of technology and the R&D intensity.

2) *Measurement of TFP*

The equation above demonstrates that in order to measure TFP, it is necessary to measure the marginal productivity of technology first. Based on Watanabe's method (Watanabe and Wakabayashi, 1997; Tarasyev and Watanabe, 1999; Watanabe et al., 2001), we measure the marginal productivity of technology by solving the simultaneous equation of the internal rate of return, price of technology service, and marginal productivity of technology.

In case using value-added (V) as production output, the marginal productivity of technology can be formulated as follows:

$$q = \frac{\partial V}{\partial T} \quad \textit{(marginal productivity of technology)} \tag{6}$$

In order to measure the marginal productivity of technology, it is requested to simultaneously solve the following equations (7), (8) and (9) in advance (see [5]):

$$P_t = (1 - gs) \cdot [(Rls \cdot Dl + Rms \cdot Dm + Res \cdot De) + Rks \cdot Dk \cdot (\overline{r} + \rho)/(1 - ct)] \tag{7}$$

$$\frac{\partial V}{\partial T} = \frac{GTC \cdot (P'_t / P_t)}{GLC + GCC + GTC \cdot (P'_t / P_t)} \cdot \frac{V}{T} \tag{8}$$

$$e^{mr} = \int_0^\infty \frac{\partial V}{\partial T} e^{-(\overline{r}+\rho)} dt = \frac{\partial V}{\partial T} / (\overline{r} + \rho) \tag{9}$$

where Pt: service price of technology; P't: capital price of technology; Rls, Rks, Rms and Res: shares of R&D expenditures for labor costs, tangible fixed assets, materials, and energy respectively; Dl, Dk, Dm and De: wage index, investment goods deflator, wholesale price indices of materials and energy respectively; gs: ratio of government financial support; ct: ratio of corporate tax; GLC: gross labor cost; GCC: gross capital cost; GTC: gross technology cost = R&D expenditure and $\bar{r}$:rate of internal return to R&D investment; m: time-lag from R&D to commercialization; and ρ: rate of obsolescence of technology.

2.2. *Trends in TFP growth rate in Japan's electric machinery firms*

We measured the marginal productivity of technology, investment ratio of R&D, and ratio of TFP growth of 15 leading Japanese electric machinery firms. Figure 6 shows the result.

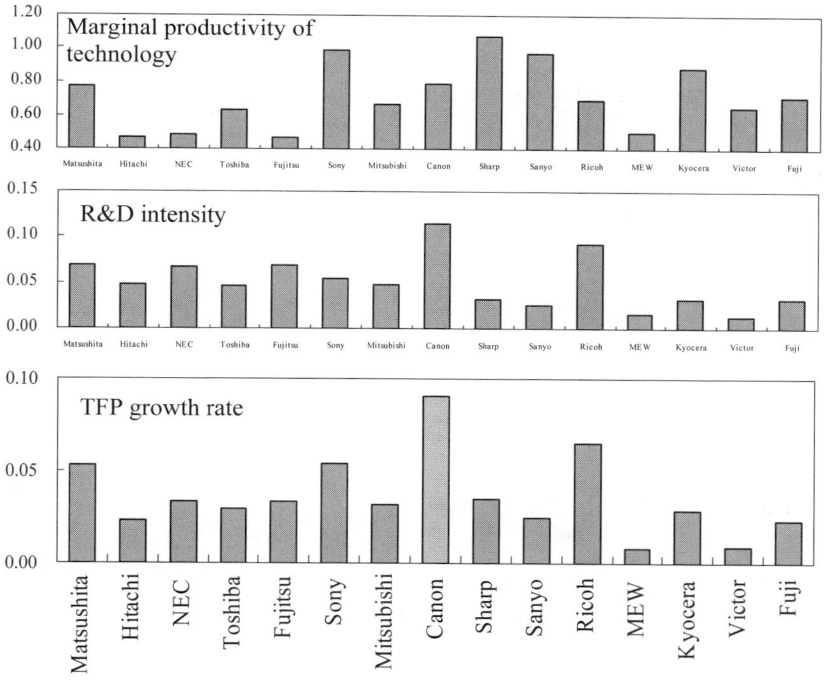

Figure 6: Comparison of TFP Growth Rate in Japan's Top 15 Electric Machinery Firms (2000–2003 average).

According to Figure 6, although large firms such as Hitachi and NEC achieved higher R&D intensity in comparison with those of medium-sized firms, including Canon, the large firms' marginal productivity of technology remains low.

This current situation of large firms in the electric machinery industry directly shows the current Japanese situation in which a high level of R&D intensity may not lead to an increase in productivity.

As in the entire Japanese economy about which the delay of structural reform in response to the globalization of the economy was pointed out, it is indicated that in large firms in the electric machinery industry, the delay in structural reform brought about a decrease of efficiency of R&D investment.

3. Dynamism of Quality of Technology and Structural Reform

3.1. *Corporate evaluation through principal component analysis*

We categorized the organizational structures of the 15 leading Japanese electric machinery firms in terms of three factors: size, financial situation, and R&D activities. As for an index of the three factors, we selected the following five items.

Size: the amount of sales and total assets

Financial Situation: Operating Income to Sales (OIS) and capital adequacy ratio

R&D Activities: R&D intensity

Using 2000-2003 data, we implemented principal component analysis for the five items above. Table 2 shows the result.

Table 2: Result of Principal Component Analysis.

	eigenvalue	proportion	cumulative
PC_1	**2.59**	**51.7%**	**51.7%**
PC_2	**1.74**	**34.8%**	**86.5%**
PC_3	0.51	10.3%	96.7%
PC_4	0.15	2.9%	99.7%
PC_5	0.02	0.3%	100.0%

	PC_1	PC_2
Sales	**0.56**	0.31
Total assets	**0.55**	0.29
OIS	-0.46	**0.46**
Capital adequacy ratio	-0.42	**0.37**
R&D intensity	0.05	**0.68**

Based on the results in Table 2, we selected PC1 and PC2 as principal components based on the Kaiser standard. We understand that PC1 and PC2 may explain about 87% of information of the object variable because of their cumulative contribution ratio. In addition, from the factor loading table, each variable that contributes to PC1 and PC2 may be summarized in the following.

PC1: the amount of sales and total assets

PC2: OIS, capital adequacy ratio and R&D intensity

We define PC1 as a principal component to represent scale factor and PC2 as a principal component to represent firms' investment efficiency including R&D activities.

Based on these two principal components, we carried out cluster analysis using the measured points of the principal components in order to evaluate and categorize firms. Figure 7 depicts the result.

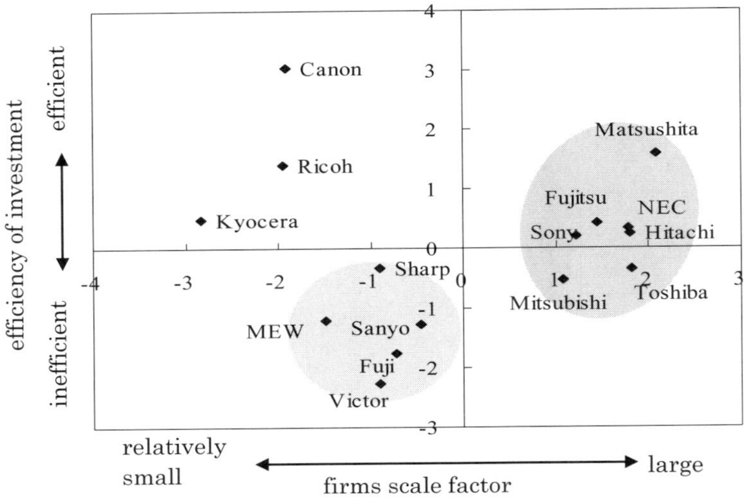

Figure 7: Scatter Diagram based on the Result of the Principal Component Analysis in Japan's Top 15 Electric Machinery Firms (2000–2003 average).

Figure 7 accords with the cluster by the scatter diagram devised based on the points of the principal components.

Consequently, the seven large firms are categorized in the same cluster not only by their size but also by the corporate investment efficiency perspective that comprehensively includes the financial situation.

3.2. *Dynamism between TFP and corporate structure*

We carried out the same analysis of firms in the 1980s and the 1990s. Based on these results, we can observe the relationship between improvement of quality of technology and firms structure in the 1980s, the 1990s, and the 2000s.

Suppose TFP is a function of PC1 and PC2 that represent firms structure, we can observe their relationship through multiple linear regression analysis. Table 3 shows the result.

Table 3: Correlation between Scale Factor, Investment Efficiency and TFP Growth Rate in Japan's Top 15 Electric Machinery Firms (1983–2003).

$$\frac{\Delta TFP}{TFP} = a + b \cdot PC_1 + c \cdot PC_2$$

	a	b	c	adj. R^2
1980s	0.107 (9.82)	0.024 (3.41)	0.026 (2.87)	0.561
1990s	0.071 (11.80)	0.011 (2.79)	0.017 (3.32)	0.545
2000s	0.036 (13.03)	-0.001 (-0.77)	0.014 (6.67)	0.755

According to the Table 3, as for the 15 leading Japanese electric machinery firms, both scale factor and investment efficiency contributed to the TFP growth rate up until the 1990s, while since the 2000s, the contribution of scale factor has been declining dramatically.

In order to examine this result further, we divided these 15 firms into two clusters based on the result of the cluster analysis: (1) Matsushita, Hitachi, NEC, Toshiba, Fujitsu, Sharp, and Mitsubishi and (2) the rest of the firms. Then, we carried out similar analysis. Table 4 shows the result.

Table 4: Correlation between Scale Factor, Investment Efficiency and TFP Growth Rate (1983–2003, divided 15 firms into two clusters).

$$\frac{\Delta TFP}{TFP} = a + b_1 \cdot D \cdot PC_1 + b_2 \cdot (1-D) \cdot PC_1 + c_1 \cdot D \cdot PC_2 + c_2 \cdot (1-D) \cdot PC_2$$

a	b_1	b_2	c_1	c_2	adj. R^2
0.060 (8.51)	-0.017 (-3.49)	0.014 (3.12)	0.016 (3.14)	0.019 (8.51)	0.871

D: dummy variable (Top 7 firms = 1, others = 0)

According to the Table 4, as for the medium-sized firms, both scale factor and investment efficiency contribute to the TFP growth rate. In contrast, as for the seven large firms, scale factor contributes to the TFP growth rate in a negative. This result shows that the group of large firms that we selected this time has become such big organizations that their inefficiency of scale (organizational inertia) diminished the contribution of investment efficiency to technological innovation.

3.3. *Market evaluation of quality of technology*

The quality of technology of firms promotes further R&D investment through its contribution to the increase of aggregate market price that represents the market evaluation. Table 5 shows this relationship.

Table 5: Correlation between Aggregate Market Price and Sales, Capital Adequacy Ratio and TFP Growth Rate in Japan's Top 15 Electric Machinery Firms (1983–2003)

$$\ln AMV = a + b\ln S + c\ln CAR + d\ln TFP$$

	a	b	c	d	$adj.\ R^2$
1980s	-0.84 (-0.54)	0.81 (7.06)	0.49 (2.25)	0.16 (1.90)	0.819
1990s	-0.81 (-0.45)	0.84 (6.88)	0.89 (3.65)	0.24 (2.31)	0.832
2000s	0.91 (2.43)	0.93 (6.28)	1.06 (4.06)	0.39 (2.24)	0.889

Table 5 demonstrates that in the market, not only the amount of sales and the health of financial situation but also the improvement of quality of technology represented in the TFP growth rate are highly evaluated.

This indicates that the structural reform itself may receive recognition in the market. In addition, it may affect corporate market evaluation through the improvement of quality of technology.

4. Conclusion

Based on the empirical analysis on Japan's electric machinery firms, this paper demonstrate the following implications:

1) As for the relatively smaller firms, both their scale factor and investment efficiency contribute to the TFP growth rate, while for the large scale firms, their scale factor has been a negative to TFP in recent years.

2) Those results suggest that an inefficiency of scale disturb the contribution of investment efficiency to technological innovation in case of too large firms.
3) And the market evaluates not only the amount of sales and the strength of financial situation but also the quality of technology.
4) In the market, reforming financial structure is highly evaluated on both side of the increase the strength of financial situation and improvement of quality of technology.
5) In conclusion, dynamism between improvement of quality of technology through efficient R&D investment and market evaluation can be forming a virtuous circle depending not on quantity but on the corporate institution that promotes structural reform.

References

Japanese Electrical Electronic & Information Union, Monthly Report (2003–2006).

Tarasyev, A. and Watanabe, C. (1999). Optimal Control of R&D Investment in a Techno-Metabolic System. International Institute for Applied Systems Analysis (IIASA) Interim Report, IR-99-01.

Watanabe, C. and Kondo, R. (2003). Institutional Elasticity towards IT Waves for Japan's Survival. *Technovation*, 23(4), pp. 307–320.

Watanabe, C. and Wakabayashi, K. (1997). The Perspective of Techno-metabolism and its Insight into National Strategies. *Research Evaluation,* 6(2), pp. 69-76.

Watanabe, C., Zhu, B., and Tou, Y. (2001). Theoretical Analysis and Empirical Demonstration of Optimal R&D Investment Trajectory Control. *The Journal of Science Policy and Research Management*, 16(1/2), pp. 83–101.

Chapter 6

Evaluation of Nuclear Knowledge Management for the Light Water Reactor and Fusion Reactor: A Case Study of Japan Atomic Energy Research Institute (JAERI)

Kazuaki Yanagisawa

Japan Atomic Energy Agency (JAEA), 1233 Watanuki, Takasaki, Gunma, 370-1292, Japan

Tel. +81-27-346-9325, Fax. +81-27-346-9480

E-mail: yanagisawa.kazuaki@jaea.go.jp

Funds invested in a 45-year study of the light water reactor (LWR) technology in JAERI totaled 4.2b$ (billion dollars) for research including human resources of 34,718 man × years. The benefits to taxpayers from this JAERI work was estimated to be about 6.3 b$, resulting in a favorable cost-benefit ratio of 1.5 (6.3/4.2). Funds invested in the 32-year study of FR were 5.4b$ for R&D and 0.6b$ (6,331 man × years) for personnel. Estimates are that after commercialization in 2050, the share of FR will reach 23.4% in year 2100 and that a commercialized FR will generate revenue from electricity as high as 1,687b$ during the period 2050-2100.

1. Introduction

This paper describes the result of case studies of long-term nuclear knowledge management (NKM) with respect to the light water reactor (LWR) and the fusion reactor (FR). We assess the potential benefits of these alternatives under conditions of substantial uncertainty.[1]

2. Purpose and Method

2.1. *Nuclear knowledge management (NKM)*

Knowledge Management (KM) is defined by the International Atomic Energy Agency (IAEA, 2005) as an integrated, systematic approach to identifying, managing and sharing an organization's knowledge, and enabling persons to create new knowledge collectively and thereby helping achieve the objectives. NKM identifies, optimizes, and actively manages intellectual assets either in the form of explicit knowledge held in intangible products or tacit knowledge possessed by individuals or communities in the nuclear fields (Snowden, 1988).

2.2. *Purpose*

In the present study the author wish not only to show the validity of long-term NKM as a key factor of INES -- namely LWR and FR, but also to assess their hypothetical benefits through the year 2100 under conditions of substantial uncertainty. It should be stressed that those factors are important intellectual assets of JAERI developed to date. Additionally, in the Framework for Nuclear Energy Policy constructed up by the Japan Atomic Energy Commission, a LWR, a fast breeder reactor (FBR), a high temperature gas-cooled reactor (HTGR), and an FR are all defined as eligible and prominent candidates for long-term nuclear energy sources.

---

[1] In October 2005, JAERI was reorganized and renamed the Japan Atomic Energy Agency (JAEA). The topic of this paper addresses research activities done within JAERI. For this reason, the fast breeder reactor (FBR) technology is omitted.

3. Results and Discussion

In the following discussion, the rate of currency is constant for all years; 1 US $ = 121 yen. The values are expressed in constant, 1995 denominations.

3.1. *Light water reactor (LWR)*

3.1.1. *Economic impact of LWR on the electricity market*

In 1997, gross electricity production totaled 3,494 bkWh (billion kilo-watt-hour) and 865 bkWh in the United States (US) (USDOE, 1997), and Japan, respectively. The US level was four times greater than that of Japan. Gross electricity generated in the US is highly biased to the use of fossil fuels such as coal, whereas in the case of Japan most can be attributed to nuclear energy. The US produced 629 bkWh of nuclear-based electricity, which sold for 39 b$ (billion dollars), while Japan produced 311 bkWh, which sold for 47 b$; the difference in value being some 21%.

In terms of economics, the explicit beneficiary of LWR in Japan is ambiguous because those technologies were originated in the US: the Federal R&D expenditures from 1950 (before the Atoms for Peace Program) to 1962 were 1.9 b$ and those from 1963 to 1975 were 1.3 b$ (2003 base), totaling 3.2 b$ (Bezdek and Wending, 2006) during the 1950–1975 period. Because of that initial investment, revenues of electricity generated by the 52 LWR units in Japan is 47 b$/year and 11 b$ for relevant reactor components (1999 base).

3.1.2. *A share of JAERI*

The following points are discussed for a better understanding of the economic share for JAERI. Market Creation Effect (MCE) is hereinafter defined by

MCE = MCP (Market Creation Product) × A rate of value added to the new products (from I-O) table (Ministry of Economy, Trade and Industry, 1995) (1)

MCP = Revenue from products born in a newly created market induced partly or fully affected by JAERI outputs $\times$ a ratio of contribution by R&D performance to a total amount of sales revenue $\times$ a ratio of R&D performance contributed by JAERI (2)

Then cost benefit effect (CBE) can be determined by

CBE = MCE / Total amounts of investment (research and personnel) (3)

LWR market

The revenue from electricity sales during 1970-2000 (Ministry of Finance, 2000) in Japan was 760 b$. For the facilities, the revenue accruing from the upstream of the fuel cycle to the downstream of the fuel cycle covering the period from 1977 to 2000 was estimated (Japan Atomic Industrial Forum, 2001) to total 248 b$. All revenues obtained in individual years were deflated to 1995 denominations.

Revenue of LWR in total was; 760+248=1,008b$ (4)

R&D ratio of LWR and JAERI

As discussed in the previous paper (Yanagisawa, 2006), two R&D ratios were defined as follows:

R&D percent of LWR on average from 1978 to 1999 = 6.2% (5)

R&D percent of JAERI = 20% (6)

MCP and MCE

In the nuclear market, the MCP of JAERI is given by

$$MCP_{electricity} = 760b\$ \times 0.062 \times 0.2 = 9.4\ b\$$$

$$MCP_{facilities} = 248b\$ \times 0.062 \times 0.2 = 3.1\ b\$$$

By using the I-O table, MCE of JAERI for LWR is given by

$$MCE_{JAERI} = 9.4 \times 0.542 + 3.1 \times 0.386 = 6.3b\$ \quad (7)$$

This is the research output (results) of JAERI.

Invested amounts of JAERI

Invested amount (i.e., income) of JAERI
= 4,194 M$ or 4.2b$ including personnel cost (8)

Cost Benefit Effect (CBE)

For JAERI, funds invested in the 45-year study of LWR were 4.2 b$, including human resources of 34, 718 man×years. A large part of the funds consisted of supporting the construction and operation of JPDR (Japan Power Demonstration Reactor), JMTR (Japan Materials Testing Reactor) and NSRR (Nuclear Safety Research Reactor). The positive return from JAERI to the tax payers is about 6.3 b$, as shown in equation (Eq. 7). Long-term robust NKM can possibly result in a CBE_{JAERI} attributed to JAERI for LWR of 1.5 (6.3/4.2). Thus, comparing the indexed income of 1.0, the outcome of JAERI is 1.5 (>1). JAERI is a national research institute and this figure may be regarded as sufficiently high because many high risk and complex tasks were completed successfully.

3.2. *Fusion Reactor (FR)*

The JAERI investment in FR program through 2000 was about 6 b$, a value almost 40% of the total JAERI budget (15 b$ for research). This consists of 5.4 b$ for R&D cost and 0.6 b$ for personnel (6,331 man×years) cost. When assessing long-term NKM, one must examine the cost reduction of electricity due to commercialized FR and the CBE due to the creation of an FR market. Under the design-based scenario, the commercialization of FR will be started in 2050 and operations will progress through 2100.

3.2.1. *A cost reduction of electricity due to commercialized FR*

3.2.1.1. Assumptions necessary to make the estimation

The following six assumptions are made for subsequent estimations.

a1 An estimation will be done on all electricity markets in the world
a2 The DSE model -- a current LDNE2 (Lions, 2000) model installed in the earth regeneration program promoted by RITE (2001) is used.
a3 The energy scenario is in conformity with IPCC92a (ARIES, 1992).
a4 By assuming that the CO_2 concentration in 2100 is capped below 550 ppm, an infrastructure of power source installations is determined as the optimal solution for minimizing total energy costs.
a5 After commercialization of FR in the nuclear market in 2050, the facility availability factor and the operating costs will be modified gradually. Incorporating these, generation costs are decreased to 2.3% /year at the beginning of commercialization and 0.25%/year after 25 years.
a6 The cost of fossil fuel is determined as a function of the accumulated amount of consumption.

3.2.1.2. Results of estimation

We can state the following;

- The world-wide power generation of FR is about 30%; i.e., 25×10^{12} kWh of the gross generated power. During this period, the generation cost of FR is gradually decreased in accordance with the rate indexed by the aforementioned item (a.5).
- If FR is not commercialized until 2100, electricity generation costs may potentially be increased as much as an additional 6.6¢/kWh. This is attributable to the depletion of the fossil source and the cost increase of the alternatives due to environmental problems.
- After 2050, the cost reduction effect of world-wide power generation due to the commercialization of FR may result in a discounted value of the “amount of cost reduction due to FR commercialization” multiplied by “generated electricity by commercialized FR”. Therefore, an accumulation of savings during 2050–2100 using a discount rate of 2% is as follows:

The total cost reduction may be approximately 5,785 b$ in the world (2000 base) (9)

The cost reduction by 2100 is 306b$/year (2000 base) (10)

3.2.1.3. The contribution by Japan to the total cost reduction in the world

The expenditure on R&D for commercialized FR in Japan is about 1/3 of that in the world. A contribution of Japan to the total cost reduction is given by

$$5{,}785\text{b\$} \times 0.3 = 1{,}736\text{b\$} \tag{11}$$

where 0.3 is the ratio of Japanese expenditures to those of the world.

3.2.1.4. The contribution of JAERI to domestic cost reduction

A ratio of R&D

With respect to the effect of cost reduction, the R&D ratio corresponds closely to the gross profit excluding the prime cost, which is estimated to be 20%.

A ratio of the JAERI contribution

In the previous study (Yanagisawa, 2002), the ratio of JAERI contributions to the R&D of LWR in Japan was estimated to be 20%. Because the time span of R&D for FR is twice that of LWR (40 years), a ratio of the JAERI contribution to the R&D of FR is 20% x 2, that is, 40%.

Results of the JAERI efforts for reducing domestic cost

The contribution of JAERI to the amount of domestic cost reduction (1,736b$) is given by

$$1{,}736\text{b\$} \times 0.2 \text{ (R\&D ratio)} \times 0.4 \text{ (JAERI contribution)} = 139\text{b\$} \tag{12}$$

3.2.2. *A CBE during creation of the FR market*

The creation of the FR market will produce two types of research impacts: One is the retail sale of electricity generated by commercialized FR and the other is the retail sales due to the construction and operations of FR facilities and equipments. In the present study the author wishes to show the result of the former only.

3.2.2.1. Retail sale of electricity generated by commercialized FR

Assumptions

To estimate the size of the electricity market created by FR, the following assumptions are made:

- The growth rate of demand and supply of energy (DSE) after 2000 will be changed as follows;
  - Through 2010; 1.2%
  - 2010–2050; 0.7%
  - 2050–2100; 0.5%

  (Decrease from 0.7% to 0.5% is due to a population reduction expected to occur after 2050)
- The share of commercialized FR is assumed to be 12.6% in 2070 and 23.4% in 2100.
- Currently, there is no index of cost for FR electricity sales cost. Therefore, the author indexes the value from the LWR series. In 1997, the generated electricity from 52 LWR units was 0.3007×10^{12} kWh and the total revenue of those was 46.8b$, resulting in a nominal unit cost of 15.5¢/kWh/year. This value is cited to commercialized FR.
- Electricity generated by commercialized FR is increased linearly during 2050–2070 and during 2070–2100.

All assumptions are summarized in Table 1.

Table 1: Electricity output generated by commercialized FR.

Year of Evaluation	Growth rate of demand and supply of electricity	Gross electricity generated (GkWh)	Share of FR (%)	Electricity output by FR (GkWh)
1999	-	917.6	0	0
2010	1.2% per year	1029.2	0	0
2050	0.7% per year	1360.4	0	0
2070	0.5% per year	1503.1	12.6	189.4
2100	0.5% per year	1645.7	23.4	408.5

Note: GkWh means Giga (10^9) kWh

Gross sale of electricity generated by commercialized FR

Using the data in Figure 1, the gross sale of electricity generated by commercialized FR during 2050-2100 is determined as

Area 1:

189.4GkWh × (2070-2050)/2 × 15.5¢/kWh = 294b$

Area 2:

(189.4GkWh+408.5GkWh) × (2100-2070) /2 × 15.5¢/kWh = 1,393b$

Gross sale of electricity
= area 1 + area 2
= 294b$+1,393b$=1,687b$ (13)

According to this result, the FR market is estimated to be 34b$/year (1,687b$ / 50 years). The discounted cost from 2100 to the present is 63b$/year. This estimate is to show that the FR market is actually greater than that of LWR market (47b$/year).

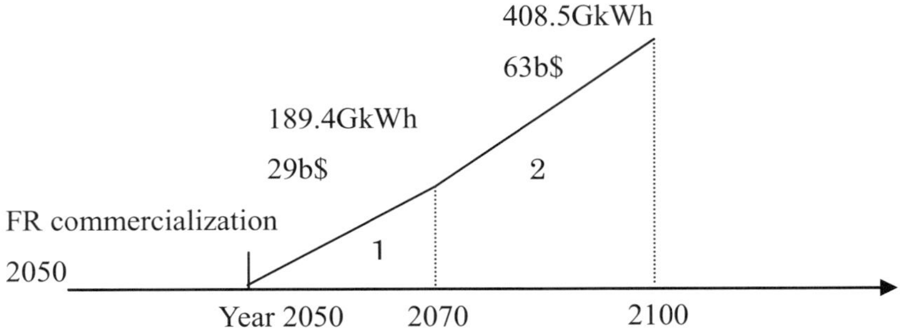

Figure 1: Gross sale of electricity generated by FR, 2050–2100.

The contribution of JAERI to the electricity generated by commercialized FR

The ratio of R&D

Japan has played an important research role in the R&D of FR from the beginning of technological development and will continue to play an active role in each stage of experimental, prototype and demonstrative fusion reactors. It is expected for this development to take more than 90 years. Electric generation by commercialized FR technology may not occur if the R&D data is not successful. This means that the role of R&D for commercialization of FR is very significant. By this reason, the author estimates 15% (10% × 1.5) to be the ratio of R&D, where 10% is the potential maximum R&D ratio at any one private business in Japan.

The ratio of the JAERI contribution

In assisting the development of commercialized FR, electric power companies, plant makers, and JAERI will all be contributing up to 2050. Then, the ratio of the JAERI contribution will be 1/3 (33%), corresponding to the primary cost of retail sale.

Impacts attributed to JAERI from FR power generation

The final impacts attributed to JAERI (MCP) from the creation of a nuclear power generating market by the commercialized FR is given by

$$MCP_{electricity} = 1{,}687\ b\$ \times 0.15\ (R\&D\ ratio) \times 0.33\ (JAERI\ contribution) = 83\ b\$ \quad (14)$$

Using the economic input-output (I-O) tables, we use the ratio of value added to the electricity, that is, 0.542.

$$MCE_{electricity} = 83b\$ \ x\ 0.542 = 45b\$ \quad (15)$$

3.2.3. *Summary of FR*

An economic effect caused by FR commercialization is as follows:

Cost reduction effect

Cost reduction effect is 1,736b$ and the share of JAERI ($MCP_{reduction}$) is 139 b$. This effect does not accompany the MCE.

Electricity market creation

Created electricity market is 1,687 b$. Then, $MCP_{electricity}$ is 83 b$. It is notable to say that MCP_{JAERI}^{FR} is 7 times than that of LWR (MCP_{JAERI}^{LWR} = 12.5 b$) (Yanagisawa, 2006). It is notable to say that MCE_{JAERI}^{FR} is 45b$, which is 7 times larger than that of LWR (MCE_{JAERI}^{LWR} = 6.3 b$) (Yanagisawa, 2006).

A total investment to JAERI at 1970-2002 is 5.4+0.6=6.0b$. Assuming a continuous investment through 2100, total amounts of investment (research and personnel) are broadly calculated as 24.4b$. The MCE of JAERI will be 45b$ for electricity, then according to the definition, CBE_{JAERI}^{FR} is 1.8 (45 / 24.4). Comparing with CBE_{JAERI}^{LWR} = 1.5, FR may have an almost equal cost performance.

4. Conclusions

LWR, and FR are important intellectual assets that have been developed by JAERI, and they are also promising technological visions for fostering the concept of INES. The author incorporated case studies for assessing the potential benefits of these technologies by means of long-term NKM. Recognizing the substantial amount of uncertainty contained in the estimates, the results obtained are as follows.

Revenue from LWR in Japan was 760 b$ during 1970-2000, and the share of JAERI (MCP) of this activity was 9.4b$ for electricity and 3.1b$ for facility. For JAERI, funds invested in the 45-year study of LWR were 4.2b$ including human resource of 34,718 man×years. The benefit from this JAERI activity to the taxpayers was estimated to be about 6.3 b$, and the cost benefit ratio of the JAERI program is thus 1.5 (6.3/4.2). JAERI is a national research institute and this figure may be regarded as sufficiently high because many high risk and complex tasks were conducted successfully.

Commercialized FR will induce a cost reduction in electric power of 1,736b$ during 2050-2100, and the share attributable to JAERI for this activity (MCP) is 139b$. Revenues generated by the gross sale of electricity by the commercialized FR will be 1,687b$, and the share attributable to JAERI for this activity (MCP) is 83 b$. Revenues generated by the domestic and foreign equipments related to FR will be 1,984 b$, and the share attributable to JAERI for this activity (MCP) is 119b$. Assuming a continuous investment through 2100, total amounts of investment (research and personnel) are broadly calculated as 24.4b$. The MCE of JAERI will be 45b$ for electricity, then according to the definition, CBE_{JAERI}^{FR} =1.8 (45 / 24.4). Comparing with CBE_{JAERI}^{LWR}=1.5, FR may have an almost equal cost performance in the future.

Acknowledgements

Grateful acknowledgment is addressed to the Mitsubishi Research Institute Inc., for their excellent contribution to the study of LWR and FR. Thanks are also addressed to Dr. R. H. Bezdek, President, Management Information Services, Inc., U. S. A. for his encouragement to this study and Dr. R. Hino, Nuclear Applied Heat Transfer Division, JAEA for his valuable comments.

References

ARIES (1992) Scenario for future energy consumption; http://aries.ucsd.edu/ARIES/MEETINGS/0601-USJ Workshop/Konishi.pdf#search='IPCC92a'

Bezdek R. H. and Wending R. M. (2006). A Half Century of Federal Incentives for Nuclear Energy: What Do the Numbers Show and What Do They Mean? Management Information Services, Inc., Washington, D. C

IAEA (2005). Nuclear knowledge management glossary. Presented at the Workshop on Managing Nuclear Knowledge, Trieste, Italy.

Japan Atomic Industrial Forum, Inc. (1965-2000).

Japan Electricity Association Newspaper Division (2000.

LIONS (2000). Luma Dependent Nonlinear image Enhancement, http://www.lions.odu.edu/org/vlsi/TSWG-FRT-ProgramReview. ODU.pdf#search='LDNE2'(2000)

Ministry of Economy, Trade and Industry (1995).

Ministry of Finance, 1970-2000.

Nuclear System Association (2001).

RITE (2001). Research Institute of Innovative Technology for the Earth http://www.rite.or.jp/

Snowden, D. (1988). A framework for creating the sustainable programme. IBM Real Business Guide on Knowledge Management. (complete reference)

USDOE (1997), http://www.eia.doe.gov/cneaf/nuclear/page/nuc_reactors/

Yanagisawa K., et al. (2002). An Economic Index regarding Market Creation of Products obtained from Utilization of Radiation and Nuclear Energy (IV)-Comparison between Japan and the U. S. A. *Journal of Nuclear Science and Technology*, 39(10), pp. 1120–1124.

Yanagisawa, K. (2006). Evaluation of nuclear knowledge management: an outcome in JAERI. *International Journal of Nuclear Knowledge Management,* 2(2), pp. 91–104.

Chapter 7

Technology Balance: Technology Valuation According to IASB's Value in Use Approach

Günther Schuh, Sascha Klapper and Christoph Haag

Fraunhofer Institute for Production Technology IPT
Department of Technology Management
Aachen, Germany

Until the early 1980s, up to 75% of the market value of a business was defined by the tangible assets that appeared on the balance sheet. In 2005, that number was less than 25%. In the course of this "intangible" dynamics, technology-related intangible assets proved to be a major value driver for many companies and thus are to be treated adequately in the financial statement. Fraunhofer IPT develops a methodology to valuate technologies according to International Financial Reporting Standards. The methodology allows for estimating a technology's future cash flows along its whole life cycle, from early stage development until late market phase. Therefore, so-called real options, that come along within the technology's remaining life cycle, have to be taken into account at each valuation point. In this chapter, the initial situation and motivation for the research project as well as the developed methodology are described in detail.

1. Introduction

While in the early 1980s, up to 75% of the market value of a company was defined by the tangible assets that appeared on the balance sheet, that number was less than 25% in 2005. Today, physical and financial assets yield at best an average return on investment. Extraordinary profits and dominant competitive positions are achieved by the sound deployment of intangibles (Lev, 2001; Kenton, 2005; Von Scheffer and Loop, 2005).

In a large number of firms, investments in technology-related intangibles (i.e., R&D) have become the major productive input, particularly for those, who operate in technology- and science-based sectors (Aboody and Lev, 2003). Consequentially, in 2005, the world's 1.000 most technology-intensive companies spent \$407 billion for research and development – representing an increase of six percent compared with the year before (Booz Allen Hamilton, 2006).

Along with the growing relevance of technology-related intangibles, executives are faced with new challenges regarding the value-based management of their companies. Major issues not only concern the internal controlling of technologies but also their external reporting in order to constitute shareholder value (Lev, 2001; Croenenberg et al., 2007).

The International Accounting Standards Board (IASB), responsible for the elaboration of the International Financial Reporting Standards (IFRS), has also become aware of the great significance of technology-related intangible assets, and has set the course for their adequate treatment in the financial statement. According to the latest IFRS, a technology arising from own development shall be valuated and recognized in the balanced sheet in case certain recognition criteria can be fulfilled (IASB, 2006).

But that again demands for suitable valuation methods, so that a reproducible and comprehensible technology valuation can be accomplished. In this regard, stakeholders in charge – like R&D controlling, technology management and accounting – still rest in its infancy. New valuation methods are to be developed (Krimpmann, 2006).

The Fraunhofer Institute for Production Technology IPT attends to this demand in its research project Technology Balance, dealing with the question, how to valuate technologies along its life cycle from early stage development until late market phase. The research objective is to design a methodology for a cash flow based technology valuation, taking the technologies' future economic benefits into account and considering the latest IFRS requirements.

The Technology Balance valuation methodology will be introduced in § 3. Before that, § 2 will give a short overview of current approaches for the valuation and measurement of intangible assets. Finally, § 4 will summarize the presented findings.

2. State-of-the-Art in Technology Valuation

In recent years, research institutes as well as consulting agencies have elaborated numerous methods that deal with the measurement and valuation of intangible assets. Some of those methods do not aim at a monetary valuation, others apply to intangible assets as a whole and are rather ineligible for the valuation of specific technologies and only few of them fulfill the requirements of IFRS (Günther et al., 2004; Andriessen, 2004; Arthur D. Little, 2005).

In § 2.1, we present an overview about four basic IASB-approved valuation categories, which exhibit the guardrails for any IFRS-conform technology valuation method. Approaches that are suitable for the Technology Balance will be outlined in § 2.2. Before that.

2.1. *IASB-approved valuation categories*

In the IFRS framework, which sets out the concepts that underlie the preparation and presentation of financial statements, IASB approves four general approaches for the valuation of assets (IASB, 2005):

Historical Costs: Assets are recorded at the amount of cash or cash equivalents paid or the fair value of the consideration given to acquire them at the time of their acquisition.

Current Costs: Assets are carried at the amount of cash or cash equivalents that would have to be paid if the same or an equivalent asset was acquired currently.

Fair Value: Assets are carried at the amount of cash or cash equivalents that could currently be obtained by selling the asset in an orderly disposal.

Value in Use: Assets are carried at the present discounted value of the future net cash inflows that the item is expected to generate in the normal course of business.

Cost-based valuation approaches are easily to apply, when the necessary data – e.g. from R&D controlling – is available. However, they rarely provide veridical results, as the future economic benefit of an asset can differ radically from its original costs, especially in case of technologies (Woodward, 2003; Smith and Parr, 2005). Thus, cost-based approaches are not suitable for technology assessment.

Fair value is the preferred valuation approach of the IASB (Lienau and Zülch, 2006; Woodward, 2003; IDW, 2004). When applicable, it reflects the true market value of an asset. But its application demands for an active market for the valued asset or at least an analogue market where similar assets are traded. Considering technologies, these requirements are scarcest accomplished (Krimpmann, 2006; Woodward 2003).

Hence, for technologies, Value in Use is the most appropriate valuation approach. The IASB recommends various scenarios with different occurrence probabilities to estimate the associated cash flows. This proceedure is known as the Expected Cash Flow Approach (Woodward, 2003; IDW, 2004; Deloitte, 2004; FASB, 2006). This approach will provide the basis for the Technology Balance methodology.

2.2. *Current valuation methods for intangible assets*

In the following, some selected measurement and valuation methods by research institutes and consulting agencies and their contribution to the Technology Balance methodology will be outlined:

2.2.1. *Multiple methods ("Rules of Thumb")*

Multiple methods are based on the assumption that a constant figure (e.g. earnings per share) can be related to the value of intangible resources. Consequently, a multiple must be identified based on empirical data, which describes how much revenue has been caused by the investment in the relevant intangible (Günther et al., 2004; Razgaitis, 2003). Multiple methods are also known as "Rules of Thumb".

Such concepts—like the Interbrand Approach (Haig and Perrier, 1997)—can serve as a useful guideline for quick decision-making (Razgaitis, 2003). However, it is rather inappropriate for technology valuation.

2.2.2. *Excess return approaches*

Excess return approaches like RAVETM or EVATM (Strack and Villis, 2001) assume that intangible resources are earning a premium return in addition to an average return, e.g. the average return within a certain branch (Günther et al., 2004). Similar to multiple methods, these concepts are able to valuate a company's intangibles as a whole, but are not appropriate to valuate individual technologies.

2.2.3. *Indicator approaches*

Due to problems concerning the monetary valuation of intangible resources, indicator methods concentrate on the creation of complex measurement systems behind monetary terms (Günther et al., 2004). Methods like the Skandia Navigator (Edvinson and Malone, 1997) or the Value Chain Scoreboard (Lev, 2001) do not allow the explicit calculation of a particular monetary figure, but rather give a qualitative survey of a company's situation concerning intangibles. Hence these concepts are not suitable for technology valuation and balancing according to IFRS.

2.2.4. *Market derived measures*

Capital market derived measures like the market-to-book-value ratio (Andriessen, 2004) or the Tobin's Q (Stewart, 1997) also valuate a company's intangibles as a whole. Such concepts assume that the difference between the book value and e.g. the market value of a company results from its intangibles.

The approach is relatively simple to apply and efficient in its conclusion. However, by itself it is not suitable for the valuation of individual intangibles like technologies (Günther, Kichner-Khairy and Zurwehme, 2004).

2.2.5. *Discounted cash flow methods*

One of the most common concepts used to determine the value of an intangible is the discounted cash flow approach. Methods based on this approach assume that an intangible's value can be measured by the present value of the net economic benefit over its life. According to this, periodic cash flows—resulting from revenues minus cost —are projected. Capitalization of these cash flows, using a risk-adjusted discount rate, arrives at the net present value of the intangible (Günther et al., 2004; Razgaitis, 2003; Smith and Parr, 2005). Different methods like the Brand Money Value (Högel et al., 2002) or the Customer Lifetime Value (Homburg and Daum, 1997) have been developed and are all based on this concept.

2.2.6. *Real option approach*

Option-based valuation approaches have grown in importance in recent years. In order to display the flexibility of R&D projects, the real option approach takes different management opportunities, that occur in the course of an R&D project, into account and uses the analogy with the financial options theory to calculate the value of those opportunities. The approach can be used as an improved discounted cash flow valuation (Günther et al., 2004; Andriessen, 2004).

3. Development of the Technology Balance

As seen previously, there are numerous methods which aim at the measurement of intangibles or at least make a contribution to it. However, none of them exhibits a holistic methodology for a monetary technology valuation, but some qualify for a partial adoption in the Technology Balance methodology.

Now we present the new methodology. The adopted approaches from the state-of-the-art will be pointed out in § 3.1 After that, § 3.2 will introduce the concept of the new Technology Balance methodology.

3.1. *State-of-the-art approaches*

The discounted cash flow concept can be assigned to IASB's Value in Use approach. In its modified form, the expected cash flow approach, where different scenarios are taken into account for the estimation of future cash flows, the IASB recommends its application in order to calculate the value in use of intangible assets. Thus, it will be adopted for the new methodology and exhibit the basic structure of the Technology Balance.

In addition to the expected cash flow approach, the real option approach will serve as the second core idea of the Technology Balance. The technologies' life cycles will be considered as sequences of numerous decision-making points, where each point offers various real options to choose from.

3.2. *Concept of the technology balance methodology*

The concept of the Technology Balance is outlined in the following. Figure 1 shows the methodology on a macro level. Its five subsystems aim at a reproducible and comprehensible technology valuation:

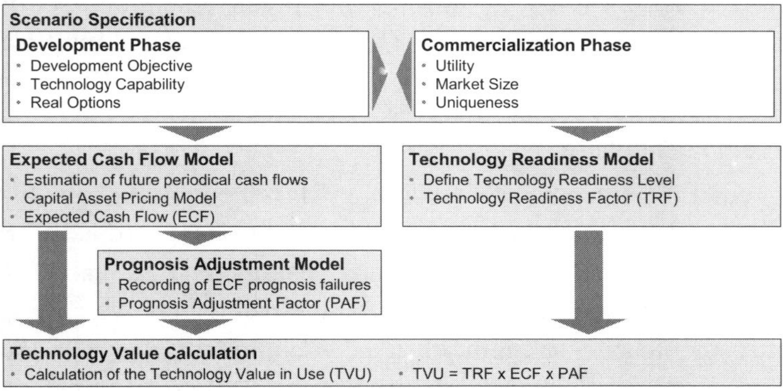

Figure 1: Concept of the Technology Balance.

3.2.1. *Scenario specification*

The Scenario Specification aims at the foresight of future cash flow scenarios and its occurrence probability based on the outcome of the technology development as well as the technology commercialization phase. The scenarios are generated using a simplified version of the well established scenario technique. The scenario technique is based on the two basic principles of Cross-linked Thinking and Multiple Future, which are illustrated in Figure 2.

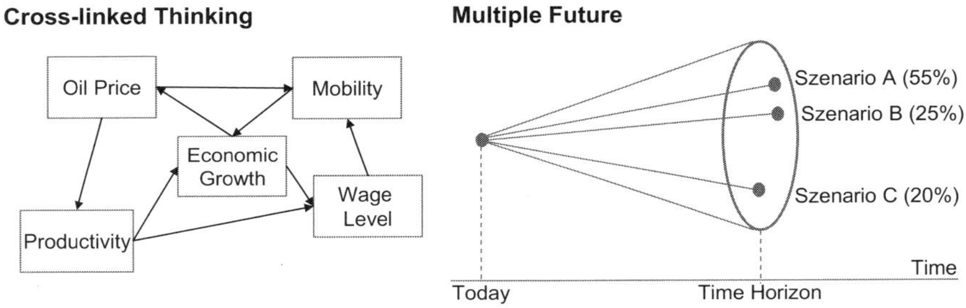

Figure 2: Basic Principles of the Scenario Technique (Gausemeier et al., 2001).

The future is thought of in complex pictures, as it is not enough to consider a company's environment as a simple system. In fact, it is essential to support the systematic foresight into the future by means of cross-linked thinking.

The scenario technique allows multiple possibilities how the future might develop. It assumes that the future is not predictable, but a set of different future scenarios is supposable. This approach is called multiple future (Gausemeier et al., 2001).

3.2.1.1. Development phase

In order to predict the outcome of a technology development process, the development objectives have to be specified at first, using ordinally or cardinally scaled technology performance parameters. Usually, a technology can be characterized by three to five parameters which e.g. might include the tensile strength of a new aluminum alloy or the scrubbing speed of a milling technology. Regarding these parameters, other possible outcome scenarios – due to different real options at certain decision-making points during the development process and due to external influence factors, that can effect the technology development progress – are identified and their occurrence probabilities are assessed.

Subsequently, the technology capability is estimated which indicates the superiority of the new technology in comparison to another existing state-of-the-art technology. This superiority is calculated based on the technology performance parameters.

Furthermore, the required resources to complete the technology development have to be displayed in the scenarios. Therefore, the real option approach is applied in order to predict possible cost scenarios and their occurrence probabilities. Real options may for example consist in the extension of the development budget in case of an unexpected positive intermediary development result. Contrariwise the project budget might also be reduced or activities can be put on hold when strived results could not be accomplished.

3.2.1.2. Commercialization phase

Commercialization Scenarios are generated in respect of three criteria: technology utility, technology market size and technology uniqueness. First of all, the utility of an employed technology can be expressed by eight basic effects as shown in Figure 3.

	Product-induced	**Process-induced**
Input (Resource Effect)	Economization (1) / Expansion (5)	Economization (2) / Expansion (6)
Output (Customer Effect)	Enhancement (3) / Impairment (7)	Enhancement (4) / Impairment (8)

Figure 3: Utility Effects of an Employed Technology.

For different technologies, those basic effects can appear in different combinations. In the following, some examples will clarify the idea.

A new technology focusing on the functionality of a product might enhance its customer effect (3 in Figure 3) but at the same time cause an expansion of necessary resources (5) as the product costs increase. If a new technology aims at the downgrading of a product to adjust it to lower customer needs, a technology might cause an impairment of the customer effect (7) but contribute to product (1) and production economization (2). A third potential case is that a new process technology enhances a product's customer effect (3), for example, due to a better surface quality, whereas at the same time an expansion of necessary process resources (6) is caused. To assess the technology utility, all possible effects have to be considered and monetary quantified. In order to predict customer effects, functional analysis of the product are applied to estimate the impact on the customer's appreciation of the implemented new technology as a function owner of the product. On the other hand, trade-off analysis (e.g., between the viscosity of a new casting material technology and processing times) estimates the resource effect of a new technology.

The technology market size reflects the amount of products and customers as well as the market cycle duration for the new technology to exert its utility. Both the technology utility (where is this defined?) as well as the technology market size have to be specified in terms of scenarios with respective occurrence probabilities and then to be combined to holistic commercialization scenarios.

In addition, the technology uniqueness has to be evaluated using the St. Gallen Escalation-Steps-Model (Müller-Stewens, Lechner, 2001). The question, to what extent the technology contributes to a sustainable unique selling point has to be answered. When the technology uniqueness cannot be fully guaranteed, the commercialization scenarios will be endued with a restriction factor.

3.2.2. *Technology readiness model*

In order to valuate technologies in early stages of their development, it becomes necessary to determine to what extent the technology is completed at the time of the valuation. Therefore, technology readiness levels are defined in the forefront of the technology development project.

If a certain level is accomplished at the time of the valuation–i.e., when certain level criteria are proven to be fulfilled–this level enters into the technology valuation with a respective Technology Readiness Factor (TRF) as shown in Figure 4.

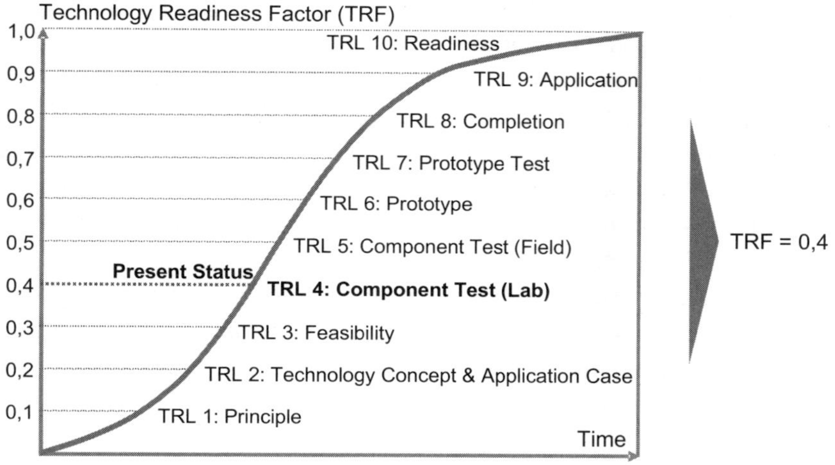

Figure 4: Determination of the Technology Readiness Factor (TRF).

3.2.3. *Expected cash flow model*

The Commercialization Scenarios and Development Scenarios provide information about expected periodical payments-in and payments-out. On basis of these information, the Expected Cash Flow (ECF) can be calculated and discounted per period. In this context it is necessary to discount the payments-in with a higher interest rate than the payments-out, due to the higher implicit risk. Interest rates will be determined using the Capital Asset Pricing Model. In compliance with the IFRS, expected cash flows are to be estimated for a time horizon of five years (Lienau and Zülch, 2006). For the subsequent years, an infinite series of annuities is assumed and determined by extrapolating the previous cash flows according to the Gordon-Growth-Model (Gordon and Shapiro, 1957).

3.2.4. *Prognosis adjustment model*

The estimation of expected cash flows will be based on scenarios, i.e. assumptions about the future. Those assumptions are by nature afflicted with errors, which can only be revealed by the course of time. Those errors shall be kept in mind and provide a learning effect for subsequent technology valuations: On the one hand, predicted scenarios have to be revised on basis of new learning. On the other hand, over- or underestimations of technology-implicated cash flows shall result in an immediate adjustment of future technology valuations. Thus, companies will also be prevented from manipulating their financial statement reporting by continuously over- or under-valuating their technology-based intangible assets.

The Prognosis Adjustment Factor (PAF) will record the prognosis errors of the past and thus adjust the current technology value in regard to those errors. In doing so, errors which have recently been committed, are weighted higher than such that are long ago.

3.2.5. *Technology value calculation*

Within the closing Technology Value Calculation the three components TRF, ECF and PAF are multiplied to assess the Technology Value in Use (TVU). Figure 5 shows conceivable curve progressions for the four parameters.

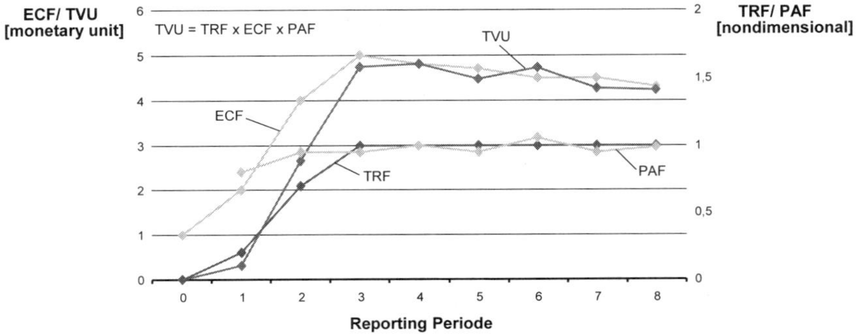

Figure 5: Exemplary Curve Progressions of Technology Value in Use (TVU), Technology Readiness Factor (TRF), Expected Cash Flow (ECF) and Prognosis Adjustment Factor (PAF).

TRF by nature increases until technology completion. After that, its value constantly equals "1". ECF also rapidly increases until technology completion due to the yearly reduction of future development costs. Subsequently it tends to decrease due to the decrease of commercialization time. PAF varies around "1" depending on if rather an over-estimation (PAF < 1) or an under-estimation (PAF > 1) has been dominant in the past.

4. Summary

The introduced concept of the Technology Balance methodology allows for the valuation of technology-related intangibles according to IASB's expected cash flow approach. Companies are enabled to determine and monitor the value of a technology along its whole life cycle, from early stage development until late market phase. By considering different scenarios and real options in the technology's development and market phase, the methodology incorporates the flexibility of technology decisions but also the dependence of a technology's real value on external factors.

By integrating an automated prognosis adjustment mechanism into the calculation on basis of empirical learning, estimation errors of the past are immediately taken into account for subsequent technology valuations.

All things considered, the Technology Balance methodology could act as a powerful tool for the purpose of value-based management. It enables companies' executives to keep track of their technology-related intangibles and to give investors a better view on their companies real value.

References

Aboody, D. and Lev, B. (2003). Information Asymmetry, R&D, and Insider Gains. In: Intangible Assets (Hand, J. R. M. and Lev, B., ds.), pp. 366–386, Oxford University Press, Oxford.

Andriessen, D. (2004). Making Sense of Intellectual Capital. Elsevier Butterworth-Heinemann, Oxford-Burlington.

Arthur D. Little (2005). Intellectual Capital Management and Reporting – Documentation of Survey Results. www.adl.com [05 February 2007].

Coenenberg, A.G. and Salfeld, R. (2007). Wertorientierte Unternehmensführung. Vom Strategieentwurf zur Implementierung. Schäffer-Poeschel Verlag, Stuttgart.

Deloitte (2004). IFRS 3: Unternehmenszusammenschlüsse sowie Änderungen der IAS 36 (Wertminderung von Vermögenswerten) und IAS 38 (Immaterielle Vermögenswerte). www.deloitte.com/de [28 October 2005].

Edvinson, L. and Malone, S. (1997). Intellectual Capital – Realizing your company's true value by finding its hidden brainpower. Harper Collins Publishers, New York.

FASB (2006). Statement of Financial Accounting Standards No. 157—Fair Value Measurements. Financial Accounting Series. Financial Accounting Foundation, Norwalk (USA).

Gausemeier, J., Ebbesmeyer, P. and Kallmeyer, F. (2001). Produktinnovation - Strategische Planung und Entwicklung der Produkte von morgen. Carl Hanser Verlag, München.

Gordon, M. J. and Shapiro, E. (1956). Capital Equipment Analysis: The Required Rate of Profit. *Management Science,* 3(1), pp. 102–110.

Günther, T., Kichner-Khairy, S. and Zurwehme, A. (2004). Measuring Intangible Resources for Managerial Accounting Purposes. In: Intangibles in der Unternehmenssteuerung (Horvath, P. and Möller, K., eds.). Verlag Vahlen, München.

Haig, D. and Perrier, R. (1997). Valuation of Trade Marks and Brand Names. In: Brand Valuation, 3rd edition (Perrier, R. and Stobart, P., eds.), pp. 25–34. Premier Books, London.

Högel, S., Hupp, O., Maul, K. H. and Sattler, H. (2002). Der Geldwert der Marke als Erfolgsfaktor für Marketing und Kommunikation. In: Der Geldwert der Marke. 7th special edition (Gesamtverband Kommunikationsagentur, ed.), pp. 37–80. Horizont Productions, Frankfurt am Main.

Homburg, C. and Daum, D. (2002). Marktorientiertes Kostenmanagement: Kosteneffizienz und Kundennähe verbinden. Frankfurter Allgemeine Zeitung Verlag, Frankfurt am Main.

IASB (2005). Framework.

IASB (2006). IAS/ IFRS

IDW (2004). Entwurf IDW Stellungnahme zur Rechnungslegung: Bewertungen bei der Abbildung von Unternehmenserwerben und bei Werthaltigkeitsprüfungen nach IFRS. Institut der deutschen Wirtschaftsprüfer e.V., Düsseldorf.

Jaruzelski, B., Dehoff, K. and Bordia, R. (2006). Smart Spenders: The Global Innovation 1000. *strategy+business*, issue 45, New York.

Kenton, C. (2005). Value Beyond the Balance Sheet. *BusinessWeek Online.* www.businessweek.com [01 February 2005].

Krimpmann, A. (2006). Fair Value – Welcher Wert ist fair? *VDMA Nachrichten*, issue 12/2006, pp. 29–31. VDMA Verlag, Frankfurt am Main.

Lev, B. (2001). Intangibles – Management, Measurement, and Reporting. Brookings Institution Press, Washington, D.C.

Lienau, A. and Zülch, H. (2006) Die Ermittlung des value in use nach IFRS. *Zeitschrift für internationale und kapitalmarktorientierte Rechnungslegung KoR*, issue 05/2006, pp. 319–329. Verlagsgruppe Handelsblatt, Düsseldorf.

Müller-Stewens, G. and Lechner, C. (2001). Strategisches Management. Schäffer-Poeschel Verlag, Stuttgart.

Razgaitis, R. (2003). Valuation and Pricing of Technology-based Intellectual Property. John Wiley & Sons Inc., Hoboken, New Jersey.

Smith, G.V. and Parr, R. L. (2005). Intellectual Property – Valuation, Exploitation and Infringement Damages. John Wiley & Sons Inc., Hoboken, New Jersey.

Stewart, T.A. (1997). Intellectual Capital: The New Wealth of Organizations. Doubleday/Currency, New York.

Strack, R. and Villis, U. (2001). Die nächste Generation im Shareholder Value Management. *Zeitschrift für Betriebswirtschaft*, issue 1/2001, pp. 67–84.

Von Scheffer, G. and Loop, D. (2005). Finanzierung mit Patenten: Patentbewertung für die Praxis. *Ifo Schnelldienst* , issue 7/2005, pp. 21–25.

Woodward, C. (2003). Valuation of Intellectual Property. PriceWaterhouseCoopers, London.

Section III

The Economic Value of Green Technologies and Sustainable Development

Chapter 8

A Modeling Framework for the Diffusion of Green Technologies

*Mitsutaka Matsumoto**, *Shinsuke Kondoh**, *Jun Fujimoto*** *and Keijiro Masui**

**Environmentally Conscious Design and Manufacturing Group*
Advanced Manufacturing Research Institute
National Institute of Advanced Industrial Science and Technology (AIST), Japan
***University of Tokyo, Japan*

We propose a modeling framework based on multi-agent models for analyzing the effects of environmental measures that depend on environmental technological innovations. In the present paper, we first outline existing approaches and clarify our objective in formulating the proposed framework. Previous major approaches have been 1) economic rationalistic models and 2) logistic curve models. We augmented logistic curve models by applying multi-agent models and setting the parameters of technological advancement and parameters of consumer preferences in the multi-agent framework. In addition, as an example of application of the proposed framework, we estimate the diffusion of hybrid electric vehicles (HV) in Japan until 2030 and the corresponding effect on CO2 emission levels.

1. Objective

Many environmental measures depend on technological developments, such as the development of energy efficient products and renewable energies. In such measures, in addition to technological development, we must promote the diffusion of such technologies.

Our research group has been creating a model to simulate and analyze the diffusion of green technologies. Figure 1 shows an image of our model. The model consists of the data of the target product, consumers, the market, and the social environment. The model simulates the diffusion of green technologies based on the above listed data and evaluates the effects of mitigating environmental load. Users can set scenarios, such as future changes in product prices and changes in tax systems. The model simulates diffusions under these scenarios.

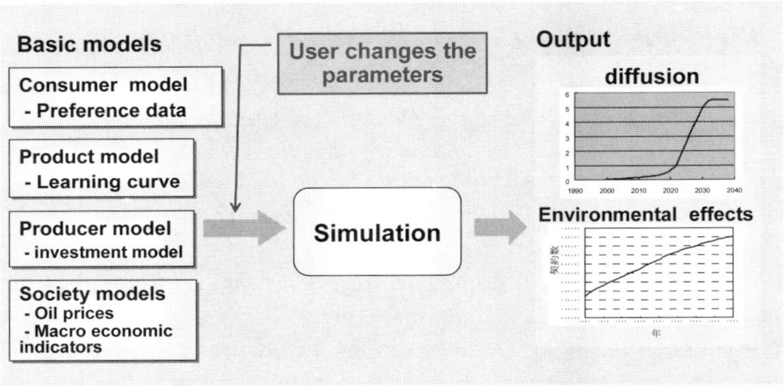

Figure 1: Outline of the model for diffusion analysis.

The clean energy vehicle (CEV) is an example of green technology. Clean energy vehicles include hybrid electric vehicles (HVs), fuel cell vehicles, electric vehicles, and diesel engine vehicles, which are more fuel efficient than existing normal gasoline vehicles (GV). In the present paper, we present an application of our model to the diffusion of CEVs in Japan. In Japan, HVs are gaining popular. From 1997 to 2005, a total of 258 thousand HVs were sold, which corresponds to approximately 0.5% of all automobiles (passenger cars) in Japan. Figure 2 outlines some features of HVs and GVs. Hybrid electric vehicles are more expensive with respect to initial price (approximately \$2,000~\$4,000), but have better mileage performance (70~100% better than GVs).

In creating our model, we ask the following questions:

To what extent and how rapidly could HVs be diffused in Japan in the future?

To what extent will HVs be diffused in 2010? What about in 2020? From the viewpoint of environmental policy, it is also important to ask: What effect does this diffusion have on reducing the CO_2 emissions of the country?

We also analyze factors influencing diffusion, which may include: 1) initial car price, 2) gasoline price, 3) mileage performance, and 4) consumer preferences. We simulate the influence of changes on diffusion.

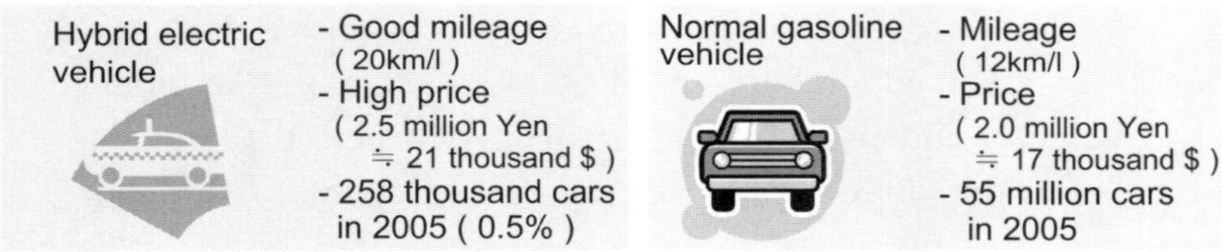

Figure 2: Features of HV (hybrid electric vehicles) and GV (gasoline vehicles) in Japan.

In developing the model, we focused on the consumer model. Product diffusion is influenced by a number of factors including consumer preferences, corporate strategies, technological development, governmental policies, and natural resource constraints. However, in many cases, it is adequate to assume that consumer preference is the most influential factor. Many of the existing diffusion models are based on the consumer model. In the following sections, we present existing diffusion models and then describe our diffusion model.

2. Existing Models

The diffusion of innovative technologies has been defined as the process by which that innovation "is communicated through certain channels over time among the members of a social system" (Rogers 1983). As such, the diffusion process consists of four key elements: innovation, communication channels, time, and the social system. The logistic curve model, or often called Bass model, is a major diffusion model. The model mainly focuses on communication channels, which are the means by which information about an innovation is transmitted to or within the social system. Among the diffusion models, we describe the economic rationalistic model which has also often been used in evaluating diffusion of green technologies, and the logistic curve model.

2.1. *Economic rationalistic model*

The economic rationalistic model assumes the economic rationality of consumers in purchasing. AIM end-use model (Kainuma, Matsuoka and Morita 2003) is a representative model in the approach. In the model, when there are some options in purchasing (e.g. either HV or GV), consumers are assumed to choose products whose sum of the initial costs (price of vehicle) and running costs in pay-back period (e.g. costs for gas, tax, etc., for three years) are more reasonable. Although the assumption is suitable when the purchaser is a corporation, it is not suitable to capture general consumer behaviors. For example, in the case of consumers' purchase of automobiles which we consider below, consumers consider, not only costs, but also designs, brand images, comfort, safety performance, and so on.

2.2. *Logistic curve model*

The logistic curve model is a major model for diffusions (Bass 1969). Temporal transition of diffusions often show *S*-shaped curves (Figure 3). In early stage of diffusion, small number of consumers adopts the product. As diffusion proceeds, more consumers adopt it. After spreading, diffusion saturates.

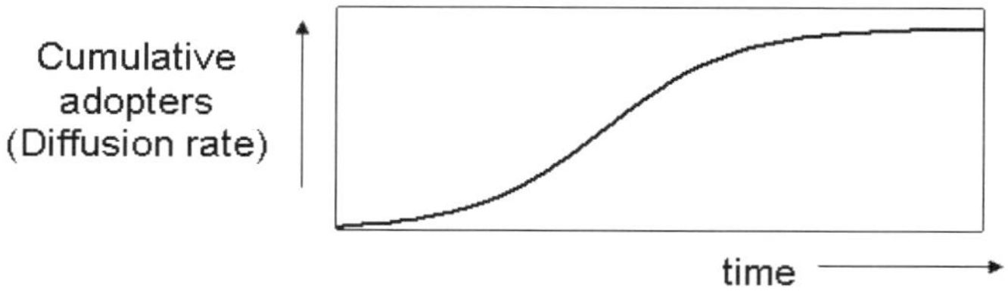

Figure 3: A logistic curve.

The process is formulated as follows (Mahajan and Peterson 1983):

$$\frac{dN(t)}{dt} = g(t)\left[\tilde{N} - N(t)\right] \tag{1.1}$$

where *N(t)* represents cumulative adopters in time *t*. $\tilde{N}$ represents number of potential adopters, that is sum of adopters and non-adopters in time *t*.

$$N(t) = \int n(t)dt \tag{1.2}$$

n(*t*) represents number of consumers who adopt in time *t* (not cumulative number). *dN*(*t*)/*dt* is diffusion pace in time *t*. *g*(*t*) is coefficient of diffusion. *g*(*t*) could be function of time or function of cumulative number of adopted consumers. When *g*(*t*) is a constant *a*, Eq. (1.1) is

$$\frac{dN(t)}{dt} = a\left[\tilde{N} - N(t)\right] \tag{1.3}$$

This model is called external influential model.

When *g*(*t*) is proportional to *N(t)*, the equation is

$$\frac{dN(t)}{dt} = bN(t)\left[\tilde{N} - N(t)\right] \tag{1.4}$$

This model is called internal influential model.

When *g(t)* is *a+bN(t)*, the model is called *mixed influential model.*

The *internal influential model* assumes that diffusions occur through communications between adopted consumers (*N(t)*) and non-adopted consumers ($\left[\tilde{N} - N(t)\right]$). The model is also called *pure imitation diffusion model.* Eq. (1.4) is

$$N(t) = \frac{\tilde{N}}{1 + \frac{\tilde{N} - N_0}{N_0}\exp\left(-b \cdot \tilde{N} \cdot (t - t_0)\right)} \tag{1.5}$$

The boundary condition is $N(t) = N_0$ for $t = t_0$. The equation represents logistic curves like Figure 3. Setting *a, b,* N_0 and $\tilde{N}$ determines the curve.

2.3. *Multi-agent model*

Multi-agent model is a framework in which many agents are set in a computer program and their decision rules are defined as micro level rules. The macro level phenomenon that is resulted from the micro-level rules is observed. An advantage of using multi-agent framework is the richness and flexibility of description, especially the description of micro-level rules. By using the framework in diffusion model, we could formulate consumer preferences in detail.

3. Model Formulation

The logistic curve model is a major diffusion model. However, in this model, the consumer preferences cannot be modeled in detail, because the diversity of consumer preferences is not modeled in the formulation. Accordingly, the influences of changes in factors such as initial price and mileage performance on HV diffusion could not be analyzed in the logistic curve model.

The procedure for analyzing innovation diffusion using multi-agent model is described in Figure 4.

We formulate agent-behaviour as follows. We define agents' preference as follows:

$$u_{ij} = \sum_{k} w_{i,k} \cdot t_{j,k} + \lambda_{i,j} \cdot share_j \tag{1.6}$$

where u_{ij} represents agent *i*'s preference for product *j*, i.e., $j \in$ {gasoline vehicle, hybrid electric vehicle}. *k* represents preference factors such as $k \in$ {price, design, performance, ...}. w_{ik} represents agent *i*'s weight with respect to *k*, referred to herein as "consumer parameters". $\lambda_{i,j}$ presents agent *i*'s weight with respect to market share ($share_j$) and is referred to herein as the "neighbour effect" term. $t_{j,k}$ represents the strength of product *j* with respect to *k*. Since the strength is affected by technological development, these terms are referred to "technology parameters." We define that agent *I* purchases product *J*, if $J = \arg\max u_{I,j}$. (Another method of formulation is to define agents as choosing a product with a probability function of u_{ij}.) We set the parameters using questionnaires completed by consumers and statistics of past transitions of diffusion. We performed scenario analysis by shifting values of parameters.

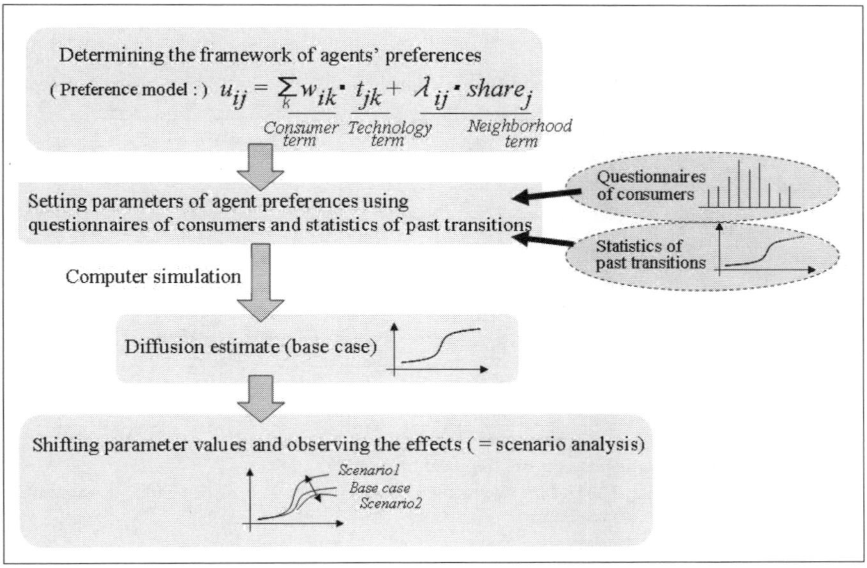

Figure 4: Analysis procedures.

4. Analysis of Diffusion of Clean Energy Vehicles Using the Proposed Model

As an example of diffusion analysis using the proposed framework, we present our analysis of the diffusion of CEVs in Japan. Estimating the effects of CEV diffusion on the abatement of CO_2 emissions in a country is important to environmental policies. We are presently creating a model to simulate the diffusion of CEVs and CO_2 emissions of the automobile-transportation-sector in Japan. Clean energy vehicles include hybrid electric vehicles (HVs), diesel engine vehicles, fuel-cell vehicles, and electric vehicles. Since the majority of CEVs in Japan for coming decades will be HVs, we considered the diffusion of HVs.

In Japan, the fist HV was released by Toyota Co. in 1997. By the end of 2005, 258 thousand HVs had been sold (Table 1). Since the number of automobiles in Japan is approximately 55 million, HVs presently make up 0.5% of all automobiles. We analyzed the future diffusion of HVs in Japan following the procedure presented in the previous section.

Table 1: Diffusion of Hybrid electric vehicles in Japan (1997-2005) in thousand of units.

Year	Number of cars sold (*1)	Cumulative sales (*2)
1997	-	3.7
1998	-	22.5
1999	-	37.4
2000	12.9	50.4
2001	25.1	74.6
2002	15.5	91.0
2003	42.2	133.2
2004	65.3	198.4
2005	59.8	258.2

Source: *1 JAMA (Japan Automobile Manufacturers Association, In.c.)
http://www.jama.or.jp/eco/eco_car/shipment/index.html
*2 (Year 1997-2002) JAMA http://www.meti.go.jp/report/downloadfiles/g40224b100j.pdf (p.8)

4.1. *Model setting*

We set the model as follows.

Step 1 Setting the consumer agents
There were 55 million automobiles in Japan in 2004. We set 55 thousand automobile-user agents, which means that we set 1/1000 miniature automobile-user society. For simplicity, we assumed that the total number of automobile does not change for the future. We assumed that automobile-user-agents purchase new cars every 10 years, which is automobile's average life-span in Japan. Agents determine whether they purchase HV or GV (gasoline vehicles).

Step 2 Setting the consumer preferences
In purchase of automobiles, consumers consider automobile's performances such as traveling performance, safety performance, riding

comfort, designs, gasoline mileage, price, eco-image; and also they are affected by other consumers. The influence from other consumers is the basis of logistic curve model. Among the performance factors, difference between HV and GV appears in price, mileage and eco-image. We defined agents' consumer-preference-function with these three factors and a neighbor-effect-factor. A neighbor-effect-factor represents the influence of adopters to non-adopters on motivating purchase of new products (HV). We define u_{ij} (consumer i's preference to automobile j (j is HV or GV)) as follows.

$$u_{ij} = w_{i,price} \cdot t_{j,price} + w_{i,mileage} \cdot t_{j,mileage} + w_{i,eco} \cdot t_{j,eco} + \lambda_{i,j} \cdot share_j \quad (1.7)$$

where $w_{i,price}$ represents agent i's weight of price on automobile preferences, and $t_{j,price}$ represents strength of j in price factor. $w_{i,eco}$ represents agent i's environmental consciousness and $t_{j,eco}$ represents environmental performance of j. $\lambda_{i,j}$ is weight of neighbour-effects for agent i in j. In our current model, we set $\lambda_{i,j}$ to be 0 for j=GV, because we regard that reputation are influencing only for new technologies. We define $share_j$ as market share of j. Defining that agent i's preference for HV to be $u_{i,HV}$ and preference for GV to be $u_{i,GV}$, agent i purchases HV when

$$u_{i,HV} - u_{i,GV} > thres \quad (1.8)$$

and otherwise, agent i purchases GV. The constant '*thres*' is the threshold. We set the value of *thres* so that the diffusion of HV in model is consistent with the actual diffusion of HV in Japanese automobile market from 1999 to 2004.

Step 3 Setting the technology parameters ($t_{j,k}$)
$t_{j,k}$ is technology parameter. In the current model, we set $t_{j,k}$ to be either 1 or 0 --- $t_{j1,k}$ =1 and $t_{j2,k}$ =0 if j_1 is stronger than j_2 with respect to k. The values are in Table 2.

Table 2: Technology parameters ($t_{j,k}$).

j \ k	price	mileage	eco-image
GV	1	0	0
HV	0	1	1

Step 4 Setting the consumer parameters ($w_{i,k}$)
We set the distribution of consumer preferences, or consumers' weights for price, gasoline mileage and environmental consciousness. We set the parameters using the result of our questionnaire survey.

i) Base scenario

We carried out a questionnaire survey to ask consumers the weights of the factors on their purchase of automobiles. We obtained answers from 1,070 respondents. Table 3 shows a part of the results.

Table 3: Weights of consumer preferences.

	Very important	Important	Fairly important	Not so important	Total
weights	5	3	1	0.1	
Price	577	393	82	17	1,070
Mileage	348	420	235	67	1,070
Eco-image	128	403	383	156	1,070

We set the weight values 5, 3, 1 and 0.1 to the answers "very important", "important", "fairly important" and "not so important". We set the values of w_{ij} for 55,000 agents in the model at the same rate with Table 1.3. For example, (577/1070)*55000 agents' $w_{i,price}$ is 5, (393/1070)*55000 agents' $w_{i,\, price}$ is 3, and so on.

ii) Scenarios

We set scenarios of rise of oil prices. Since HVs are better in gasoline mileage than GVs, rise of oil prices promotes diffusion of HV. Oil prices rise more than twice in past six years (2000~2006). We estimated the effect of oil price rise. We set scenarios in which oil price rises by 1.7 times and 3.4 times, respectively, from 1998 to 2030. We set the scenarios by rising the weights of "mileage" of all consumer agents' preferences (wi ,mileage) by 1.7 times and 3.4 times.

Step 5 Consumer preferences in the scenarios

We calculate the consumer agents' preferences in scenarios using the technology parameters (set in Step 3) and the consumer parameters (set in Step 4). Distribution of consumer preferences follows.

i) Base scenario

Figure 5 presents the distribution of consumer agents' preference (without neighbour effects) in base scenario, that is [($w_{i,\ price}\ t_{j,price} + w_{i,\ mileage}\ t_{j,\ mileage} + w_{i,eco}\ t_{j,\ eco}$) for *j*=GV, minus ($w_{i,\ price}\ t_{j,price} + w_{i,\ mileage}\ t_{j,\ mileage} + w_{i,eco}\ t_{j,\ eco}$) for *j*=HV] with the parameters described in Table 1.1 and Table 1.2. The mean and variance of distribution in Figure 5 are -0.8 and 2.6, respectively. We approximated the distribution with normal distribution.

We assumed that potential HV adopters are the agents whose preference value (in Figure 5) is negative. The ratio is 65.1%. We assumed that the number of HVs owned in Japan in future is 36 million (65.1% of 55 million).

We set the parameters λ_j for HV in Eq. (1.7) and *thres* in Ineq. (1.8) so that simulated diffusions are consistent with the actual diffusion of HVs in Japan from 1999 to 2004. The calculated values were λ_j=140 and *thres* =7.

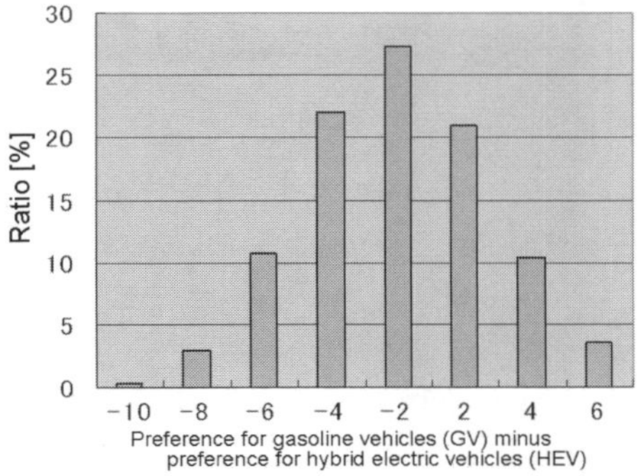

Figure 5: Distribution of preferences.

ii) Oil price rise scenarios

Figures 6 and 7 present the distribution of consumer preferences when oil price rises by 1.7 times and 3.4 times, respectively. Ratios of consumer agents who prefer HV to GV are 77% and 89%, respectively (65.1% in base scenario).

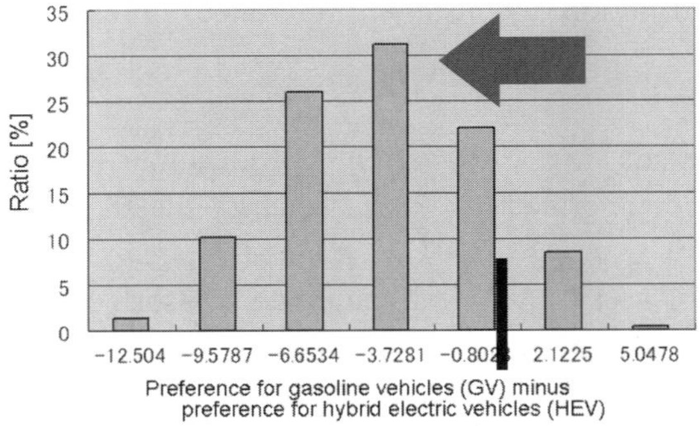

Figure 6: Distribution of preferences.

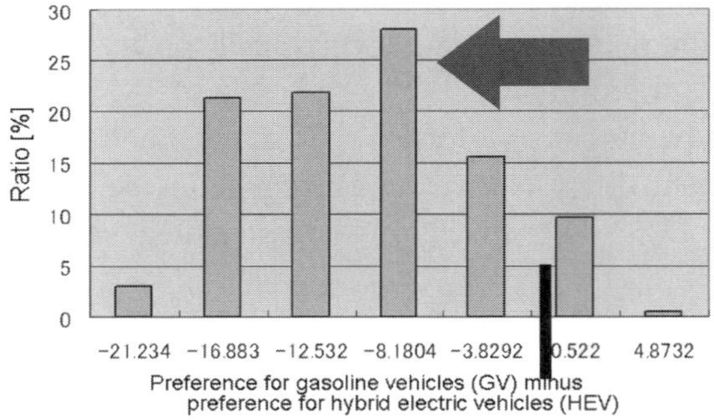

Figure 7: Distribution of preferences when oil price rises by 3.4 times.

4.2. *Simulation results*

i) Base case scenario

Setting parameters as above and following the procedure in Figure 8, we carried out a simulation. Table 4 and Figure 9 (the curve in the lowest) present the result for base scenario. The estimated diffusion during 1999 and 2004 is different from the actual diffusion during the term, because the actual diffusion curve does not strictly follow a logistic curve. The diffusions of HVs in 2010, 2020 and 2030 were estimated to be 1.3, 33.1 and 36.0 millions, respectively. To estimate the effect of the abatement of CO2 emissions, we assumed that the ratio of gasoline mileage of HVs and GVs is 1.48:1 on average. The abatement effect includes only automobile uses by consumers (it does not include business use and cargo transports). In 2030, 36 million consumers own HVs and the effect of the abatement of CO2 emissions is 27 million ton-CO2, which is 2% of total CO2 emissions in Japan in 2000.

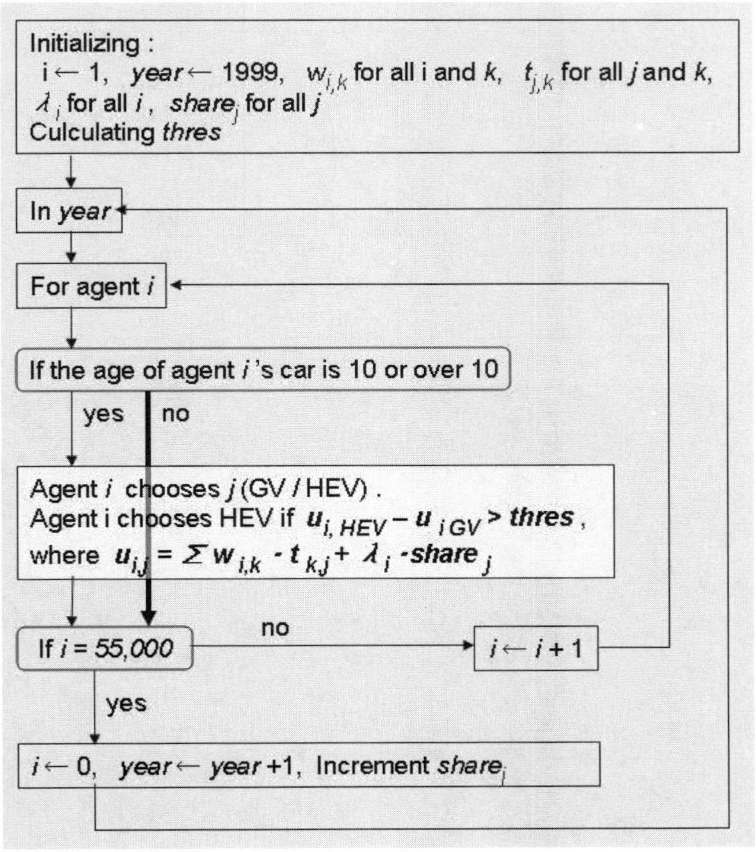

Figure 8: Simulation flowchart.

ii) Oil price rise scenarios

We set scenarios in which the consumer agents' preference weights for gasoline mileage ($w_{i,\ mileage}$) rise to 1.7 times (and 3.4 times) from 1999 to 2030 linearly. Figure 9 presents the results. Eventual diffusion number of HV increased by 1.2 times and 1.4 times for oil price rise by 1.7 times and 3.4 times, respectively.

Table 4: Estimated diffusion of HV (unit: thousand) and effects of abatement of CO_2 emissions (unit: 10 thousand ton-CO_2).

year	Actual diffusion	Estimate diffusion	Abatement of CO_2 emissions
1999	37	23	1.7
2000	50	56	4.3
2001	75	88	6.7
2002	91	120	9.1
2003	133	162	12.3
2004	197	218	16.6
2010	-	1,269	96.4
2020	-	33,082	2514.2
2030	-	36,000	2736.0

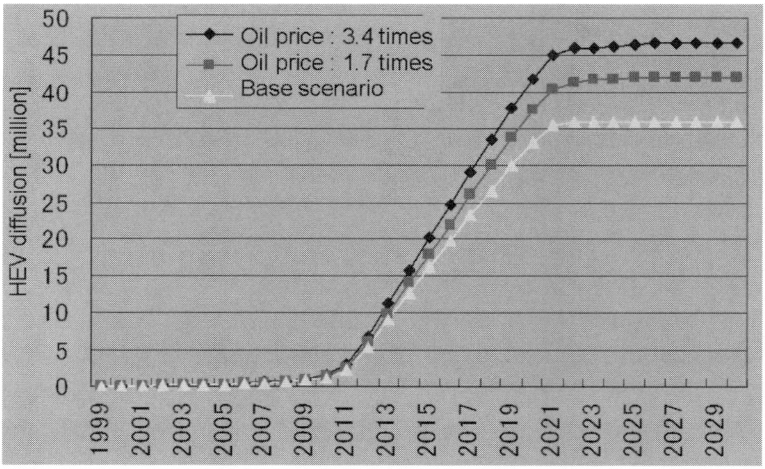

Figure 9: Diffusion curves of HV.

5. Conclusion and Future Research

We proposed a framework to estimate and analyze the diffusion of innovative technologies. The main approaches used to estimate the effects of innovation diffusion on national CO_2 abatement have been the economic rationalistic model and the logistic curve model. Neither model could describe "technological development" or "consumer preferences" explicitly. We used a multi-agent modeling framework and defined the agents' preferences as Function (1.6). With this formulation, both "technological development" and "consumer preferences" are explicitly described. Therefore, the effects of the development of specific technologies in products, or the effects of changes in consumer preferences, on the diffusion of products are analyzed. In addition, the "variety of consumer preferences" is represented in the model. In the economic rationalistic model, for example, consumer preferences are defined uniformly. The multi-agent model can reflect differences in preferences by considering the questionnaire data in the model.

Our current and future research areas include:

1) refining consumer preference parameters
2) refining technological development parameters
3) formulating suppliers (producers) in the model
4) formulating policies in the model
5) formulating interactions among parameters

First, in the example of HVs, we set consumer preference parameters ($w_{i,k}$) to be 5, 3, 1 and 0.1, which correspond to the questionnaire answers. The parameters are weights assigned by consumers to factors (price, mileage, and eco-image). The weights could be defined more elaborately. We could use marketing analysis methods such as conjoint analysis. We then set technological development parameters as shown in Table 1.1. The parameters should be set more accurately by quantifying the factors (price, mileage and eco-image) based on actual performances. In addition, the temporal changes of the factors should be modeled. Next, we input suppliers into the model. The diffusion of products (in the case of HVs) depends on the marketing strategies of suppliers as well as consumer preferences. The framework of modeling governmental policies such as research investment, preferential treatment, and taxation, should also be provided. Finally, formulation of the interactions among parameters in the model is an area for future study. For example, technological development is motivated by product diffusion, or sometimes by consumer preferences, which indicates that technological development parameters ($t_{j,k}$) are functions of diffusion rates ($share_j$) or consumer preferences ($w_{i,k}$).

Acknowledgement

The work was financially supported by Ministry of Economy, Trade and Industry (METI), Japan.

References

Bass, F. M. (1969). A new product growth model for consumer durables, *Management Science*, 15, pp. 215–227.

Hafele, W. (eds.) (1981). Energy in a Finite World. Ballinger Pub. Co., Cambridge.

Kainuma, M, Matsuoka, Y. and Morita, T. (eds.) (2003). Climate Policy Assessment: Asia-Pacific Integrated Modeling, Springer-Verlag, Tokyo.

Mahajan, V, Muller, E. and Wind Y. (eds.) (2000). New Product Diffusion Models. Kluwer Academic Publishers, Boston.

Mahajan, V and Peterson, R.A. (1985). Models for Innovation Diffusion. Sage University Press, Beverly Hills.

Rogers, E. M. (1983). Diffusion of Innovations. 3rd ed. The Free Press, New York.

Tsuchiya, H. (1999). Learning curve cost analysis for model building of renewable energy in Japan, Experience curve for policy making. *Proc. the IEA Workshop*, pp 10–11, Stuttgart, Germany.

Chapter 9

A Green Operations Framework and Its Application in the Automotive Industry

Breno Nunes and David Bennett

Aston University, United Kingdom

This chapter describes a framework for Green Operations to improve the environmental decisions of organizations in the automotive industry. The model draws on three major fields of research: environmental management, operations management, and automotive production. The framework should lead to a better understanding of environmental issues among researchers and practitioners that are interested in sustainability, especially for the automotive industry. Its relevance derives from the combination of concepts, strategies and knowledge of sustainability and the automotive industry to provide an integrated approach to sustainability in car makers' operations.

1. Introduction

This chapter proposes a framework that unites the various environmental practices for operations with the goal of achieving a sustainable competitive advantage. Some decision makers, within both management and engineering, argue that it is extremely difficult to ascertain today whether a product could be considered environmentally friendly in ten years' time (Mildenberger and Khare 2000). Nevertheless, the question for companies has become not whether to commit to a strong environmental, health, and safety record, but how to do so in the most cost-effective manner (Kleindorfer et al., 2005).

Wells and Orsato (2005) point out that vehicle manufacturers have responded to regulatory and market pressures, but the technological paradigm of car design and manufacture substantially limits the alternatives available to them. The authors also affirm that most of actions of manufacturers have been designed to reduce costs by reaching greater economies of scale in car production and sales. As a result, most of the gains in efficiency and fuel economy were offset by the increase of the car fleet and engine power. Complex products like automobiles need a broad discussion of their environmental impacts. For instance, car use has received the most attention due to its large consumption of fossil fuels and, therefore, greenhouse gas emissions. Moreover, other environmental impacts from car use are congestion, noise, deaths from accidents and the disposal of used cars. In manufacturing, the use of raw materials, contributes to the depletion of natural resources while painting ingredients are a significant source health hazards. Furthermore, energy used to transport components and final products to and from large assembly plants are also discussed by some authors. The Green Operations approach can help address all these concerns by including environmental awareness in operations management.

2. Green Operations

Among all functions of a manufacturer (i.e., finance, marketing and operations), operations has the greatest environmental impact because it is the main source of harmful products. Therefore, it is most important to find opportunities to reduce the environmental burdens from the organisation's activities through environmental management programmes (Gupta, 1995). Life-Cycle Analysis (LCA) is a technique that represents a holistic approach to avoid the transfer of environmental impact from one phase of the life-cycle (of a product or process) to another. Nevertheless, while in principle it is a straightforward concept, the life-cycle approach can be swamped by complexity and the minutiae of data requirements and collection for an accurate analysis (Orsato and Wells, 2007). Even so, there is a coherence behind seeking to identify all aspects of the impact of a product, not least to ensure that 'improvements' made in one area do not conflict with ones in others.

The discussion about how to integrate environmental issues into the company's activities needs continuous reflection. An important approach is to consider Environmental Management as a strategic tool. In this way, at a corporate level, the environmental SWOT analysis can help to identify external threats (e.g., competitors gain market shares with green products) and opportunities (e.g., offering an "environmental-friendly" product), and relate them to internal strengths (e.g., research and development capabilities for "clean" processes and "green" products) and weaknesses (e.g., hazardous wastes) of a company (Gupta and Sharma, 1996).

Identifying environmental threats and opportunities or setting sustainability challenges may contribute to the development of a sustainable competitive advantage for a company. It is important to note that there already exist some practices that might facilitate the company's path towards a sustainable future. The following sections provide a brief explanation of these practices: Green Buildings, Eco-Design, Green Supply Chains, Green Manufacturing, and Reverse Logistics.

3. Practices for Green Operations

3.1. *Green buildings*

Environmental issues should be taken into account from the very beginning of the operations cycle, i.e., from the planning and construction of an industrial plant.

Paumgartten (2003) presents figures about the USA, where non-residential buildings are responsible for 30-40% of all the nation's energy consumption, 30-40% of atmospheric emissions, 60% of all electricity use, 25% of all water use, 35-40% of the municipal solid waste stream, 25-30% of all use of wood and materials. Analysing the whole life of a building, construction costs account for only 11% of its total cost, whereas operation and alteration costs amount to 75%. The core idea of Green Buildings is to reduce operational costs through better environmental performance. For example, the criteria used by the "LEED" methodology of the US Green Building Council to assess environmental performance of a building consist of the sustainability of the site (e.g., using green or brown fields, proximity of markets or suppliers, etc.), water efficiency (e.g., use of rainwater), energy use (e.g., intelligent control systems.), resources and materials (e.g., use of certified wood, etc), indoor environmental quality (e.g., low emitting materials of volatile organic compounds) and finally, innovation and design process (e.g., use of innovative techniques). Ries et al. (2006) affirm that there are five major areas of improvement: (1) gains in worker productivity, (2) reductions in health and safety costs, (3) improvements in indoor environmental quality, (4) reduction in maintenance costs and, (5) energy and water savings.

This practice is also important to non-manufacturing activities (sales centres, offices, etc.) where energy and water may be the main environmental concerns. Another concern related to companies' facilities is the end-of-life of their buildings and the residual financial value.

3.2. *Eco-design*

This practice refers to the environmental concerns in product or process design and, as such, it influences the entire life-cycle of a product. For example, Mildenberger and Khare (2000) say about the long decision lead time in the automotive industry: "According to the German car-maker BMW, it takes about 3-4 years to design a car, 7-8 years to manufacture it, and it would be in use for about 10-12 years; thus, in all, a decision taken today will have its effect for about a quarter of a century if it is not victim of irresponsible disposal of waste." Moreover, product design is often complicated by uncertainty inherent in the evolution of environmental trends and regulations (Kleindorfer et al., 2005).

Among characteristics like functionality, product safety, comfort, efficiency and aesthetics, R&D teams are also required to consider the impact of the product/process on the natural environment. Karlsson and Luttropp (2006) argue that Eco-design also ought to include concepts such as sustainable consumption, reduction of the volume of "desire" and aims to enable human satisfaction in concert with a positive role in sustainable product development. As an example, power and speed have been differentiators for some car automakers, although their customers are not usually allowed to drive their vehicles at high speeds due to the traffic law.

For Sarkis (1998) Eco-design has a number of functional sub-components: design for reusability, recyclability and remanufacturing; design for disassembly; and design for disposal. To achieve successful implementation of Eco-design practices it is essential to have a combination of customer needs and desires with respect to the environment, aimed at bringing satisfaction and functionality. Last, but not least, communication throughout the organisations' functions (marketing, engineering, etc.), avoiding concentration and isolation of R&D teams was noted by Boks (2006). The interaction, communication and commitment of people outside R&D can avoid a non-feasible idea from going to implementation causing further and higher costs.

3.3. *Green supply chains*

The green supply chains (GSC) concept brings environmental and sustainability issues to the buyer-supplier relationship. According to Gilbert (2001), greening the supply chain is the process of incorporating environmental criteria or concerns into organizational purchasing decisions and long-term relationships with suppliers.

Between 1996 and 2002, the cost of materials bought by the 100 biggest US manufacturers increased by almost 12% according to Purchasing Magazine's estimates (Liker and Choi, 2004). Moreover, in 2001, the Malcolm Baldrige National Quality Award Committee made "key supplier and customer partnering and communication mechanisms" a separate category on which it would judge the best companies in the United States. These facts show how important it is to promote a solid and sincere relationship with suppliers to build competitive advantages. Indeed, there are three approaches involved in the development of a GSC: environmental, strategic and logistical (Nunes, 2004). Considering the environmental sphere, a company could consider the use of cleaner fuels for transportation, green purchasing and also transfer of environmental technologies to suppliers. As GSC also involves long term relationships with suppliers, there is a strategic approach involving the selection of suppliers, partners, and the knowledge or technology to be transferred. These are strategic issues that might affect the factors of adaptability, agility and alignment of objectives in the supply chain. The third approach relates to logistics. Thus, GSC would also need to be considered taking into account costs, feasibility to move, store and distribute material, personal training, etc.

Gilbert (2001) supports the idea of two categories of initiative to stimulate greener supply chains. The first involves better coordination with suppliers on environmental efforts to enable development of greener or more environmentally friendly products. The second is demanding improved environmental performance at suppliers' operating facilities, such as the requirement to obtain ISO 14000 certification or achieve a set standard of performance. In the US, the Environmental Protection Agency (EPA) shows that proactive management of supplier environmental performance can lead to product and process simplification, more efficient resource utilization, quality improvement, liability avoidance, and enhanced leadership image (EPA, 2000).

3.4. *Green manufacturing*

For the countries of the OECD (Organisation for Economic Cooperation and Development) manufacturing accounts for 40% of sulphur dioxide emissions (the precursor of acid rain), 60% of water pollution, 75% of non-hazardous waste and 90% of hazardous waste (OECD, 1995). As a result, research spanning the engineering, natural sciences, public policy, economics and business literature has proposed various strategies and actions to improve the environmental performance of manufacturing (Jones and Klassen, 2001).

The concept of green manufacturing is based on the philosophy of industrial ecology. The basic aims of green manufacturing are to increase efficiency continuously and therefore, "Reducing" the use of inputs through a reduction of waste. "Re-use, "Re-manufacturing" and "Recycling" complete the fours "Rs" of industrial ecology. It is important to highlight the level of energy required to reduce, re-use, re-manufacture and recycle. "Reducing" is the most important practice from the point of view of energy conservation. On the other hand, Re-using, Re-manufacturing and Recycling practices are important to avoid the final disposal of valuable or harmful components as landfill. In addition to this philosophy of 4 Rs, a Green Manufacturing policy must consider regulatory compliance and employee healthy and safety.

Major benefits are basically: cost reduction, energy and water conservation, minimization of overall output of waste, and in terms of financial, ecological and health costs. Moreover, it is also possible to identify new products and business opportunities through the use of wasted materials (Martin, 2001). Total Quality Environmental Management and Environmental Management Systems may play a very important role in the establishment, implementation and monitoring the environmental performance of manufacturing function.

3.5. *Reverse logistics*

Reverse Logistics is considered in many publications as part of green supply chain management but in this work it is treated as a separate tool from green supply chain practices. The main reason for this is because different skills and ways of thinking are needed to run a programme of Reverse Logistics. De Brito (2003) noted that traditional logistics uses "forward" thinking; nevertheless, product recovery approach is a vital condition for the future. Fleischmann et al. (1997) give some examples of how several countries have enforced environmental legislation, charging producers with responsibility for the whole product life cycle. Nevertheless, companies can make Reverse Logistics a profitable activity by recognising the value of some components in the products discarded at their end-of-life, by avoiding landfill costs or environmental liabilities because of fines. Shrivastava (1995) note that in some end-of-life items, such as used tyres, 95 percent of what is discarded as waste is usable energy. Similarly, discarded automobiles have many reusable components and materials, but they are simply scrapped because currently it is too expensive for them to be recovered. Thus, a set of criteria is needed to identify what can be reused. Fleischmann et al. (1997) suggest a number of criteria based on (1) reuse motivation, (2) type of recovered items, (3) form of reuse, and (4) involved actors.

According to Fleischmann et al. (2000), product recovery networks comprise the following functions: (1) collection; (2) inspection/ separation; (3) re-processing; (4) disposal and, (5) re-distribution. Dowlatshahi (2005) considers the strategic factors for Reverse Logistics, explaining the implications of costs, quality, customer services, environmental concerns and finally, political/legal concerns. For instance, does the cost of Reverse Logistics increases significantly the overall cost of the products? What are the costs of not doing it? How does an illegal action affect the company's image and financial performance? Moreover, is the customer keen to buy a product that has reused components? All of these questions show how Reverse Logistics must be addressed as a strategic weapon.

Therefore, a Reverse Logistics system should take into consideration the internal processes of an organisation. Better integration of vendors and after sales policies will be required. For example, most automakers have no contact with their product after the warranty services finish, so one strategy adopted by Toyota has been to extend the warranty time. The results were that the customers felt more confident about their vehicles and Toyota sold more after sales services.

Like other Green Operations practices, Reverse Logistics is also strongly influenced by the product design. The more product and process development include valuable components in their composition, the more Reverse Logistics can become a less cost-sensitive activity, i.e., the final disposal having a high value; reuse, remanufacture and recycle would become a viable alternative as they could pay off the costs of collecting, separating and treating scrap instead of buying new raw materials.

4. A Generic Framework for Green Operations

This section presents a generic framework for Green Operations, providing the idea that there are green practices that could be included in every phase of the operations function.

Figure 1 shows the activities of the operations functions organised into six main blocks: production capacity (planning), process and product development, supplier relationship (and selection), production, in-bound and out-bound logistics, and after sales.

The activities of operations appear in the centre of the framework, where the Green Operations practices also reside. They are represented by "existing environmental practices for manufacturing". The box labelled "innovation" shows that Green Operations may require the development of new practices, technologies and production systems.

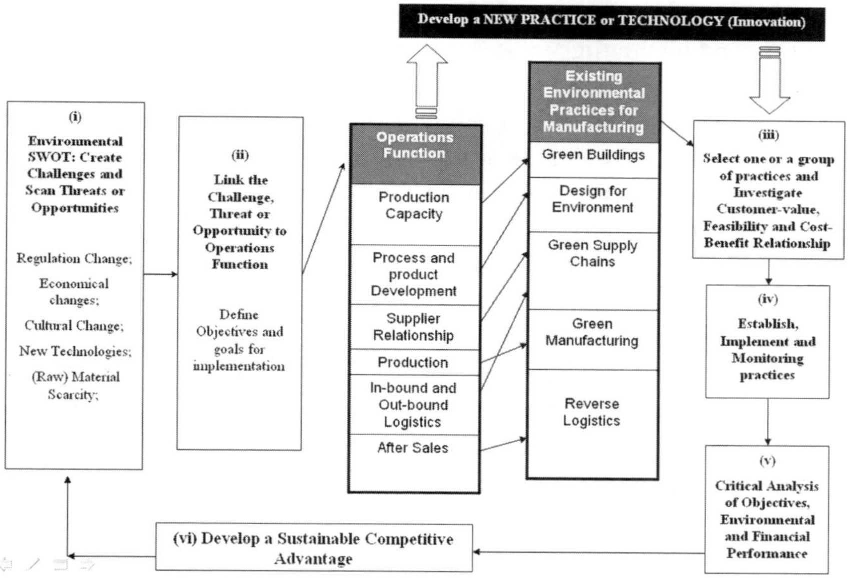

Figure 1: General framework for Green Operations.

There are two basic strategies: incremental innovation through successive improvements of the current production system or more extensive changes to the production system. In each case, there is a need to analyse the impact of the proposed production system on the company's environmental performance.

As well as analyzing the impact of the proposed changes in the production systems, it is fundamental for successful development of the framework and the implementation of the Green Operations Practices (GOPs) to find out the effectiveness of this framework and the impact of the 5 GOPs on the company's environmental performance, which does not have any clear criteria yet. As an example, Rothenberg et al. (2005)

suggest an Environmental Benchmarking for automakers, which involves four categories: regulatory compliance, gross emissions, efficiency and life-cycle analysis. Therefore, an analysis of the contribution of GOPs to each of those four categories could help companies to take greater advantage of this current framework and the future model that will be developed from it.

The framework seeks a dual benefit by using a strategy technique (SWOT) and integrating environmental issues in the corporative strategic agenda. Using an environmental SWOT may enable the company to have better results from the investments of its environmental policy. On the other hand, understanding the green opportunities and practices related to the broad processes of operations may motivate the top administration to build up a more sustainable corporate strategy.

Step (i) is to run an environmental SWOT analysis. In this phase the company should run sessions to identify its strengths and weaknesses regarding environmental issues as well as possible threats and opportunities. Taking into consideration the life-cycle approach is important to visualise external factors that might affect the company. By using the environmental SWOT companies can contextualise the framework and green their operations based upon local and global scenarios and align the environmental business practices to the corporate and operations strategy.

Step (ii) is to link the challenges, threat or opportunities to the operations function. Managers need to associate the most important and urgent challenges, threats and opportunities to the value chain of their organisations. According to their processes and internal analysis (strengths and weaknesses), they can set up environmental objectives and goals, methods for implementation and measurement.

Step (iii) involves the selection of one, or a group, of practices and the investigation of customer-value, feasibility and/or cost-benefit relationships: At this time, it must be assured that the environmental objectives and goals have enough support from top administration, the employees or suppliers are competent to implement and monitor and, finally, the forecast addresses for a positive cost/benefit relationship. The value of these actions for customers is measured;

however, even if there is no customer support, some of the practices should be implemented for avoiding future problems with regulation.

Step (iv) is to establish, implement and monitor practices related to daily procedures and instructions of those green operations practices. Having an environmental management system is recommended as well as running periodic PDCA (plan; do; check; act) cycles. These may enhance the chances of continuous improvement; which is particularly important after radical innovations.

Step (v) is the critical analysis of objectives, environmental and financial performance. The definition of indicators and benchmarking methods to analyse the objectives, environmental and financial performance is essential to organisational learning in environmental management. The benefits sometimes can come from a different area where the practice was implemented (e.g., design for eco-efficiency: implementation is in design, however, the results are in the product use). Therefore, detailed and systemic analysis should be combined to a better assessment of the organisation's progress towards the development of a sustainable competitive advantage.

Finally, step (vi) is to promote the development of a sustainable competitive advantage, which consists of achieving a privileged position in the long-term, ahead of competitors, without damaging the ethical image of the company.

The next section provides a contextualisation of the Green Operations framework for the automotive industry.

5. Application of the Green Operations Framework in the Automotive Industry

The application of the framework to the automotive industry follows below. It is important to highlight that item 5.1 employs the Environmental SWOT, associating the threats to the automotive industry to the broad processes of a company's operations function. Moreover, the current example of the framework application considers the end-of-life of the vehicles as the most urgent and important threat to take actions against. If these premises change it could affect the following steps: the link to operations function, investigation of customer-value, feasibility and cost-benefit relationship, monitoring, the critical analysis, and the development of a sustainable advantage.

5.1. *Environmental SWOT and its link to the operations function (Steps i and ii)*

For the purpose of this application of the framework, only the threats and their respective implications have been taken into account. Even though the challenges and opportunities may vary according to the local context, nevertheless, it is possible to think of some global threats that are likely to affect the car industry world wide.

(1) Zero Emission Vehicle Regulation: California is leading the drive towards zero emission regulation. Should the policy be followed in other places, companies that are already engaged in using cleaner fuels would have an advantage. The implications for the operations function would be in the design and manufacturing activities. The supplier relationship could be also affected; however, the predominant concerns are focused on the ability to produce hybrid or electric cars.

(2) Scarcity or price increase for oil and raw materials: "Given an estimate that China will produce over 6 million vehicle units in 2005, it is expected that there will be significant increases in imports of metals. Moreover, the large transportation system in China is based on gasoline and diesel fuels, which would dramatically increase China's dependence on oil imports. "A sobering fact is that if China's vehicles per capita were the same as the United States, the oil demand in China would exceed worldwide oil production by 18%" (Zhu et al. 2007).

The use of different material. (plastics or magnesium) may have a positive impact on companies' environmental and financial performance. Moreover, a shortage of oil could lead to a price increase, and therefore would make other alternative fuels viable. The use of renewable fuels such as ethanol or biodiesel will probably be related to the strategy for the national energy matrix (e.g., flexible fleet fuelling and biodiesel from mamona seeds in Brazil). The environmental benefits of using electricity or hydrogen to run cars is dependent on the way they will be generated. Besides product design, manufacturers will need to produce reliable cars using these new technologies; thus, manufacturing should also be on the environmental strategy agenda.

(3) Tighter regulations on gross plant emissions: Some companies have moved their operations to developing countries for various reasons such as low labour costs, fiscal incentives and less strict environmental regulations. However, developing countries are starting to increase the pressure to reduce gross emissions from manufacturing plants. Therefore, the impact might be on the construction and operation of these plants.

(4) Final Disposal Regulation: Another forthcoming threat for the automotive industry arises from the final disposal of cars. There is increasing pressure on car automakers to assume responsibility for their scrapped vehicles. In this case, the threat links to their after sales activities; thus, a structure for disassembly car will probably be needed.

5.2. *Investigating customer-value, feasibility and/or cost-benefit relationship (Step iii)*

Taking a hypothetical example, let us suppose the higher pressure for a company involves the need for final disposal of cars. Thus, if manufacturers are required to deal with this aspect they should select Reverse Logistics and try to excel in it. As a consequence, companies that recognise the residual value of some discarded components and have efficient collection networks will be able to apply the reuse obligation strategically and even commercialise the scrap, thus turning the threat into a sustainable competitive advantage.

Although Reverse Logistics might be the core competence, other environmental practices (mainly Eco-design) can play an important role. This happens because Eco-design techniques could make a car use less harmful substances, become easier to disassembly, and help the process of recovering the scrap.

5.3. *Implementation and monitoring (Step iv)*

Once practices are selected and there is administrative, technical and financial support to implement the Reverse Logistic programmes, the key partners, suppliers and customers must be involved to enhance the possibilities of collecting end-of-life products and providing a better destination to its components (reuse, remanufacture or recycle). The first decision regarding the implementation of those programmes may be to "buy" or to "make". Should the car manufacturer concentrate on its core competence and outsource end-of-life collection, separation, disassembly to other companies? Should a disassembly plant follow the pattern of a few large and centralised plants, or should it be various small and decentralised plants? Regardless of the choices, monitoring the recovery processes may still be the responsibility of the car manufacturer. Incorrect use of a product recovery network inhibits the car assembler from taking advantage of valuable components in the scrap and, what is worse, may affect negatively the image of the company if the final destination is not appropriate.

For these reasons, the sales centre could be used as an information provider for collection or even as a collection point. Key customers, such as rental companies and the public sector could return their end-of-use cars to the manufacturer or a responsible third part in the recovery process.

5.4. *Critical analysis of objectives, environmental and financial performance (Step v)*

For a Reverse Logistics system environmental goals may be established by the number of recycled cars, average cost of collecting a car, time of disassembly, etc. Internal and external benchmarking are vital to the company's analysis of its performance. The definition of indicators that allow the company to undertake a critical analysis need to be carefully scrutinised because the goals might need to follow legislation or be correlated to the current production as it has been set for other products, such as tyres in Brazil. Costs and time of recovery may be higher for new markets where the infrastructure for disassembly could be non-existent or the collection more difficult.

Having relevant indicators would give the company a better direction for managing the backward flow of products and materials.

5.5. *Develop a sustainable competitive advantage (Step vi)*

It is expected that first movers can develop expertise in Reverse Logistics and a better use of the returned components. Learning or experience curves will probably make undertake the Reverse Logistics process in a cost effective manner, allowing the company to have a superior environmental and financial performance.

6. Final Considerations and Conclusions

Since 1996, when the ISO 14000 standard series were launched, Environmental Management has been evolving continuously and steadily. Practitioners have been recognising its importance and academia is organising the knowledge to be better digested by companies. As happened earlier with other management technologies, Environmental Management is being discussed more actively and moving towards its maturity. The general framework that has been presented here is an attempt to consider specific strengths or weaknesses from the context of a given company. Although further empirical research is necessary to better understand the value of these practices and the applicability of the framework in different companies and sectors, it is believed that the union of a SWOT analysis to Green Operations Practices into a framework using a PDCA cycle is progress towards higher levels of sustainability.

This is one of the limitations of the authors' work, which was developed against a theoretical background. However, as work in progress, this study has value for bringing Green Buildings into the discussion of operations management, which was rarely involved previously. Moreover, this chapter accomplishes two out of the four phases of the wider research project: the literature review and framework development. Obviously, both might be further developed during the progress of the other stages, which are the empirical research and model validation. The authors propose that the framework will be further developed to become a significant tool to help organisations in their path to sustainable competitive advantage.

Acknowledgements

This research has been supported by the Operations and Information Management Research Group of Aston Business School and Alban, the European Union Programme of High Level Scholarships for Latin America, under the scholarship Nº E06D103633BR. The authors are grateful to both for their support.

References

Boks, C. (2006). The soft side of ecodesign. *J. of Cleaner Production*, 14(15-16), pp. 1346-1356.

De Brito, MP (2003) Managing Reverse Logistics or Reversing Logistics Management? Doctoral Dissertation, Erasmus University Rotterdam, Netherlands.

Dowlatshahi, S. (2005). A strategic framework for the design and implementation of remanufacturing operations in reverse logistics. *International J. of Production Research*, 43(16), pp. 3455–3480.

EPA (2000). The Lean and Green Supply Chain: A practical guide for materials managers and supply chain managers to reduce costs and improve environmental performance. Washington, D.C., United States Environmental Protection Agency.

Fleischmann, M. Bloemhof-Ruwaard, J.M. Dekker, R. van der Laan, E. van Nunen, J. A.E.E. Van Wassenhove, L. N. (1997). Quantitative methods for reverse logistics: A review. *European Journal of Operational Research*, 103, pp. 1–17.

Fleischmann, M. Krikke, H. R. Dekker, R. Flapper, S. D. P. (2000). A characterisation of logistics networks for product recovery. *Omega*, 28, pp. 653–666.

Gilbert, S. (2001). Greening Supply Chain: Enhancing Competitiveness Through Green Productivity. Asian Productivity Association. Tapei, Taiwan.

Gupta, M. and Sharma, K. (1996). Environmental Operations Management: An Opportunity for Improvement. *Production and Inventory Management Journal*, 37(3), pp. 40–46.

Gupta, M. C. (1995). Environmental Management and its impact on the operations function. *International Journal of Operations and Production Management*, 15(8), pp. 34–51.

Jones, N. and Klassen, R. D. (2001). Management of Pollution Prevention: Integrating environmental technologies in manufacturing. In: Greener Manufacturing and Operations: From Design to Delivery and Back (Sarkis, J., ed.), pp. 56–70, Greenleaf Publishing, Sheffield, UK.

Karlsson, R. and Luttropp, C. (2006). EcoDesign: what's happening? An overview of the subject area of EcoDesign and of the papers in this special issue. *Journal of Cleaner Production*, 14 (15-16), pp. 1291–1298.

Kleindorfer, P. R., Singhal, K. And Wassenhove, L. N. V. (2005). Sustainable Operations Management. *Production and Operations Management*, 14 (4), pp. 482–492.

Liker, J. K. and Choi, T. Y. (2004). Building Deep Supplier Relationships. *Harvard Business Review*, 82(12), pp. 104–113.

Martin, M. (2001). Implementing the industrial ecology approach with reverse logistics. In: Greener Manufacturing and Operations: From Design to Delivery and Back (Sarkis, J., ed.), pp. 24–35, Greenleaf Publishing, Sheffield, UK.

Mildenberger, U. and Khare, A. (2000). Planning for an environment-friendly car. *Technovation*, 20, pp. 205–214.

Nunes, B., Marques Jr., S. and Ramos, R. (2004). Green supply chain management: an overview of techniques and practices. Proc. X International Conference on Industrial Engineering and Operations Management. Florianopolis, Brazil.

OECD (1995). Technologies for Cleaner Production and Products. Organisation for Economic Co-operation and Development, Paris, OECD.

Orsato, R. and Wells, P. (2007). The Automobile Industry and Sustainability. *Journal of Cleaner Production*, 9 (11-12), pp. 989–993.

Paumgartten, P. V. (2003). The business case for high-performance green buildings: Sustainability and Its Financial Impact. *Journal of Facilities Management*, 2(1), pp. 26–34.

Ries, R., Bilec, M. M., Gokhan, N. M. and Needy, K. L. (2006). The economic benefits of green buildings: a comprehensive case study. *The Engineering Economist*, 51, pp. 259–295.

Rothenberg, S., Schenck, B. and Maxwell, J. (2005). Lessons from benchmarking environmental performance at automobile assembly plants. *Benchmarking: An International Journal*, 12(1), pp. 5–15.

Sarkis, J. (1998). Evaluating environmentally conscious business practices. *European Journal of Operational Research*, 107(1), pp. 159–174.

Shrivastava, P. (1995). Environmental Technologies and Competitive Advantage. *Strategic Management Journal*, 16, pp. 183–200..

Wells P and Orsato R J (2005). Redesigning the Industrial Ecology of the Automobile, *Journal of Industrial Ecology*, 9(3), pp. 15–30.

Zhu, Q., Sarkis, J. and Lai, K.-H. (2007). Green supply chain management: pressures, practices and performance within the Chinese automobile industry. *Journal of Cleaner Production*, 9(11-12), pp. 1041–1052.

Chapter 10

Creating Value with Forest-Based Biomass – Traditional Industries Seeking New Business Opportunities

Satu Pätäri, Ari Jantunen**, and Jaana Sandström***

**Technology Business Research Center, Lappeenranta University of Technology, P.O. Box 20, Lappeenranta, FIN-53851, Finland*
Tel.: +358 400 178126. E-mail: satu.patari@lut.fi
***Lappeenranta School of Business, Lappeenranta University of Technology, P.O. Box 20, Lappeenranta, FIN-53851, Finland*
E-mail: ari.jantunen@lut.fi
E-mail: jaana.sandstrom@lut.fi

This chapter explores the possible challenges, threats and opportunities that face the forest and energy industries should they engage in the exploitation of renewable raw materials for bioenergy. A prospective analysis focuses on the future development of the biomass-intensive energy using a three-round Delphi study, which was performed at the end of 2006 in Finland, complemented with expert interviews. Based on the empirical data, we also attempt to identify what resources are needed in the future, and the most promising value creation possibilities at the interface of these two traditional industries.

1. Introduction

Although the energy and forest sectors are traditional and conservative industries, the current dynamics in the pulp and paper industry (PPI) indicate that it is on the verge of a significant industrial revolution (Toivanen, 2004). For example, due to global changes, the operational environment of the Finnish forest sector has transformed considerably during the past 10 to 15 years. However, Finland is still the country most dependent on the forest sector (Paperiteollisuus, 2006). Finnish global PPI players are drastically cutting costs as a first step in increasing value creation; however, other efforts are needed because cost cutting seldom plays a crucial role in creating sustainable competitive advantage (Hayhurst et al., 2004). One of the most fruitful options available is related to the collaboration between the forest and energy industries to integrate the exploitation of the renewable biomass with pulp and paper production (Hetemäki and Verkasalo, 2006).

With bioenergy, we mean energy derived from biomass which, according to Lucia et al. (2006), is an organic resource available on a renewable basis. According to the White Paper on Renewable Sources of Energy (EU, 1997), biomass is a widespread resource that comprises woody biomass, residues of the wood working industry, energy crops, agricultural residues, and agrofood effluents, manures as well as the organic fraction of municipal solid waste or source, separated household waste and sewage sludge.

After presenting the background of the evolving energy business, the next section will present how the empirical data was collected. The following section provides an overview of the bioenergy value chain, and then we explore the value creation possibilities in the interface of the energy and forest sectors. The final section summarizes the study.

2. The Evolving Energy Business with Emphasis on Bioenergy

The global interest towards renewable energy options has grown markedly in response to concerns about greenhouse gas emissions from fossil fuels. The 1997 Kyoto Protocol calls for the industrialized countries to reduce their overall greenhouse gas emissions by at least 5.2% below their 1990 levels by the years 2008–2012 (UNFCC, 1998). Furthermore, the price and availability of non-renewable energy sources have increased the attractiveness of forest-based biomass (de Vries et al., 2007).

By 2030, the world primary energy demand is projected to increase more than by half. Fossil fuels, namely oil, coal and gas, will continue to dominate the energy supplies, oil remaining the single largest fuel as is the case at the moment. Also, the center of gravity of global energy demand is shifting, with the developing countries replacing the industrialized world as the largest group of energy consumers, according to the International Energy Agency (IEA, 2006a; 2006b). These concerns are increasing global interest for renewable energy sources and pushing for changing the policies of energy consumption (Dorian et al., 2006). All these trends open up new and promising business opportunities for the Finnish forest sector, but also pose many challenges to the global forest industry which is highly dependent on energy.

2.1. *Role of renewable energy sources in the energy business*

In 2004, renewable energy sources accounted for approximately 13% of the total primary energy supply in the world. The main renewable energy sources include combustible renewables and waste as well as hydro power. In the OECD countries, the renewable energy sources accounted for only about 6% of the total primary energy supply (IEA, 2006a). Bioenergy is the largest contributor to the renewable energy supply in the IEA member countries (55% in 2001) (IEA, 2004a).

Most of the biomass-based electricity is generated in the OECD countries, accounting for between 1% and 3% of the electricity production, and the situation has not changed significantly since 1999 (IEA, 2004b; Roos et al., 1999). In some individual countries, bioenergy has, however, a significantly higher share in the energy production (IEA, 2004b). For instance, in Finland bioenergy covers about 20% of the total primary energy supply making it one of the leading bioenergy-using countries in the world (Ericsson et al., 2004). During the next three decades, the world biomass-fuelled electricity generation is projected to triple, with most of the growth coming from the OECD Europe (IEA, 2004b). Novel energy technologies, for one, are now broadening the possibilities to use biomass well beyond the simple combustion for heating (Lucia et al., 2006).

2.2. *Role of forest-based biomass in the PPI*

For the Finnish forest industry, forest-based biomass seems to offer a value creation potential as it holds a strong position among the industrialized countries in the use of forest fuel, as well as in the related technology and methods (Björheden, 2006). Traditionally, the forest industry has regarded energy as a by-product of the pulping process to be used in the energy-intensive paper manufacturing. Now the forest-based biomass is collected carefully and used in pilot plants for heat and energy generation. Some of the underlying reasons include heavy taxation of competing fossil fuels, investment subsidies, price development of fossil fuel emission trading, energy self-sufficiency objectives, diminution of production costs, industrial, employment and regional policies, objectives set by the EU, as well as common concern about the climate change. Moreover, geographical, industrial and technical structures, together with relevant actors in the forestry and energy sectors, have created favorable conditions for bioenergy especially in Finland (Ericsson et al., 2004; Hetemäki et al., 2006).

At the moment, the Finnish forest sector is both the main provider and the main user of forest-based energy, so that it is able to collect forest biomass at a competitive price (TEKES, 2004). There are, however, many uncertainties and unanswered questions that can hinder the increased use of forest-based biomass (Hetemäki et al., 2006). For instance, how does it affect the supply and prices of raw wood material? Also, is the value-added larger if the biomass is converted into energy rather than to forestry products? Generally speaking, the main obstacle to the increased use of forest-based biomass for energy is its poor price-competitiveness compared to other fuels (TEKES, 2004). Figure 1 illustrates the wood consumption in the energy and forest sectors in Finland. Note, however, that the figure illustrates only the key inputs and outputs of the forest and energy sectors which are relevant from the perspective of this study.

Approximately 59% of the forest sector's own energy needs are met by the use of internal wood-derived by-products (Ericsson et al., 2004). Overall, roundwood and wood residues that were used for energy generation together accounted for about 14.6 million m^3. Additionally, about 5.2 million m^3 of roundwood was used as a fuelwood in dwellings. The roundwood consumption was altogether around 74.9 million m^3 in Finland in 2005 (FAO, 2004; Statistical Yearbook of Forestry, 2006).

3. Empirical Data Collection

A three-round Delphi study, performed at the end of 2006, was complemented with four expert interviews. The main objective of the Delphi inquiries was to gather expert opinions of what the value creation possibilities are in the interfaces of the forest and energy industries. That is to say, our analysis focused mainly on studying and identifying the most interesting and controversial issues that affect the future development of this business opportunity between the biomass-intensive forest industry and the evolving energy industry.

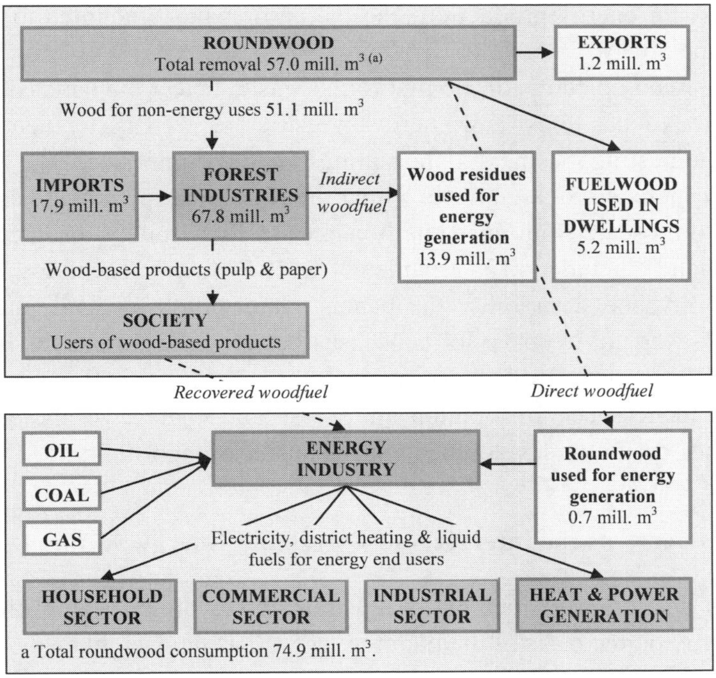

Figure 1: Wood consumption in 2005 in Finland (adapted from FAO, 2004; Statistical Yearbook of Forestry, 2006).

The time scale in the Delphi survey spanned till the year 2015, and it was conducted from the perspective of Finland although global trends were also covered. The Delphi technique was chosen in order to bring the specialists of the industries together. The panel of experts included business managers and executives, and representatives from universities as well as from joint industrial organizations of the industries. The Delphi rounds were performed as online inquiries. Unlike the other two rounds, the first Delphi inquiry took a wider perspective on the interface between the two industries under scope. In all, 36 panelists answered the first Delphi inquiry. They included representatives both from the energy and forest sectors but also from the information and communication technologies (ICT) sector.[1] Of the panelists, 17 were business managers or executives, 8 were actors of universities, and 11 respondents represented joint industrial organizations of the two industries. The second and third rounds focused on the utilization of forest-based biomass for energy production, and the participants amounted to 10 and 11, respectively.

The main objective of the first round was to detect the core assets and activities of the forest and energy industries as well as to identify the most interesting business opportunities in their interface. The second inquiry included 33 arguments which the panel of experts were asked to assess on a scale from one (totally agree) to four (totally disagree). The last round included 12 arguments and 5 future scenarios whose probability and desirability the panels evaluated. In general, the main purpose was not to strive for consensus between the experts, as is usual in Delphi studies. Instead we attempted to bring forth the issues that invoke differences of opinion the most and that have the greatest influence on future development of this business opportunity.

4. Bioenergy as the Interfaces of Forest and Energy Sectors

In this section, we attempt to illustrate the evolving value chain of bioenergy on the basis of the literature as well as the Delphi study.

---

[1] Representatives of the ICT sector were included in the first Delphi inquiry to compare the conservative forest or energy sections with an innovative and fast changing industry.

Berndes et al. (2003) highlight that most authors who have studied the future potential of bioenergy have focused on either the demand side or the supply side. *Supply side* refers to a resource-focused analysis that centers on the competition between different uses of the forest-based biomass and the total bioenergy resource base. By *demand side* we refer to a demand-driven analysis that focuses on the competition between alternative energy technologies and primary energy sources (Berndes et al., 2003). Figure 2 depicts both the supply and the demand sides and their effects on the utilization of forest-based biomass and its effectiveness. Also the main parties that are more or less interested in the exploitation of forest-based biomass, one of them being the forest sector, are illustrated in the figure (Berndes et al., 2003; Radetzki, 1997).

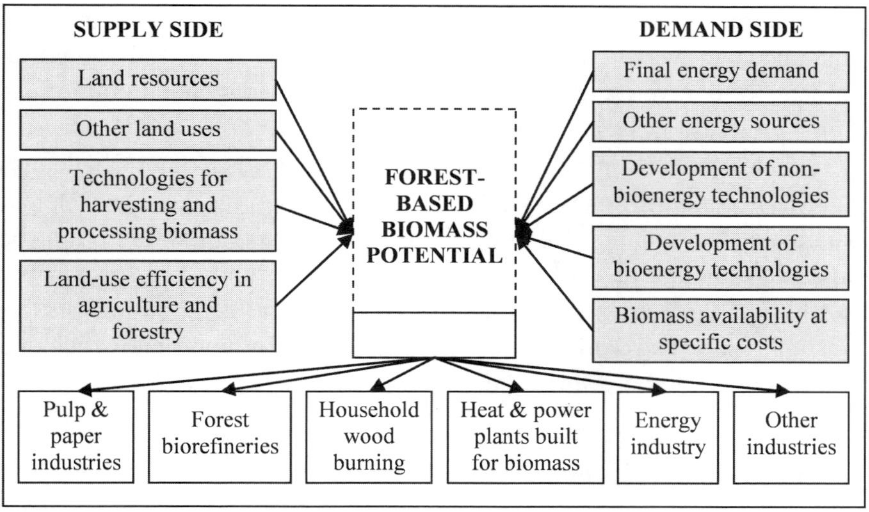

Figure 2: Factors affecting the potential use of forest-based biomass and the main actors operating in the bioenergy field (adapted from Berndes et al., 2003; Radetzki, 1997).

A recent construction called *a forest biorefinery* is a promising opportunity to make higher value-added products and to gain sustainable competitive advantage (Van Heiningen, 2007; Thorp, 2005). The biorefinery can be conceptualized as a mill whose outputs are more numerous compared to the traditional pulp mills (Lucia et al., 2006). Besides traditional wood, pulp and paper products, the forest biorefinery is able to produce, for example,. bio-based fuels and energy as well as chemicals and polymers. Principally, the PPI must maximize the value from the resources that the mills are already utilizing (Ericsson et al., 2004; Van Heiningen, 2007; Thorp, 2005).

It seems that the forest industry companies have unique positions to develop the existing pulp mills into forest biorefineries (Van Heiningen, 2006b). Although the forest sector has profound expertise in the upstream actions of the bioenergy value chain, it lacks knowledge on processing the material into energy or biofuels and delivering the products to end customers. For one, these are the fields of knowledge that the energy industry possesses. It can thus complement the capabilities of the forest industries in the evolving energy business. As one Delphi participant stated, it seems thus fruitful for the two industries to collaborate. One example of such collaboration has already been announced in March 2007 when Stora Enso and Neste Oil signed an agreement to develop technology for producing new-generation biofuels from wood residues (Stora Enso, 2007).

Besides partnering, the forest industry faces also other challenges such as technology, new skill sets and financing. And as one Delphi panelist stated, commercializing of the new products may turn out to be difficult, although he regarded the business opportunity very promising. In addition, there are no guarantees of how it will affect the forest sector. Changing such as commodity-oriented strategy to a more innovative one requires thus a lot of effort (Thorp, 2005).

5. Forest-Based Biomass as a Source of Energy

According to Radetzki (1997), the growing political and public interest towards biomass is founded largely on a supposition that the external costs (e.g. environmental damage and climate change) related to biomass are much smaller than the external costs of fossil fuels, such as coal, oil and gas. Besides, the external benefits (e.g. socially valuable employment generation) with respect to biomass are regarded as higher than those of fossil fuels. These conceptions of bioenergy are not, however, founded on any real estimates of its financial viability or cost-effectiveness on a large scale. For instance, some financial estimates were made a decade ago by Sedjo (1997) and Radetzki (1997), and the majority of the Delphi experts agreed with their arguments:

"In the future also, the added value [for wood] is certainly larger in pulp production [than in energy production]."

According to Sedjo (1997), the most promising opportunities to improve the financial viability of woodfuel energy are through innovations that would enable efficient but relatively small-scale wood power facilities. In these niche markets the wood resource would be free from competition from the industrial wood sector. Also, some experts of the Delphi study argued that the utilization of biomass in small units could turn out to be efficient.

Like Sedjo (1997) and Radetzki (1997), van Heiningen (2006a, 2006b) estimates that the total value of a mill that only produces transportation fluids is smaller than the total value of an existing kraft pulp mill that only produces pulp. But, when a mill co-produces pulp and transportation fluids, the total value-added increases by approximately 66% compared to the traditional kraft mill. The value will be almost tripled in a biorefinery that produces pulp, structural and diesel fuel products (van Heiningen 2006a; 2006b). One must, however, bear in mind that these simple estimates rest on many presuppositions and extrapolations and therefore should be regarded only as suggestive. For instance, investment costs or transportability of biomass are not taken into account in these projections. Many Delphi experts also argued that the sole production of biomass-based products does not seem profitable, and that utilization of biomass should be connected with pulp production.

To summarize, when discussing about the value creation possibilities of the energy technology of biomass production, several factors have to be taken into account. As a consequence, expert opinion with respect to the extensive utilization of biomass into energy and bio-based fuels is divided (see Table 1).

Table 1: Expert opinions about the value creation possibilities and potential.

VALUE CREATION POSSIBILITIES AND POTENTIAL
For extensive utilization of biomass:
"Can even revolutionize traditional wood processing industry. Utilization of biomass in the energy industry will be a remarkable global area of business in the future." "Highly potential." "Cooperation would diminish overlap and the resources would be directed more efficiently." "A biorefinery as well as compounds and constituents of wood have a lot of potential for also chemical industry and energy products." "So-called biorefining, that is extensive utilization of biomass for different products, is definitely coming up due to international targets and high prices of fossil fuels." "Extremely promising. Development is hindered by tax authorities' willingness to keep fuel taxation stringent also as regards renewable fuels." "The meaning of biochemicals and bioenergy is increasing. Revaluation and development of products of pulp production processes towards more efficient utilization of by-products [is needed]."
Against extensive utilization of biomass:
"I do not see any special new prospects [in the exploitation of biomass], large investments are far from core business." "In Finland, the supply of renewable raw material is sufficient to cover only a very small share of the energy demand. The most influential would be if energy efficiency of the forest industry would grow so much that a share of the bioenergy could be transmitted in the surrounding communities' favor." "At least not yet there are no guarantees for long-term business activities." "Utilization and use of biomass is already now unsustainable so it will not be the solution."

6. Conclusions

In this explorative study we have examined the emerging interface between the forest and energy sectors and the new business opportunities arising in response to growing global interest towards renewable energy options. The main focus was on Finland which has vast and not fully exploited biomass resources from forests and industrial and technical structures, making it an excellent candidate for bioenergy production. The technology can provide both the forest and energy sectors a sustainable competitive advantage.

The results of the Delphi inquiries indicate a consensus that processing biomass in rather small local units close to its origin is the most practical first step. Among the panel of experts, collaboration with the energy industry was regarded as extremely fruitful due to the industries' complementary capabilities. It thus appears that the co-production of pulp together with bioenergy and bio-based fuels would outweigh the sole production of pulp or bio-based fuels and energy. Two guiding principles in the successful adoption of the biorefinery construct were pointed out in the study: First, maximization of the value from the resources that are economically available to the mills is essential, and second, the new, novel products and services must meet the customer needs.

The success of the collaboration between the forest and energy sectors in the area of bio-energy depends on the development of new technological and marketing skills. But, as the collaboration is still in its beginnings, further research is needed. In future, it would be thus fruitful to further study this interface by looking at, for instance, the existing actors and potential new entrants that hold interest towards this area of business. Moreover, we still lack knowledge on what is really required for orchestrating globally the businesses with multitude end uses which are based on the efficient use of the forest-based biomass.

References

Berndes, G., Hoogwijk, M. and van den Broek, R. (2003). The contribution of biomass in the future global energy supply: a review of 17 studies, *Biomass and Bioenergy*, 25(1), pp. 1–28.

Björheden, R. (2006). Drivers behind the development of forest energy in Sweden, *Biomass and Bioenergy*, 30(4), pp. 289–295.

Dorian, J. P., Franssen, H. T. and Simbeck, D. R. (2006). Global challenges in energy, *Energy Policy*, 34(15), pp. 1984–1991.

Ericsson, K., Huttunen, S., Nilsson, L. J. and Svenningsson, P. (2004). Bioenergy policy and market development in Finland and Sweden, *Energy Policy*, 32(15), pp. 1707–1721.

EU (European Union). (1997). Communication from the Commission – Energy for the future: Renewable sources of energy, White Paper for a Community Strategy and Action Plan, COM(97)599 of 26.11.1997, [retrieved January 2007]. From: http://europa.eu/documents/comm/white_papers/index_en.htm.

FAO (2004). Unified Bioenergy Terminology, UBET, Food and Agriculture Organization of the United Nations, Forestry Department, Wood Energy Programme, [retrieved February 2007]. From: http://www.fao.org/documents/.

Hayhurst, D., Doak-Dunelly, T., Goodison, A., Purola, J. and Kekki, T. (2004). Profitability throughout the cycle—On demand in the forest, paper and packaging industry, IBM Business Consulting Services, Thailand, [retrieved January 2007]. From: http://www-935.ibm.com/services/th/index.wss/executivebrief/igs/x1019496?cntxt=x1019702.

Hetemäki, L., Harstela, P., Hynynen, J., Ilvesniemi, H. and Uusivuori, J. (ed.). (2006). Suomen metsiin perustuva hyvinvointi 2015, Katsaus Suomen metsäalan kehitykseen ja tulevaisuuden vaihtoehtoihin [Prosperity founded on Finnish forests, A review on the development of Finnish forestry and visions of the future], Working Papers of the Finnish Forest Research Institute, 26, [retrieved January 2007]. From: http://www.metla.fi/julkaisut/workingpapers/index.htm.

Hetemäki, L. and Verkasalo, E. (2006). Puunjalostuksen uudet tuotteet ja kehitys Suomessa [New products of wood processing and development in Finland]. In Suomen metsiin perustuva hyvinvointi 2015, Katsaus Suomen metsäalan kehitykseen ja tulevaisuuden vaihtoehtoihin (Hetemäki, L., Harstela, P., Hynynen, J., Ilvesniemi, H. and Uusivuori, J., eds.), Working Papers of the Finnish Forest Research Institute, 26, pp. 199–213, [retrieved January 2007]. From: http://www.metla.fi/julkaisut/workingpapers/index.htm.

IEA (2004a). Renewable Energy—Market & Policy Trends in IEA Countries, International Energy Agency of Paris, France, [retrieved January 2007]. From: http://www.iea.org/Textbase/publications/index.asp.

IEA (2004b). World Energy Outlook 2004, International Energy Agency of Paris, France, [retrieved January 2007]. From: http://www.iea.org/Textbase/publications/index.asp.

IEA (2006a). Key World Energy Statistics 2006, International Energy Agency of Paris, France, [retrieved January 2007]. From: http://www.iea.org/Textbase/publications/index.asp.

IEA (2006b). World Energy Outlook 2006, Summary and Conclusions, International Energy Agency of Paris, France, [retrieved January 2007]. From: http://www.iea.org/Textbase/publications/index.asp.

Lucia, L. A., Argyropoulos, D. S., Adamopoulos, L. and Gaspar, A. R. (2006). Chemicals and energy from biomass, *Canadian Journal of Chemistry*, 84(7), pp. 960–970.

Paperiteollisuus (2006). Toimialan tilanne ja tulevaisuuden haasteet [State of the paper industry and challenges of the future], Paperiteollisuuden tulevaisuustyöryhmän raportti, Finnish Forest Industries Federation and The Finnish Paper Workers' Union, 12.06.2006.

Radetzki, M. (1997). The economics of biomass in industrialized countries: an overview, *Energy Policy*, 25(6), pp. 545–554.

Roos, A., Graham, R. L., Hektor, B. and Rakos, C. (1999). Critical factors to bioenergy implementation, *Biomass and Bioenergy*, 17(2), pp. 113–126.

Sedjo, R. A. (1997). The economics of forest-based biomass supply, *Energy Policy*, 25(6), pp. 559–566.

Statistical Yearbook of Forestry. (2006). Finnish Forest Research Institute (METLA), [retrieved July 2007]. From: http://www.metla.fi/julkaisut/index-en.htm.

Stora Enso (2007). Stora Enso and Neste Oil to join forces in biofuel development, Press Release of 16.03.2007, [retrieved August 2007]. From: http://www.storaenso.com/.

TEKES (2004). Growing Power, Renewable solutions by bioenergy technology from Finland, Finnish Funding Agency for Technology and Innovation (TEKES), [retrieved February 2007]. From: http://www.tekes.fi/eng/publications/verkkojulkaisuarkisto.asp.

Thorp, B. (2005). Biorefinery Offers Industry Leaders Business Model for Major Change, *Pulp & Paper*, 79(11), pp. 35–39.

Toivanen, H. (2004). Learning and corporate strategy: the dynamic evolution of the North American pulp and paper industry, 1860-1960, PhD Thesis, Georgia Institute of Technology, School of History, Technology and Society.

UNFCC (1998). Kyoto Protocol to the United Nations Framework Convention on Climate Change, [retrieved July 2007]. From: http://unfccc.int/2860.php.

Van Heiningen, A. (2006a). Biorefinery for Ligno-Cellulosics [presentation], Maine Biomass and Biofuels Conference – "Towards energy Independence for Maine", 20-22 September 2006, Bangor, Maine, [retrieved January 2007]. From: http://www.maineswcds.org/biomass.htm.

Van Heiningen, A. (2006b). Converting a kraft pulp mill into an Integrated Forest BioRefinery (IFBR), Paper presented at the World Renewable Energy Congress IX, 19-25 August 2006, Florence, Italy, [retrieved January 2007]. From: http://www.forestbioproducts.umaine.edu/index.htm.

Van Heiningen, A. (2007). A Concept and Analysis for Converting a Kraft Mill into a Forest Biorefinery [presentation], Finnish Paper Research Community Serving Europe, 23 January 2007, Helsinki, Finland.

de Vries, B. J. M., van Vuuren, D. P. and Hoogwijk, M. M. (2007). Renewable energy sources: Their global potential for the first-half of the 21st century at a global level: An integrated approach, *Energy Policy*, 35(4), pp. 2590–2610.

Chapter 11

Innovation and Sustainable Development in Wood Furniture Design

Olivier Chéry and Elise Marcandella

Equipe de Recherche sur les Processus Innovatifs
ENSGSI – 8, rue Bastien Lepage – B.P. 90647 – 54010 NANCY CEDEX – France
Olivier.Chery@ensgsi.inpl-nancy.fr
Elise.Marcandella@ensgsi.inpl-nancy.fr

What kind of methodology would take into account the sustainability of a product? How to link all the dimensions and protagonists of sustainable development that are traditionally studied separately?
Our work takes place in a "critical emancipatory action research" framework (McKerman, 1991) because of the integration of the actor's point of view. The aim of this paper is to present a methodology to evaluate the sustainability degree during the pre-design step (early stage) of a new product. We propose a three-steps approach. First, we characterize the sustainability by both interviewing the actors concerned by the product and reviewing regulations and standards applicated to it. Next, we build a method to evaluate sustainability and in the third step, we evaluate a product in terms of its sustainability. A model based on Life Cycle Assessment that we call Sustainable Life Cycle Assessment (SLCA) illustrates our approach. Finally, we reveal an application of our methodology in the sector of wood furniture design in the East of France.
The target audience of this study are managers of SMEs: with our methodology, they will be able to evaluate product innovation in a way that favours sustainable development.

1. Introduction

Our research is on technological innovations for sustainable development. We elaborate models and work methods that take into account the concepts of sustainable development in the front end of the New Product Development (NPD). As shown in Figure 1, this front end includes product strategy formulation and communication, opportunity identification and assessment, idea generation, product definition, project planning and executive reviews (Bacon et al., 1994).

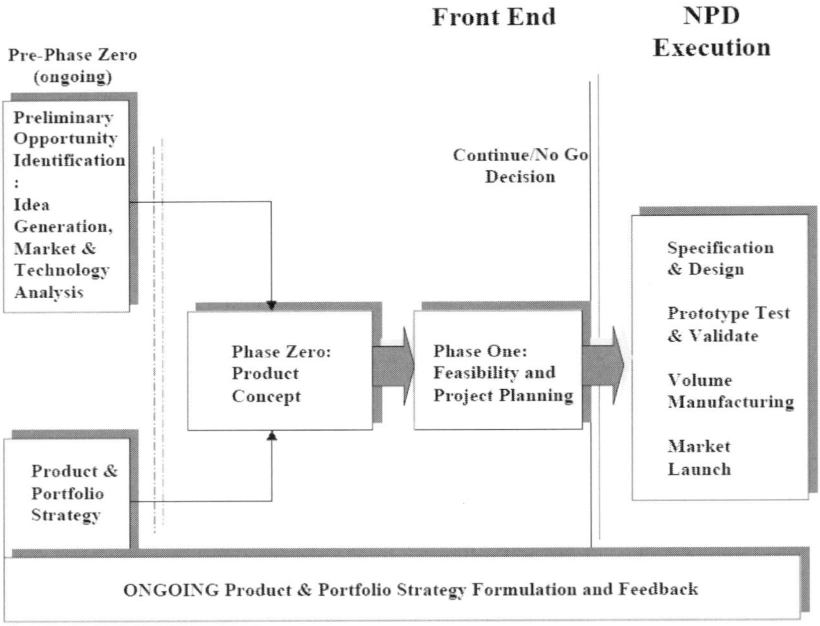

Figure 1: A model of the Front End of New Product Development (Khurana and Rosenthal, 1998).

We assume that it is possible to evaluate the sustainability of a new product just before the "go/no go" decision point. We call this phase the "pre-design step." With this evaluation, the decision-maker (manager, company head, ...) will be able to select the innovation scenario corresponding to his objectives in sustainable development.

In this paper, we present how to evaluate the sustainability of a product and present some initial results regarding the sustainability in the design of wood furniture.

2. The Methodology

2.1. *What is sustainable development?*

A commonly quoted definition of sustainable development it that it is a development that "meets the needs of the present without compromising the ability of future generations to meet their own needs" (Brüntland, 1987). Article 5 of the 2002 Johannesburg Declaration on Sustainable Development is also very explicit: "Accordingly, we assume a collective responsibility to advance and strengthen the interdependent and mutually reinforcing pillars of sustainable development - economic development, social development and environmental protection - at the local, national, regional and global levels" (UNO, 2002). As we can see, both definitions stress the links among:

All the dimensions (social, environmental economic),
All the scales (temporal, spatial),
All the actors (man, society, institutions, companies, etc.) that are encountered in the sustainable development.

Brodagh (2003) highlights that the implementation of sustainable development "is also the result of a network diplomacy in which the associations, the scientists, the companies, the labour unions and the local authorities (…) are involved with the agents of the State in international considerations and negotiations."

2.2. *The proposed methodology*

We chose the three-phased approach shown in Figure 2. First (step 1), we find a definition validated by all the stakeholders concerned about the product's sustainability. Then (step 2), we seek methods for evaluating each area associated with the sustainable development (environment, human and societal aspects and economics). Finally (step 3), we produce criteria to evaluate product sustainability. These criteria will be used in any design method capable of including the social dimension of sustainable, such as Quality Functional Deployment (QFD) or Functional Needs Analysis (FNA). These three phases can be considered simultaneously and not sequentially.

The sustainability index will be designed from the results of the steps 1 to 3. Our goal is to build a decision-making aid tool with the participation of the actors; thus, its design will depend on their contribution.

To visualize the proposed approach, we use a Life Cycle Assessment (LCA) model. In the LCA, only the environmental aspects are covered. Moreover, in our approach, economic and social aspects are included. We call this new model " Sustainable Life Cycle Assessment (SLCA)." As shown in Figure 3, this SLCA takes into account the potential impacts of the product during its whole life cycle (from birth to death) on the Environmental, Human, Societal and Economical (EHSE) targets. This representation enable to highlight the different steps to be taken into account as well as the concerned actors and the set of criteria needed to evaluate sustainability.

In the following Sections, we review the above three steps.

3. Step 1: Characterization of the Sustainability

3.1. *What is the sustainability of a product?*

Our hypothesis is that any definition of sustainability is product-dependent in terms of its economical, political and normative environment. Accordingly, all the stakeholders have to be interviewed to build a common reference framework for the sustainability of the product under consideration. We present further the application of this research through our results in wood furniture design.

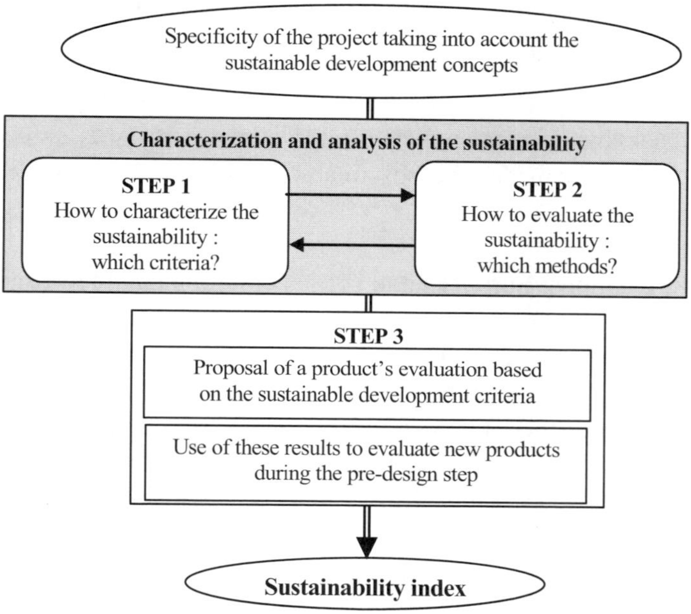

Figure 2: A global approach to design a sustainability evaluation tool in the pre-design phase.

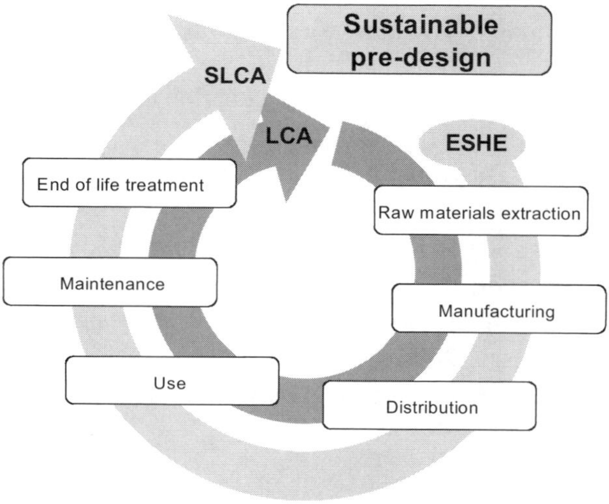

Figure 3: Proposed Product Sustainable Life Cycle Assessment.

3.2. *Regulatory, political and economical framework*

We review the regulations, standards, economical and political measures that can influence the definition on the sustainability of a product. In our case, French Institutional policies, international standards and methods can be used to consider the sustainability of a product during its life cycle, such as:

(1) Regulations: European and French environmental laws and labour laws
(2) Economical measures and management guides: labels, certification standards: ISO 9001 (Quality), ISO 14001 (Environment), OHSAS 18001 (ISO standard in Health and Safety), guides such as SD 21000 (AFNOR[1] standard in Sustainable Development)...(Delchet, 2006).
(3) Political measures: environmental taxation, competitive bidding with a sustainable development demand, Agendas 21, subsidy to buy eco products, etc. (Boutaud, 2005).

With these, we can characterize the sustainability of a product on the basis of the criteria that are appropriate of the case at hand.

4. Step 2: Methods to Evaluate Sustainability

The main methods traditionally used in situation analysis are listed here.

(1) Life Cycle Assessment (Janin, 2000; Lavallee *et al*, 2005; Le Van, 1995),
(2) Economic Sector Assessment (Floriot, 1985; Batalha, 1993; Duteurtre *et al*, 2000),
(3) Integration of the ergonomics in the pre-design phase (Marsot, 2002; Chitescu, 2005),
(4) Risk Assessment concerning the workstations (Durand, 2002),

[1] AFNOR : Association Française de Normalisation (French Standards Association)

(5) Natural and Industrial Risk Assessment: MOSAR (Organised and Systemic Method of Risk Assessment) (Perilhon, 2003), Danger Sciences Approach (Verdel, 2000), (Laurent, 2003),

(6) Prospective approach: structural analysis such as MIC-MAC, role playing games as MACTOR, SMIC Prob-expert (Godet, 2004),

(7) Data Aggregation methods, multi criteria analysis to design the "sustainability index" of a product (Brun *et al*, 2005; DGITIP, 2002; Caillet, 2003; Gauthier, 2005).

We can organize these methods to create a methodology to evaluate the sustainability of a product.

5. Step 3: Sustainability Evaluation in the Pre-Design Phase

After Steps 1 and 2 in Figure 2 have been established, we can measure the sustainability degree of a new product during the pre-design step. This approach can be visualized as a "logical circle" similar to Life Cycle Assessment (Figure 3). As a result of the steps 1 and 2 (Figure 2), we obtain a set of sustainable criteria. We can now use the new product development methods.

We choose to present two methods that readily integrate human factors and ergonomics in the pre-design phase. According to Marsot (Marsot, 2002), the ergonomic aspect is one of the parameters along the social dimension of sustainable development. These methods are Quality Function Deployment (QFD) and Functional Needs Analysis (FNA).

QFD for Total Quality Management (TQM) is a systematic approach to include the required quality level during the design. This method is relevant to our case because it allows the inclusion of the client's criteria in a bottom-up way. A consultant helps the client to formulate their demands in terms of the "what" and the "how" of the product and these are translated into design objectives and key-points in the design itself (Marsot, 2004).

FNA is established by the French standard (AFNOR X50H) (AFNOR, 1991). Its advantages are that it allows for the inclusion of end-users' expectations from a new product or a service. It should be noted, that there is a major difference between a client and an end user of a product. This distinction is important in our demonstration because the angle of view could be opposite. A user can expect a product to be functional and secured whereas a client expects to be satisfied. The approach requires the intervention of a trained leader in a workshop that gathers the various stakeholders, with the intent of making the end-user central to the design. With this technique, it is possible to consider the functionalities of the final product from the start and to integrate the environment into the product's life cycle (Puyou, 1999).

As stated earlier, both tools can integrate ergonomics as a parameter. Other aspects of sustainable development could be easily added to both of the methods as needed.

6. Application to the Wood Furniture Sector in Lorraine

6.1. *The study's context*

We have studied the entire wood furniture sector in Lorraine. Historically, Lorraine was an industrial region based on coal and iron mining until 1979. Today, most companies are small and medium enterprises (SME) and Lorraine belongs to a Development Pole that focuses on three traditional industries: textile, paper-cardboard and wood. The aim of this Pole is to design new products from renewable fibres and result in new functions.

The wood furniture sector is also of interest because it has the incentives for an eco-design approach. The companies of the sector are suffering from the Asian and East European competition. With the integration of the sustainable development, these companies would be more competitive and as well as "sustainable." Thus, our research can contribute to help managers in this sector making product development decisions.

The workflow in the common wood furniture sector in France is shown in Figure 4. Each step is presented from the forest management to the elimination of the wood furniture after using. We can also see the impact of transportation (and the associated pollution) on this sector ("T" in circles on the Figure 4).

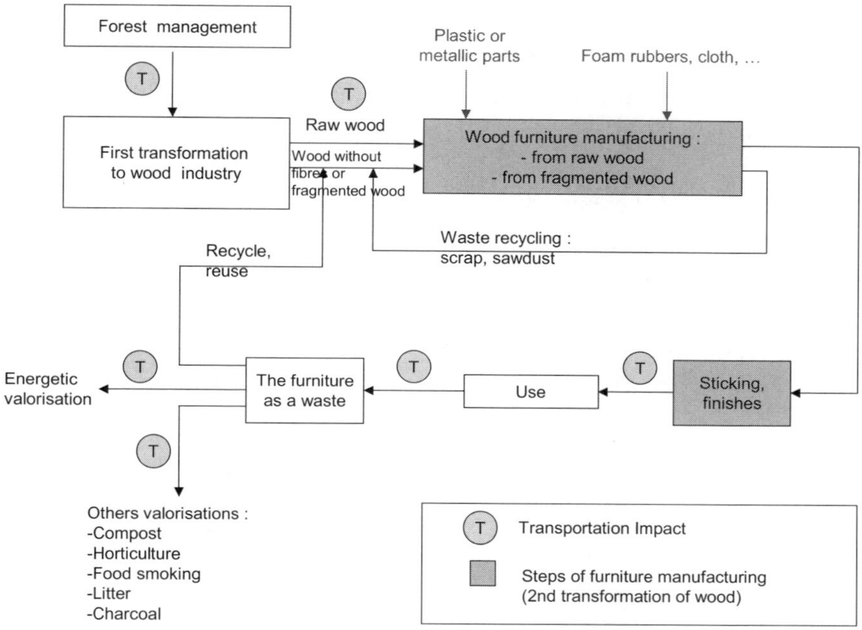

Figure 4: The wood furniture life cycle.

6.2. *Our methodology applied to the wood furniture sector*

6.2.1. *Step 1: Characterization of the sustainability — Definition of a sustainable furniture*

Step1a: Stakeholders and observed impacts

The stakeholders of this "sustainable sector" have been identified in order to define the notion of "sustainable furniture". These are

(1) Companies active in this sector
(2) Professional associations
(3) Technical and design centres (Research valorisation centres)
(4) Worker representatives and user health safety groups
(5) Research laboratories
(6) Public environmental agencies.

After discussions with the concerned stakeholders, we identified the EHSE impacts to take into account (Figure 5).

At this stage of the study, the characteristics of sustainable furniture are as follows:

(1) Respect of the environment during its life cycle,
(2) Respect of the health of involving participants of the "product's life cycle" (workers, users, carriers, etc...),
(3) Stimulate the sustained development of Lorraine through socially responsible firms.

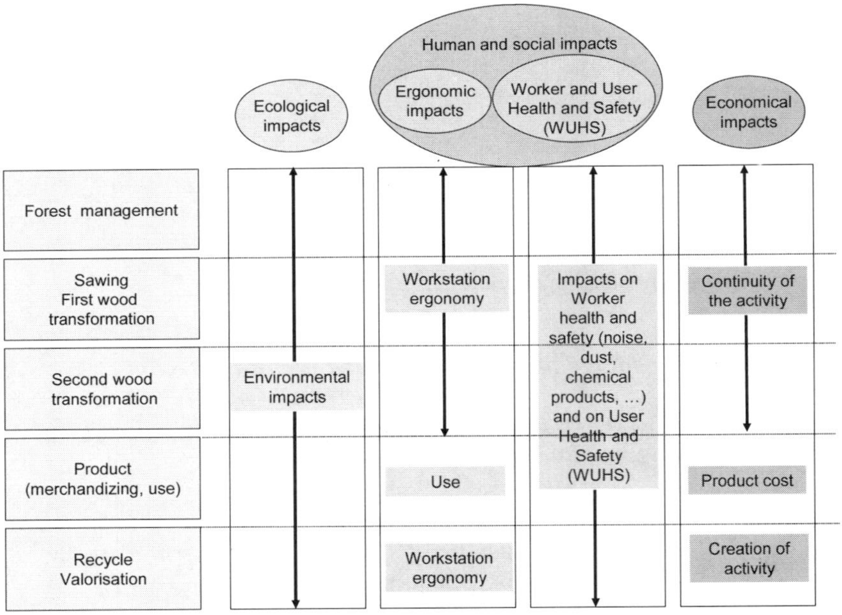

Figure 5: Environmental, Human, Social and Economical (EHSE) impacts taken into account in the wood industry.

6.2.2. *Step 1b: Regulation, political and economical framework*

In our application, the standards and laws that regulate the local wood furniture-based sector have been identified.

Figure 6 presents the specific standards and regulations used to take into account the EHSE impacts in this sector. We can see on Figure 6 that for most of the impacts presented in Figure 5, there exist norms and standards in order to regulate them. Studying these related texts, we extract criteria enable us to define criteria of sustainable furniture.

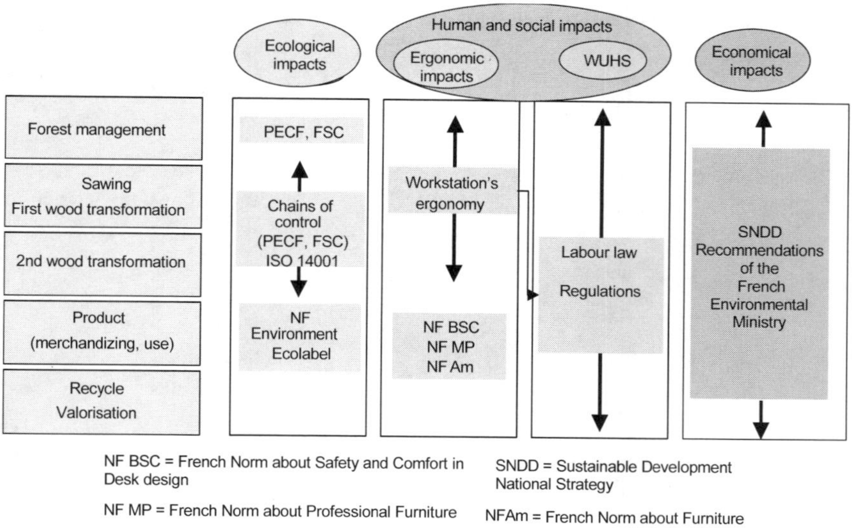

Figure 6: Economical, political and regular standards used in the wood industry.

6.2.3. *Step 2: Methods to evaluate the sustainable development*

The main methods are presented Table 1. They are used to define the system and to evaluate its environmental, social or economical characteristics. It also details when these methods are able to build evolution scenarios (i.e., decision-making aids).

6.2.4. *Step 3: Construction of the sustainability index*

By performing the third step for the wood furniture sector in Lorraine, we can construct the relevant sustainability index, This step is currently being conducted and will be the subject of a future paper.

Table 1: Methods for evaluating the sustainability of a system.

Methods	Criteria	System's Description	Evolution scenarios Decision aid making	WFBS*
Life Cycle Assessment	Environment	According to LCA	Comparison – LCA software	Yes
Economic Sector Assessment	Economical Organisational	Meso economical approach	Economical intelligence	Yes
Design and ergonomics	EHSE Use	Numerical chains	Comparison	To be defined
Risk assessment	Work health and safety	Workstation conception	Checklists	Yes
Risk assessment	Industrial risks Natural risks	Systemic tools – MADS** Danger sciences	Complex scenarios elaboration - MOSAR	To be defined
Prospective approach	Transversal approach	Structural Analysis Role playing games SMIC Prob-expert	Formulation of technological scenarios Technological benchmarking	To be defined
Data Aggregation methods-multi criteria analysis	Multi criteria analysis	Systemic tools	Complex scenarios elaboration	To be defined

* WFBS: Wood Furniture Based Sector. ** MADS: Analysis Method of System Failures

7. Conclusion and Final Perspectives

In performing this study, our goal was to propose a methodological approach to evaluate a sustainable development design.

This methodology could be a decision support tool for SME managers in order to improve new product design process when confronted by several possible innovation scenarios. Our approach will be the basis for constructing a sustainable index. Indeed, the manager will be able to choose the best scenario in function of his strategical positioning regarding sustainable development. Moreover, he will be able of measuring the evolution of the index during all the life cycle steps of the designed product.

The wood furniture sector in Lorraine is a good illustration that can be extended to other countries or others economical sectors.

Acknowledgement

We would like to thank Catherine Aubier for her help, her great ideas and her energy.

References

AFNOR (1991). Management par la valeur - Expression Fonctionnelle du Besoin et cahier des charges fonctionnel - Exigences pour l'expression et la validation du besoin à satisfaire dans le processus d'acquisition ou d'obtention d'un produit. AFNOR, Paris.

Bacon, G., Beckman, S., Mowery, D. and Wilson, E. (1994). Managing Product Definition in High-Technology Industries: A Pilot Study. *California Management Review*, Spring 1994: pp. 32–56.

Batalha, M.O. (1993). La filière comme outil d'analyse stratégique: le cas des matières grasses à tartiner au Brésil. Thèse de doctorat de l'Institut National Polytechnique de Lorraine, Nancy.

Boutaud, A. (2005). Le développement durable: penser le changement ou changer le pansement? Bilan et analyse des politiques publiques locales en matière de développement durable en France: de l'émergence d'un changement dans les modes de faire au défi d'un changement dans les modes de penser. Thèse de l'Ecole Nationale Supérieure des Mines de Saint-Etienne.

Brodhag, C. (2003). Genèse du concept de développement durable: dimensions éthiques, théoriques et pratiques. Séminaire, Développement durable et aménagement du territoire: les villes et les régions du futur. Presses Polytechniques Romandes, Lausanne, pp. 29–45.

Brun E. and Saillet, F. (2005). Étude sur l'Eco-conception—État de l'art dans le domaine de l'éco-conception. AFNOR, Paris.

Brundtland, G. H. (1987). Our common Future. Report from the UN World Commission on Environment and Development (WCED). Oxford University Press, London.

Caillet, R. (2003). Analyse multicritère: étude et comparaison des méthodes existantes en vue d'une application en analyse du cycle de vie, *Cahiers de la Série Scientifique*, Cyrano, Paris.

Chitescu, L. C. (2005). Simulation en ergonomie: facteur d'innovation dans la conception de produits—Application à la conception des systèmes de travail. Thèse de l'Institut National Polytechnique de Lorraine, Nancy.

Communication from the Commission to the Council and the European Parliament (2003)-Integrated Product Policy-Building on Environmental Life-Cycle Thinking, Com 302 final.

Delchet, K. (2006). La prise en compte du développement durable par les entreprises: entre stratégies et normalisation. Étude de la mise en œuvre des recommandations du guide Afnor SD21000 au sein d'un échantillon de PME françaises.Thèse de l'Ecole Nationale Supérieure des Mines de Saint-Etienne, Nancy.

DGITIP (2002). Indicateurs de développement durable: Comment mesurer la performance durable des entreprises? Proposition d'une grille d'indicateurs. Étude réalisée pour la Direction générale de l'Industrie des Technologies de l'information et des Postes, Paris.

Durand, E. and Lafon, D. (2002). Les services de santé au travail et l'évaluation des risques dans les petites entreprises, Symposium INRS, Grenoble, 6 juin 2002, TD 121, *Documents pour le médecin du travail*, 91, pp. 283–295.

uteurtre, G. Koussou M. O. and Leteuil. H. (2000). Une méthode d'analyse des filières—Synthèse de l'atelier du 10 - 14 avril LRVZ, N'Djamena.

Floriot, J. L. (1985). Pratique de l'analyse filière et Génie des Systèmes Industriels. Éditions Economica , Paris:

Gauthier, C. (2005). Measuring Corporate Social and Environmental Performance: The Extended Life-Cycle Assessment. *Journal of Business Ethics*, 59, pp. 199–206.

Godet, M. (2004). La boîte à outils de prospective stratégique. *Cahier du Lipsor n° 5*—Juin 2004.

Janin, M. (2000). Démarche d'éco-conception en entreprise. Un enjeu: construire la cohérence entre outils et processus. Thèse de l'Ecole Nationale Supérieure des Arts et Métiers, Chambery.

Khurana, A. and Rosenthal R. S. (1998). Towards holistic 'front ends' in new product development, *Journal of Product Innovation Management*, 15(1), pp. 57–74.

Laurent, A. (2003). Sécurité des procédés chimiques—Connaissances de base et méthodes d'analyse des risques. Lavoisier, Paris.

Lavallée, S., Normandin, D. and Sonnemann, G. (2005). L'analyse du cycle de vie des produits et services: un outil d'aide à la décision pour les décideurs publics et privés en matière de développement durable, *Liaison Energie-Francophonie*, 6, pp. 24–28.

Le Van, S. L. (1995). Life Cycle Assessment: Measuring Environmental Impact, Proceedings of the 49th Annual Meeting of the Forest Products Society, Portland, pp. 7–16.

Marcuccilli, A. et al. (1998). Evaluation des symptômes et de la fonction respiratoire en relation avec les expositions aux poussières de bois dans les ateliers de menuiserie industrielle. *Archives des maladies professionnelles et de médecine du travail*, 59(5), pp. 305–314.

Marsot, J. (2002). Conception et ergonomie—Méthodes et outils pour intégrer l'ergonomie dans le cycle de conception des outils à mains, *Les notes scientifiques et techniques de l'INRS*, (219).

Marsot, J. (2004). QFD: A methodological tool for integration of ergonomics at the design stage, *Applied Ergonomics*, 36 (2), pp. 185–192.

McKernan, J. (1991). Curriculum Action Research. A Handbook of Methods and Resources for the Reflective Pratictioner. Kogan Page, London.

Perilhon, P. (2003). Mosar—Présentation de la méthode. Sciences et Techniques pour l'Ingénieur, Traité Sécurité et Gestion des Risques, Vol. SE 4060.

Rigo, M. O. (2004). Les poussières de bois dans l'industrie du bois: conséquences physiologiques de l'inhalation d'atmosphères chargées en fines particules de bois. Etat des techniques de captation et amélioration des procédés de filtration des microparticule, *Contrat ADEME*—French Minister of Fishing and Agriculture, Final Report.

UNO (2002). Report from World Summit on Sustainable Development in Johannesburg, South Africa, from 2 to 4 September 2002, United Nations, New York.

Verdel, T. (2000). Méthodologies d'évaluation globale des risques. Applications potentielles au Génie Civil. Presses de l'Ecole Nationale des Ponts et Chaussées, Paris, pp. 23–38.

Villeneuve, C. (2003). La recherche pour le développement durable… A la recherche du développement durable? Recherche et développement durable, *Liaison Energie-Francophonie*, (61), pp. 14–19.

Chapter 12

Sustainable Development and Technology Management

Alan C. Brent and Marthinus W. Pretorius***

**Graduate School of Technology Management, University of Pretoria*
Natural Resources and the Environment, CSIR
***Graduate School of Technology Management, University of Pretoria*

The purpose of this chapter is to establish a conceptual framework for the technology management field of knowledge and identify the departure point for further research to incorporate the concept of sustainable development into the field. From a review of the literature it is concluded that sustainability aspects are not addressed adequately in technology management theories and practices. Emerging technology management practices related to sustainable development do emphasize the focus on technology strategy, selection and transfer, especially between developed and emerging economies. At the core of these issues lies technology assessment that also forms part of other technology frameworks and methodologies. It is recommended that future research to concentrate on the development of technology assessment methods, based on the modification of the Technology Balance Sheet, Income Statement and Space Map analytical techniques, that incorporate the dynamic interactions between nature and society that is researched in the newly established field of sustainability science.

1. Introduction

The World Commission on Environment and Development (WCED)'s report in 1987 is viewed as a major political turning point for the concept of sustainable development (Mebratu, 1998). Since then the influence of the concept has increased extensively in policy documents of governments and international agencies (Mebratu, 1998). The World Summit on Sustainable Development (WSSD) in 2002 highlighted this growing recognition of the concept at a global level (Labuschagne and Brent, 2005a). This need to incorporate the concept of sustainable development into decision-making, combined with the World Bank three-pillar-approach to sustainable development (World Bank, 2001), resulted in the popular business term "triple-bottom-line decision-making".

The concept of sustainability and sustainable development may be understood intuitively, but it remains difficult to express in concrete, operational terms (Briassoulis, 2001). However, many agree that sustainable development is about achieving environmental, economic, and social welfare for present as well as future generations (Azapagic and Perdan, 2000). From a governmental perspective, this can be at national and global levels (UNCSD, 2001). From an organizational perspective, this can be at project (Labuschagne *et al.*, 2005b) and technology (Brent *et al.*, 2006; 2007) levels. In some cases, stakeholders specifically require that environmental, economic, and social goals must be met across all levels of development. Sustainable development has subsequently been conceptualized as a state of dynamic equilibrium between societal demand for a preferred development and the supply of environmental and economic goods and services needed to meet this demand (Briassoulis, 2001). Systems approaches have been proposed to consider strategic sustainable development planning in different sectors (Robèrt *et al.*, 2002; Labuschagne *et al.*, 2005c). But the intricate relationships between the three dimensions of sustainable development, i.e. environmental, economic and social welfare, have been difficult to model within the concept of a clear absolute technological system (Brent *et al.*, 2006; 2007). Specifically, trade-offs between the three dimensions of sustainable development may not be possible to quantify as the benefits cannot be measured. Proposals for these trade-offs can be referred to as 'weak', i.e. indirectly indicating sustainability (Hanley *et al.*, 1997; Rennings and Wiggering, 1997; Atkinson, 2000).

Consensus on the details of how to achieve sustainable development or maintain sustainability is difficult to obtain in practice. This difficulty can be attributed to the variety of perceptions on specific socio-cultural and political contexts that change over time (Briassoulis, 2001; Brent *et al.*, 2005a). It has been also argued that sustainable development cannot be easily integrated with the practices of technology or innovation management (Coles and Peters, 2003).

From a research perspective the following main question is subsequently posed: Are sustainability aspects addressed adequately in technology management theories and practices? In other words, has technological research progressed into the field of sustainability science, as has been suggested (Kates *et al.*, 2001)? The research question focuses on mainly those large-scale technologies, i.e. technologies that can only be added in discreet sized lumps (Murto, 2000), and which are highly dependent on, or may pose risks to, the natural resource base of countries and regions (Cooney, 2004).

2. Objectives of the Chapter

The primary objective of this chapter is to establish a conceptual framework of the technology management field of knowledge, and coupled tools and methodologies, as it relates to sustainable development. The secondary objectives are to introduce a criteria framework of what sustainable development entails in different resource-based sectors where technology management occurs, e.g. the manufacturing, energy, and agricultural sectors, and to provide insight into how sustainability aspects may be measured effectively as part of technology management practices in these sectors.

From these objectives the paper aims to identify the departure point for further research in terms of incorporating the concept of sustainable development into the technology management field of knowledge, which is a specific agenda that may differ significantly from other technology management orientated research themes (Pilkington and Teichert, 2006).

3. Methods

The primary objective of the chapter was addressed by first considering the:

(i) Management of Technology (MOT) body-of-knowledge (BoK) process, which has been initiated by the International Association for Management of Technology (IAMOT, 2006), and specifically a survey on a Template for Graduate Programs and an analysis of the results of a survey of 148 Technology Management or MOT graduate programs (Portland State University, 2003).

(ii) Engineering and Technology Management Education and Research Council's identification of related research areas (ETMERC, 2006).

A search was conducted in the databases of several journals with the keywords indicated in Table 1. The IAMOT BoK survey, the ETMERC identification of related research areas, and the *Technovation* papers on 'technology management tools' and 'technology management methodology' were used to construct a mind map of the technology management field of knowledge, which is downloadable from the website of the University of Pretoria (2007; *http://www.up.ac.za/gstm*).

Mind maps are especially useful as support for intuitive-type research to highlight casual connections between different aspects (Monaghan, 2003). In the mind map overlaps between the IAMOT and ETMERC defined areas are shown with graphical links (left-hand side of the mind map). The linkages between defined technology management tools and methodologies, and associated applications (right-hand side of the mind map), and the IAMOT and ETMERC areas are shown with numeric keys. The specific linkages between the core technology management areas and sustainable development are emphasized with shadings.

The additional literature on 'technology management' identified a conceptual framework that could be improved in the context of sustainable development. The obtained literature on 'sustainable development' was used to determine how the linkages between the core technology management areas and sustainable development occur in practice.

Table 1: Journals and papers relating to technology management theories and practices, and technology management orientated sustainable development.

Journal	Keywords	References
Technovation	Technology management tools	Phaal *et al.*, 2006 Maine *et al.*, 2005 Brady *et al.*, 1997
	Technology management methodology	Liao, 2005 Jacob and Kwak, 2003
	Sustainable Development	Demaid and Quintas, 2006 Fahmy, 2005 Gerstlberger, 2004 Watanabe et al., 2003 Harris and Khare, 2002 Lambert and Boons, 2002
International Journal of Technology Transfer & Commercialisation (ABI Inform)	Sustainable Development AND Technology Management	Momaya and Ajitabh, 2005
International Journal of Services Technology and Management (CSA Illumina)		Banwet et al., 2003
International Journal of Biotechnology (CSA Illumina)		Hamilton, 2001
International Journal of Technology Management (CSA Illumina)		Bowonder and Miyake, 2000
Technological Forecasting and Social Change (CSA Illumina)		Sharif, 1992
International Journal of Technology Management (SCOPUS)		Khalil and Ezzat, 2005 Phaal *et al.*, 2004

4. Discussion

4.1. *A conceptual framework for technology management*

A conceptual framework supports understanding of an issue or area of study, provides structure, communicates relationships within a system for a defined purpose, and supports decision-making and action (Phaal *et al.*, 2004). Such a framework has been introduced at the firm level (see Figure 1) (Phaal *et al.*, 2004) to represent how technological and commercial knowledge combine to support strategy, innovation and operational processes in a firm. To use it for sustainable development, it needs to be expanded to accentuate the external-to-internal drivers of sustainable development (Labuschagne and Brent, 2005b). Furthermore, and especially for large-scale resource-oriented technologies, the system must be extended beyond the firm level, i.e. the life cycle of the technology, or asset, and the life cycle of the associated product value chain must be considered (Brent *et al.*, 2005b; 2007). Such an extended life cycle system is illustrated in Figure 2.

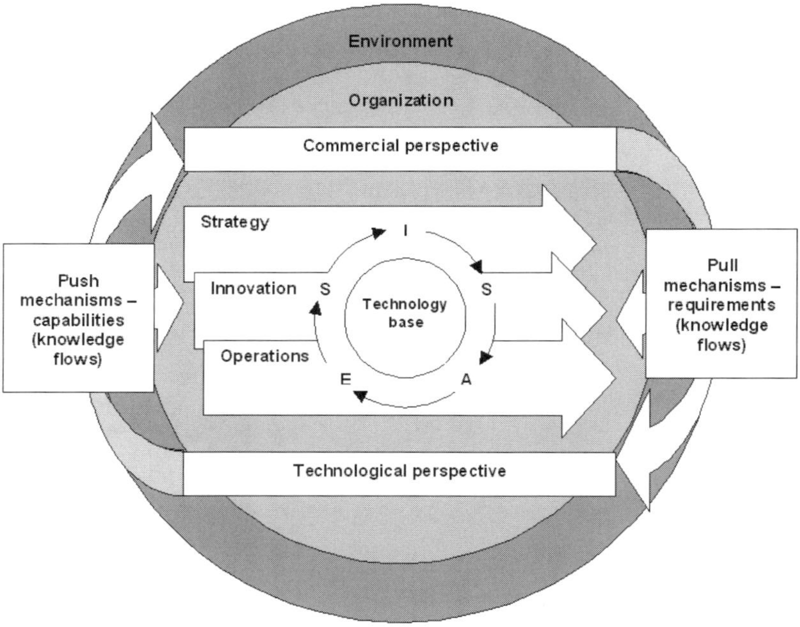

Figure 1: Conceptual technology management framework at firm level (adopted from Phaal *et al.*, 2004).

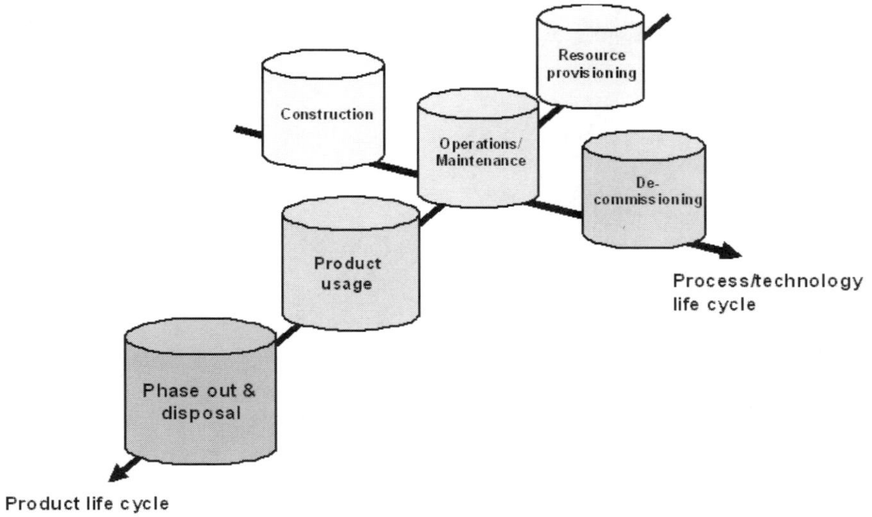

Figure 2: Life cycle system for large-scale resource-oriented technologies (adopted from Brent *et al.*, 2005b; 2007).

4.2. *Defining a conceptual technology management framework in the context of sustainable development*

The framework introduced (Labuschagne *et al.*, 2005c) focuses on large-scale resource-oriented technologies and emphasizes the operational initiatives in industry must be evaluated separately in terms of internal and external economic, social and environmental performances. The internal operational sustainability must also be ensured, e.g., technology management practices, and a fourth dimension of sustainable development has been suggested (Labuschagne *et al.*, 2005c; Mulder and Brent, 2006). Therefore, it is proposed that technology management, as it relates to sustainable development, should be conceptualized as a triangular-based pyramid (see Figure 3). The three conventional dimensions of sustainable development form the base or foundation of the pyramid, and supports sustainable technology management practices at the top of the pyramid.

The conceptual framework indicates two planes of influence. First, technology management practices (at the firm level) influence other

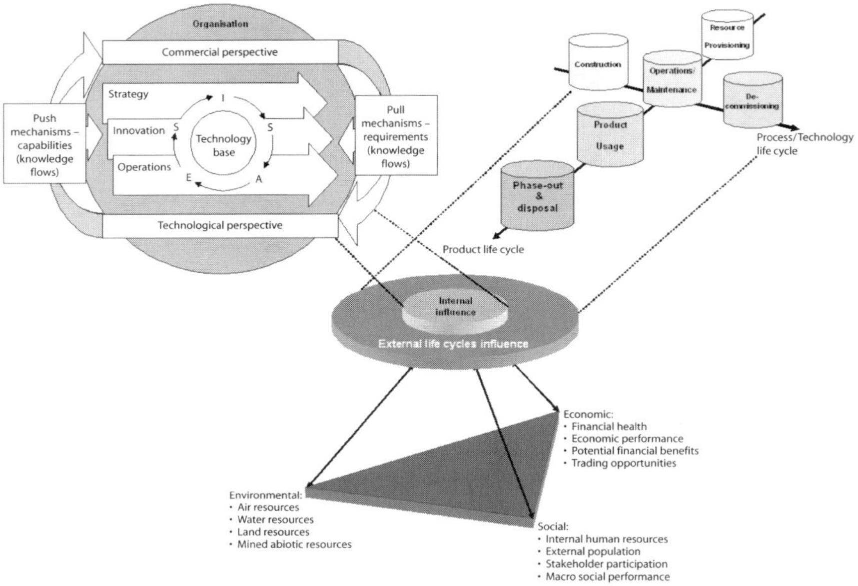

Figure 3: Conceptual framework for technology management in the sustainable development context.

internal operations, but sustainable development aspects, e.g. economic forces, natural resource constraints, and social behavior, may also influence internal operations. In turn, internal operations do exercise influence on different sustainable development aspects. Similarly, there is interaction between internal operational initiatives, the technology and product life cycle phases outside the firm level, and sustainable development aspects.

It has been stated that conceptual frameworks exist largely in the mind and require practical devices to 'interface' with the real world, in terms of both the development (induction) and application (deduction) of frameworks (Phaal *et al.*, 2004). The devices, i.e. tools and methodologies, depicted on the right of the technology management mind map (*http://www.up.ac.za/gstm*) are primarily concerned with the interfaces between two planes of the conceptual framework. This is reflected in the defined research and education focus areas of IAMOT and ETMERC.

Table 2: Current technology management research and applications in relation to sustainable development.

Reference	Description of paper focus
Demaid and Quintas, 2006	The uncertainty associated with the changing legal and ethical imperatives of sustainable development and the related additional complexity of knowledge management in a specific sector; the similarities between the fields of sustainable development and risk are specifically highlighted.
Fahmy, 2005	Technological trends in specific sectors due to sustainable development pull and push drivers with a subsequent strategic plan and policy advice for decision-makers.
Gerstlberger, 2004	Systematic design of regional innovation systems for policy support, whereby the multidimensional aspects of sustainable development aspects are considered for effective, sustainable knowledge transfer in networks.
Watanabe *et al.*, 2003	Policy options to substitute technologies in a specific sector for competitive advantage; sustainable development, from an ecosystem perspective, is used as basis to formulate an approach for competitive innovation.
Harris and Khare, 2002	Strategy development for a specific sector due to sustainability pull and push drivers; sustainable development risk are identified that decision-makers must consider for the long-term survival of the sector.
Lambert and Boons, 2002	Societal and environmental problems related to mixed industrial parks, i.e. an extension of the industrial symbiosis concept, are identified, and solutions are proposed to ensure the continuity and sustainability of these parks.
Momaya and Ajitabh, 2005	Strategic management of technology to sustain the competitiveness of organisations; sustainable development is synonymous with management performance and competitiveness in terms of productivity, growth, returns and market capitalisation.
Banw et *et al.*, 2003	Technological competitiveness must be achieved to realise sustainable development, and the internal processes and assets that derive performances are important for decision-makers; no emphasis is placed on external drivers.
Hamilton, 2001	Defining characteristics of technological trends and response firms to propose changes in management practices for effective technology transfer.
Bowonder and Miyake, 2000	Combining knowledge management and ecosystem theory concepts to sustain competitive advantage in an uncertain business context.
Sharif, 1992	Increasing international cooperation to ensure the advancement and spread of technology that is economically efficient, commercially attractive, and environmentally sound, and that leads to self-reliance; technology-oriented policies are addressed.
Khalil and Ezzat, 2005	Globalisation, competitiveness, and the risk of marginalisation of developing nations; responses in public policy are highlighted, with emphasis on human resource development.

Interfaces between the planes and the sustainable development aspects have been considered in theory, albeit to a lesser extent. Table 2 summarizes the obtained literature that deals with such interfaces. In these cases the technology management research and applications were mainly associated with the sub-areas of risk management and decision-analysis or support, and is highlighted in the technology mind map (*http://www.up.ac.za/gstm*).

4.3. *Emerging technology management practices related to sustainable development*

It has been noted that, as a research area, technology management is extremely diverse (Pilkington and Teichert, 2006). This is emphasized in the mind map (*http://www.up.ac.za/gstm*). Furthermore, in the sustainable development context, technological research is viewed as one of the four branches of sustainability science (Kates *et al.*, 2001), i.e. concentrating on the design of devices and systems to produce more social goods with less environmental harm. Sustainability science in turn can be defined as the study and integration of particular issues and aspects of radical, systemic approaches to innovation and learning for ecological and social sustainability (Struyf, 2003). The merger of these two fields has led to concepts such as Environmentally Sound Technologies (ESTs), i.e. technologies that have the potential for significantly improved environmental (and social) performance relative to other technologies (IETC, 2003a).

The European Institute for Technology and Innovation Management (EITIM, 2001) states: "technology management addresses the effective identification, selection, acquisition, development, exploitation and protection of technologies (product, process and infrastructural) needed to maintain a market position and business performance in accordance with the company's objectives."

For ESTs, the emphasis is not only on the firm level, but also on the regional, national and international levels (IETC, 2003b). This again stresses the requirement to expand the technological system that is managed, as is shown in the conceptual model (Figure 3), and an adaptation to the EITIM definition is proposed, i.e. *technology management addresses the effective identification, selection, acquisition, development, exploitation and protection of technologies (product, process and infrastructure) needed to sustain the competitive advantage of regional sectors in accordance with the sector, regional, national and international sustainable development objectives.*

A number of cases have been documented in literature that supports the proposed definition of technology management (see Table 3). The table further shows that the literature on technology management and sustainable development increasingly deals with three main issues:

(i) Integrated strategies across companies, sectors, regions, and, in some cases, across countries.

(ii) Selection of appropriate technological options across companies, sectors, regions and countries.

(iii) The transfer of technologies (and knowledge) across companies, sectors, regions and countries.

A focal point of these three issues is that of technology assessment or evaluation, which also forms part of other technology frameworks and methodologies as is shown in the technology management mind map (*http://www.up.ac.za/gstm*). Technology evaluation is a set of principles, methods and techniques or tools for effective assessing the potential value of a technology and its contribution to a company's competitiveness and profitability (Bakouros, 2005). Models (Pretorius and de Wet, 2000) and metrics (Geisler, 2002) have been introduced to assist the technology assessment process at firm level.

The following statements have been made with regards to the ongoing development of metrics (Geisler, 2002):

(i) Technology is not judged by its existence alone, nor is its mere existence a sufficient condition for successful usage.

(ii) We cannot evaluate technology unless and until we put it in the context of social (and environmental) and economic phenomena.

(iii) Technology is not defined and evaluated by what it is, but by the criteria outside itself – by its actual and potential users.

These statements support the system expansion component of the conceptual framework (Figure 3), and the notion of sustainability performance indicators that have been proposed for technology management purposes (Labuschagne *et al.*, 2005c; Brent *et al.*, 2005b; 2007).

Table 3: Emerging technology management research and applications in relation to sustainable development.

Reference	Description of paper focus
Grieve, 2004	An accepted strategy for medium- and large-scale industry sectors in less developed countries is identified as capability building for technology options based on technology transfer with the aim of achieving competitiveness in international markets; the 'intermediate technology' approach is also introduced for the clustering of small-scale developments in sectors of the third-world.
Tsoutsos and Stamboulis, 2005	A strategy is suggested that focuses on selected niches with the aim of integrating the innovation dimension into a policy for specific technology options; the growth in successful applications would lead to the development of new industry sectors in countries.
Knot *et al.*, 2001	Strategies for enhancing the flexibility of technological systems, which is increasingly required because of uncertainties and fast developments, to promote alternative technology options and change in industry sectors.
Coles and Peters, 2003	A more informed analysis of technological innovation, and associated options, is suggested for discussions about the future direction of industrial society and subsequent strategies that is required to adapt specific sectors to sustainability requirements.
Bessant and Francis, 2005	Mechanisms are explored for transferring technologies into sectors of developing countries, by first characterising technologies, and then identifying strategies for organisational development to facilitate such transfers.
Malairaja and Zawdie, 2004	Policy issues are discussed that need to be addressed to enhance the effectiveness of the transfer and innovation of specific technologies in sectors of developing countries.
Ayele, 2005	Analysis and strategy of how new technologies can be delivered in specific sectors of developing countries; specifically the transfer of knowledge between sectors and between innovation processes is addressed.
Harris and Pritchard, 2004	Adaptation of a technology transfer model for application at company, network and government level for symbiotic strategy formulation.

4.4. *Sustainability performance indicators for technology management*

General technical, economic, environmental and social indicators have been proposed for technology transfer evaluations (Dunmade, 2002). For large-scale resource-oriented technologies specific sustainability indicators have subsequently been developed, which are described in detail elsewhere (Brent and Visser, 2005; Labuschagne and Brent, 2006;

Mulder and Brent, 2006). Although the applications of these indicators do attempt to follow a holistic approach, constraints have been noted where sustainability information is required from parts of the expanded system that is not controlled by the particular technology management decision-makers. Especially in the initial research and development phases of technology management, a set of principles, methods and techniques or tools must be established for effectively assessing the potential value of a technology and its contribution to sustainable development during the market uptake phases of its life cycle (see Figure 4).

5. Conclusions

The turn of the millennium has seen increasing efforts to align technological research with the emerging field of sustainability science (Clark and Dickson, 2003). However, the field of science and technology for sustainability is in its infancy (AAAS, 2006). From the review of the literature summarized in this paper, it is concluded that sustainability aspects are not addressed adequately in technology management theories and practices.

The proposed conceptual framework is based on an existing framework for technology management sustainable development. It defines the context in which sustainable technology management should occur in practice. An expanded system perspective is required, that not only includes the respective technological, operational and business life cycles across companies, sectors, regions and countries, but also the dynamic interaction between macro, meso, and micro economies, societies at large, and the natural environment, as perceived by sustainability science. A modification to the definition of technology management has subsequently been proposed.

Emerging technology management practices related to sustainable development do emphasize the focus on technology strategy, selection and transfer, especially between developed and emerging economies. At the core of these issues lies technology assessment; this also forms part of other technology frameworks and methodologies. It is therefore recommended to concentrate on the development of technology assessment methods, as they are used in technology management practices, which incorporate the intrinsic modeling that is researched in the field of sustainability science. To this end, the modification of the available Technology Balance Sheet, Income Statement and Space Map analytical techniques are currently being investigated, with specific emphasis on the initial research and development phases of technology management.

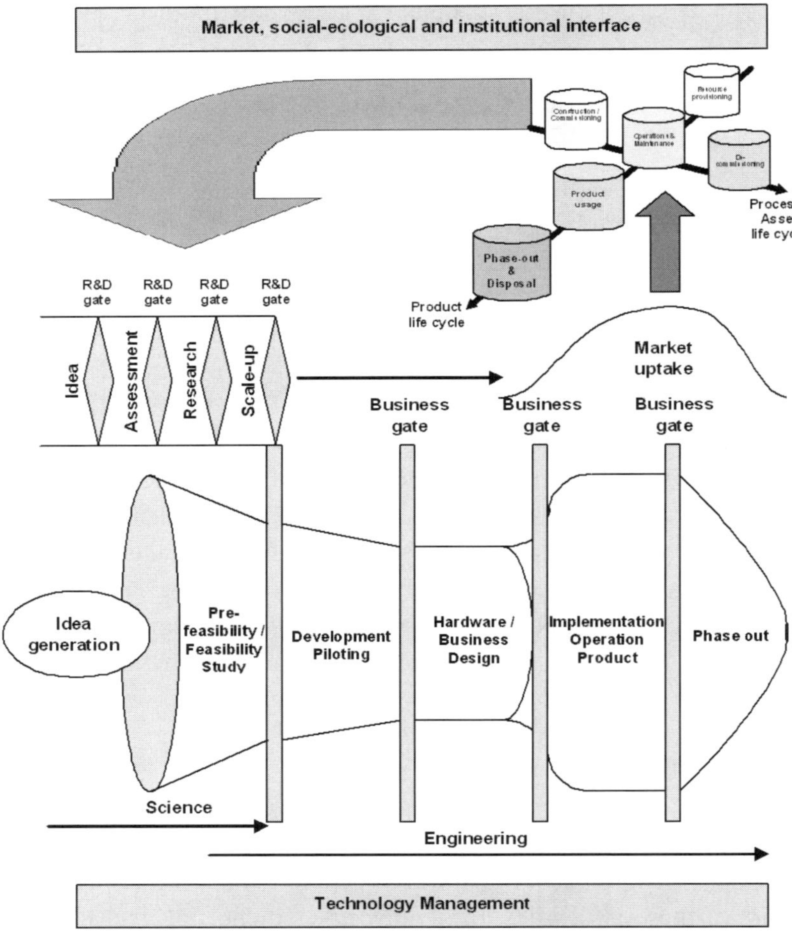

Figure 4: Technology life cycle interventions and associated evaluated systems.

Ultimately, the challenge lies in the formation and coordination of transdisciplinary research teams that are required to reach truly sustainable technology management practices.

References

American Association for the Advancement of Science (AAAS) (2006). FORUM: Science and Innovation for Sustainable Development. Website: http://sustsci.aaas.org/, accessed 20 December 2006.

Atkinson, G. (2000). Measuring corporate sustainability. *Journal of Environmental Planning and Management*, 43(2), pp. 235–252.

Ayele, S. (2005). Biotechnology generation, delivery and adoption: The case of Bt biopesticide in Eqypt. *International Journal of Technology Management and Sustainable Development,* 4(2), pp. 75–91.

Azapagic, A. and Perdan, S. (2000). Indicator of sustainable development for industry: A general framework. *Transactions of the Institution of Chemical Engineers (IchemE)*, 78(B4), pp. 243–261.

Bakouros, Y. (2005). Technology evaluation. Portland International Conference for the Management of Engineering and Technology (PICMET), Portland, Oregon.

Banwet, D. K., Momaya, K. and Shee, H.K. (2003). Competitiveness through technology management: an empirical study of the Indian software industry. *International Journal of Services Technolgy and Management*, 4(2), pp. 131–155.

Bessant, J. and Francis, D. (2005). Transferring soft technologies: Exploring adaptive theory. *International Journal of Technology Management and Sustainable Development,* 4(2), pp. 93–112.

Bowonder, B. and Miyake, T. (2000). Technology management: A knowledge ecology perspective. *International Journal of Technology Management*, 19(7), pp. 662–684.

Brady, T., Rush, H., Hobday, M., Davies, A., Probert, D. and Banerjee, S. (1997). Tools for technology management: An academic perspective. *Technovation*, 17(8), pp. 417–426.

Brent, A.C., Heuberger, R. and Manzini, D. (2005a). Evaluating projects that are potentially eligible for Clean Development Mechanism (CDM) funding in the South African context. *Environment and Development Economics*, 10(5), pp. 631–649.

Brent, A. C., van Erck, R. P. G. and Labuschagne, C. (2005b). A sustainability cost accounting methodology for technology management in the process industry. International Association for the Management of Technology (IAMOT), Vienna, Austria.

Brent, A. C. and Visser, J. K. (2005). An Environmental Performance Resource Impact Indicator for Life Cycle Management in the manufacturing industry. *J. Clean. Prod.*, 13(6), pp. 557–565.

Brent, A. C., van Erck, R. P. G. and Labuschagne, C. (2006). Sustainability Cost Accounting: Part 1—A monetary procedure to evaluate the sustainability of technologies in the South African process industry. *South African Journal of Industrial Engineering*, 17(2), pp. 35–51.

Brent, A. C., van Erck, R. P. G. and Labuschagne, C. (2007). Sustainability Cost Accounting: Part 2—A case study to demonstrate and assess the introduced monetary procedure to evaluate the sustainability of technologies in the South African process industry. *South African Journal of Industrial Engineering*, 18(1), pp. 1–17.

Briassoulis, H. (2001). Sustainable development and its indicators: Through a (planner's) glass darkly. *Journal of Environmental Planning and Management*, 44(3), pp. 409–427.

Clark, W.C. and Dickson, N.M. (2003). Sustainability science. Proceedings of the National Academy of Sciences of the United States of America, 100(14), pp. 8059-8061, website: http://www.pnas.org/cgi/content/full/100/14/8059, accessed 20 December 2006.

Coles, A.-M. and Peters, S. (2003). Sustainable development, global innovation and advanced technologies: The case of fuel cells. *International Journal of Environmental Technology and Management*, 3(3/4), pp. 278–289.

Cooney, R. (2004). The precautionary principle in biodiversity conservation and natural resource management: An issue paper for policy-makers, researchers and practitioners. IUCN Policy and Global Change Series, 2, World Conservation Union, IUCN, Cambridge, UK.

Demaid, A. and Quintas, P. (2006). Knowledge across cultures in the construction industry: Sustainability, innovation and design. *Technovation*, 26(5–6), pp. 603–610.

Dunmade, I. (2002). Indicators of sustainability: Assessing the suitability of a foreign technology for a developing economy. *Technology in Society*, 24(4), pp. 461–471.

Engineering and Technology Management Education and Research Council (ETMERC), 2006. Main research areas. Website: http://www.etmerc.org, accessed 20 December 2006.

European Institute for Technology and Innovation Management (EITM) (2001). Our purpose and mission. Website: http://www-eitm.eng.cam.ac.uk, accessed 20 December 2006.

Fahmy, Y. M. (2005). Catalysis role for sustainable industrial development in Egypt with prospective. *Technovation*, 25(6), pp. 645–655.

Geisler, E. (2002). The metrics of technology evaluation: Where we stand and where we should go from here. *International Journal of Technology Management*, 24(4), pp. 341–374.

Gerstlberger, W. (2004). Regional innovation systems and sustainability: Selected examples of international discussion. *Technovation*, 24(9), pp. 749–758.

Grieve, R.H. (2004). Appropriate technology in a globalizing world. *International Journal of Technology Management and Sustainable Development,* 3(3), pp. 173–187.

Hamilton, W. F. (2001). The biotechnology revolution: Lessons for technology management research and practice. *International Journal of Biotechnology,* 3(1–2), pp. 157–167.

Hanley, N., Shogren, J. and White, B. (1997). Environmental economics in theory and practice. Palgrave Macmillan, United Kingdom.

Harris, R. and Khare, A. (2002). Sustainable development issues and strategies for Alberta's oil industry. *Technovation*, 22(9), pp. 571–583.

Harris, S. and Pritchard, C. (2004). Industrial Ecology as a learning process in business strategy. *Progress in Industrial Ecology*, 1(1/2/3), pp. 89–111.

International Association for Management of Technology (IAMOT) (2006). IAMOT Body-of-Knowledge (BoK). Website: http://www.iamot.org/bok/, accessed 20 December 2006.

International Environmental Technology Centre (IETC) (2003a). Environmentally Sound Technologies and Sustainable Development. United Nations Environment Programme, website: http://www.unep.or.jp/ietc/knowledge, accessed 20 December 2006.

International Environmental Technology Centre (IETC) (2003b). Technology Transfer: The Seven "C"s for the Successful Transfer and Uptake of Environmentally Sound Technologies. Division of Technology, Industry and Economics, United Nations Environment Programme, website: http://www.unep.or.jp/ietc/knowledge, accessed 20 December 2006.

Jacob, W. F. and Kwak, Y. H (2003). In search of innovative techniques to evaluate pharmaceutical R&D projects. *Technovation*, 23(4), pp. 291–296.

Kates, R. W., Clark, W. C., Corell, R., Hall, J. M., Jaeger, C. C., Lowe, I., McCarthy, J. J., Schellnhuber, H. J., Bolin, B., Dickson, N. M., Faucheux, S., Gallopin, G. C., Grübler, A., Huntley, B., Jäger, J., Jodha, N. S., Kasperson, R. E., Mabogunje, A., Matson, P., Mooney, H., Moore III, B., O'Riordan, T. and Svedin, U. (2001). Sustainability Science. *Science*, 292(5517), pp. 641–642.

Khalil, T. M. and Ezzat, H. A. (2005). Management of technology and responsive policies in a new economy. *International Journal of Technology Management*, 32(1–2), pp. 88–111.

Knot, J. M. C, van den Ende, J. C. M. and Vergragt, P. J. (2001). Flexibility strategies for sustainable technology development. *Technovation*, 21(6), pp. 335–343.

Labuschagne, C. and Brent, A. C. (2005a). Sustainable Project Life Cycle Management: The need to integrate life cycles in the manufacturing sector. *International Journal of Project Management*, 23(2), pp. 159–168.

Labuschagne, C., Brent, A. C. and Claasen, S. J. (2005b). Environmental and social impact considerations for sustainable project life cycle management in the process industry. *Corporate Social Responsibility and Environmental Management*, 12(1), pp. 38–54.

Labuschagne, C., Brent, A. C. and van Erck, R. P. G. (2005c). Assessing the sustainability performances of industries. *Journal of Cleaner Production*, 13(4), pp. 373–385.

Labuschagne, C. and Brent, A. C. (2006). Social indicators for sustainable project and technology life cycle management in the process industry. *International Journal of Life Cycle Assessment*, 11(1), pp. 3–15.

Lambert, A. J. D. and Boons, F. A. (2002). Eco-industrial parks: Stimulating sustainable development in mixed industrial parks. *Technovation*, 22(8), pp. 471–484.

Liao, S.-H. (2005). Technology management methodologies and applications: A literature review from 1995 to 2003. *Technovation*, 25(4), pp. 381–393.

Maine, E., Probert, D. and Ashby, M. (2005). Investing in new materials: a tool for technology managers. *Technovation*, 25(1), pp. 15–23.

Malairaja, C. and Zawdie, G. (2004). The 'black box' syndrome in technology transfer and the challenge of innovation in developing countries: The case of international joint ventures in Malaysia. *International Journal of Technology Management and Sustainable Development*, 3(3), pp. 233–251.

Mebratu, D. (1998). Sustainability and sustainable development: Historical and conceptual review. *Environmental Impact Assessment. Review*, 18, pp. 493–520.

Momaya, K. and Ajitabh, A. (2005). Technology management and competitiveness: is there any relationship? *International Journal of Technology Transfer and Commercialisation.*, 4(4), pp. 518–524.

Monaghan, P. (2003). Interdisciplinary research design. School for New Learning, DePaul University, website: http://snl.depaul.edu, accessed 20 December 2006.

Mulder, J. and Brent, A. C. (2006). Selection of sustainable agriculture projects in South Africa: Case studies in the LandCare programme. *Journal of Sustainable Agriculture*, 28(2), pp. 55–84.

Murto, P. (2000). Competitive equilibrium and investments in a growing market: The choice between small- and large-scale electricity production. Annual Meeting of the Energy Economics Subprogram of the NERI, Reykholt, Iceland, website: http://www.ioes.hi.is/rammi4.htm, accessed 20 December 2006.

Phaal, R., Farrukh, C. J. P. and Probert, D. R. (2006). Technology management tools: Concept, development and application. *Technovation*, 26(3), pp. 336–344.

Phaal, R., Farrukh, C. J. P. and Probert, D. R. (2004). A framework for supporting the management of technological knowledge. *International Journal of Technology Management*, 27(1), pp. 1–15.

Pilkington, A. and Teichert, T. (2006). Management of Technology: Themes, concepts and relationships. *Technovation*, 26(3), pp. 288–299.

Portland State University (2003). ETM study: List of current responses. Website: http://www.etm.pdx.edu/survey/results.asp, accessed 20 December 2006.

Pretorius, M. W. and de Wet, G. (2000). A model for the assessment of new technology for the manufacturing enterprise. *Technovation*, 20(1), pp. 3–10.

Rennings, K. and Wiggering, H. (1997). Steps towards indicators of sustainable development: Linking economic and ecological concepts. *Ecological Economics*, 20(1), pp. 25–36.

Robèrt, K.-H., Schmidt-Bleek, B., Aloisi de Larderel, J., Basile, G., Jansen, J.L., Kuehr, R., Price Thomas, P., Suzuki, M., Hawken, P. and Wackernagel, M. (2002). Strategic sustainable development: Selection, design and synergies of applied tools. *Journal of Cleaner Production*, 10(3), pp. 197–214.

Sharif, N. (1992). Technological Dimensions of International Cooperation and Sustainable Development. *Technology Forecasting and Social Change*, 42(4), pp. 367–383.

Struyf, I. L. R. (2003). Inter-organizational learning for sustained competitiveness and ecological sustainability – the case of beta-testing activities of alternative-fuel, fuel cell-driven public transport buses. Doctoral research paper, Erasmus Centre for Sustainability and Management, Erasmus University, Rotterdam, the Netherlands.

Tsoutsos, T. D. and Stamboulis, Y. A. (2005). The sustainable diffusion of renewable energy technologies as an example of innovation-focused policy. *Technovation*, 25(7), pp. 753–761.

United Nations Commission on Sustainable Development (UNCSD) (2001). Indicators of sustainable development: Guidelines and methodologies. Website: http://www.un.org/esa/sustdev/isd.htm, accessed 20 December 2006.

University of Pretoria (2007). Technology Management Mind Map. Website: http://www.up.ac.za/gstm, accessed 24 August 2007.

Watanabe, C., Kondo, R. and Nagamatsu, A. (2003). Policy options for the diffusion orbit of competitive innovations: An application of Lotka–Volterra equations to Japan's transition from analog to digital TV broadcasting. *Technovation*, 23(5), pp. 437–445.

World Bank (2001). What is Sustainable Development. Website: http://www.worldbank.org/depweb/english/sd.html, accessed 20 December 2006.

Section IV

The Knowledge Chain and Value Creation

Chapter 13

Commercializing Breakthrough Technologies: Scenarios and Strategies[1]

J. Roland Ortt, Chintan M. Shah**, and Marc A. Zegveld**

**Department of Technology, Policy and Management (TPM), Delft University of Technology (TUD), Jaffalaan 5, 2628 BX Delft, the Netherlands, j.r.ortt@tudelft.nl, m.a.zegveld@tudelft.nl*
***Department of TPM, TUD, the Netherlands, and Bluewater Energy Services BV, chintanmshah@gmail.com*

In this chapter we focus on different strategies for commercialising breakthrough technologies. After the invention of their technology, pioneers might be confronted with completely different scenarios, each of which requires another strategy. These strategies are studied for the cases of the photocopier, video cassette recorder and microwave oven. Several conclusions are derived from the cases. The main actors involved in the breakthrough technologies change considerably in the course of time, and so do the customer segments and user applications. Usually, niche markets emerge first that diverge considerably from the mass market applications that emerge later. In our cases, pioneers of breakthrough technologies never create a mass market. After the pioneer, multiple entrants adopt a similar strategy but many of them have to leave the market later. Only a small number of them survives, either by consistently adopting a niche market strategy or by creating a mass market.

[1] We gratefully acknowledge the valuable comments from Dr. M. Hashem Sherif.

1. Introduction

This chapter focuses on strategies for commercialising breakthrough technologies. Breakthrough technologies represent an advance in technology that is so significant that either attainable price/performance ratios are altered dramatically or an entirely new kind of application is enabled that changes the behaviour pattern of end users (Tushman & Anderson, 1986). Examples of breakthrough technologies or products based on these technologies include dynamite, Kevlar, radar, laser, telegraphy and television.

Although developing and subsequently commercialising breakthrough technologies can be very rewarding, it is also a risky endeavour for companies. On the one hand, there are potential high gains, like achieving a competitive advantage that can contribute significantly to a firm's growth and profitability (Veryzer, 1998; Kleinschmidt & Cooper, 1991). Breakthrough technologies have been the source of new product categories, new markets and new industries (Christensen, 1997; Abernathy & Clark, 1985). The history of the American company Raytheon, for example, is intimately connected with the radar, Xerox with the photocopying machine and The Bell Company with the transistor. On the other hand, it is remarkable how many companies that are involved in the invention of breakthrough technologies lose out when the technology is applied on a large scale (Olleros, 1986; Tellis & Golder, 1996). Projects dedicated to breakthrough technologies are risky and expensive, and it usually takes several years to produce results (Leifer *et al.*, 2000). Technical, market-related and organisational uncertainties associated with these kinds of projects are much higher than with projects aimed at incremental improvement (Burgelman & Sayles, 1986).

In order to look at the causes of the risks involved and the appropriate strategies for dealing with them, we describe the process of development and diffusion of breakthrough technologies in the next section. We show that alternative scenarios can emerge after the invention of a technology. In the third section, different strategies for commercialising these technologies will be presented and their result in the scenarios will be discussed. In the fourth section, case-study results are presented for the photocopier, video cassette recorder and microwave oven – all products based on breakthrough technologies. These cases focus on the following questions: Who were the most important actors during the phases of the pattern? In which applications and customer segments did they introduce their products? What were the business models that they adopted? Conclusions and discussions are presented in the final sections.

2. The Process of Development and Diffusion of Breakthrough Technologies

In many cases, the diffusion of products based on breakthrough technologies shows a remarkably similar S-shaped pattern (Rogers, 1986; Williams, Rice & Rogers, 1988). This similarity seems to imply that it would be relatively easy to predict market results and that the risks are relatively limited. However, the S-shaped pattern does not reflect the actual risks experienced by pioneering companies. Empirical results indicate that the process of development and diffusion begins much earlier (Ortt & Schoormans, 2004) and that the S-curve is only the final phase of a much longer process that is difficult to predict.

Ortt and Schoormans (2004) distinguish three phases. In the first phase, from invention to first market introduction, the technology is developed into a marketeable product. In this phase, prototypes are developed and tested until the products are considered market-ripe. In the second phase, from the initial market introduction up to large-scale industrial production and diffusion, alternative product types incorporating the technology emerge and new applications for the technology appear. In practice, an erratic process of diffusion may occur in this phase instead of a smooth S-curve (Clark, 1985). In this phase, the diffusion is often characterized by the periodic introduction, decline and re-introduction of multiple products in multiple small-scale applications (Carey & Moss, 1985). In the third phase, from large-scale production and diffusion up to the substitution of the technology, we usually encounter the S-curve. Similar phases to the ones outlined here are distinguished by Agarwal and Bayus (2002) and Tushman and Anderson (1986).

Table 1: Milestones in the process of development and diffusion of the cases.

Cases	Invention; Market introduction; Diffusion takes off	Length first phase	Length second phase	Length of phases before large-scale production and diffusion
Photo copier	1938; 1949; 1976	11	27	38
Video recorder	1951; 1956; 1971	5	15	20
Microwave oven	1945; 1947; 1972	2	25	27

An overview of essential milestones in the process of development and diffusion of the photocopier, video cassette recorder and microwave oven is shown in Table 1.

Table 1 shows that in each of the cases, at least two decades elapsed before the diffusion took off. Compared to the usual life span of patents, approximately 18 years, this is quite a long period of time. Other authors have found similar periods (Agarwal & Bayus, 2002; Utterback & Brown, 1972).

In all three cases the first phase, from invention to first market introduction, is relatively short compared to the next phase. In general, we believe that in specific situations each phase can disappear. Our ideas are summarised in two propositions: (i) The phases can vary considerably in length. One or more phases may even disappear; (ii) The entire process can stop in each phase. These propositions suggest a more unpredictable process than the regular occurrence of the S-shaped pattern would have us believe. In practice, the actors involved in the commercialisation of breakthrough technologies may face different scenarios after the invention of their technologies.

After studying the pattern of development and diffusion of more than fifty breakthrough technologies, we distinguish three important scenarios. In Scenario 1, there is a long first phase, which means that it takes a long time before a product based on a new technology can be introduced in the market. In scenario 2, the product is ready for the market within a short time after its invention, but there is a lengthy second phase, which means that it takes a long time before the product diffuses in a mass market. In scenario 3, the first two phases are relatively short, which means that both phases almost disappear and products quickly diffuse in a mass market.

3. Strategies During the Process

There are different strategies for coping with the uncertainty that is inherent in the various scenarios. However, the real consequences of the uncertainty become clear when we look at the position of a pioneering company, *i.e.* a company that has invented a breakthrough technology and is now considering introducing this technology in the market. Suppose that such a company can choose between three main strategies: a niche market, a mass market and a wait-and-see strategy.

A niche market strategy means that the pioneer invests in small production facilities to manufacture a specific product that is tailored to the niche, and deploys specific marketing and distribution resources for this niche. Empirical research indicates that niche market strategies can be very successful, especially in the case of innovative products (Hultink *et al.*, 1997; De Bruyne *et al.*, 2002).

A mass market strategy means that the pioneer invests in large-scale production, distribution and marketing facilities. Being the first with a mass market strategy makes sense, for example, when network effects in the market enable the first mover to establish a strong position in the market and subsequent movers are confronted with potential consumers that are locked in and are therefore reluctant to switch to another type of product (Arthur, 1996). A prerequisite for this strategy is a market situation that enables a relatively fast diffusion.

Adopting a wait-and-see strategy means that the pioneer deliberately decides to become a follower. This strategy implies that the company develops the technology, prepares for its marketing and distribution and monitors the market results of competitors, waiting for the moment when the introduction becomes a commercially viable option. Tellis and Golder (1996) clearly indicate that, in the case of breakthrough technologies, followers often conquer the market and pioneers usually disappear before the technology becomes a success. Followers can decide to keep their knowledge regarding the technology up to date, and at the same time introduce their products at the commercially appropriate moment. Other followers may choose to invest in venturing or buy stakes in the pioneering companies in order to be prepared when technology and market are ripe (Chesbrough, 2003; Constantinos & Geroski, 2005).

If we combine these three strategies with the three scenarios distinguished in the previous section, it becomes clear what risks are facing the pioneer (see Table 2). A mass market strategy requires a very

Table 2: Result of three strategies in different scenarios.

Mass market strategy	Niche market strategy	Wait-and-see strategy
Scenario 1: Long first phase (between invention and first market introduction)		
The product introduction may be announced but is postponed. The technology may develop further and take on a different form. Stopping the preparations for production, marketing and distribution activities is sure to result in considerable losses.	The product introduction may be announced but postponed. Stopping the preparation for production, marketing and distribution activities is less expensive compared to the case of a mass market strategy.	The strategy requires relatively small investments to keep the knowledge of the technology up-to-date and to prepare for marketing and distribution. No losses due to a failed introduction.
Scenario 2: Short first phase and long second phase (from initial market introduction to large-scale production and diffusion)		
Sales will be disappointing. The technology and the market may develop further, albeit slowly. Adapting the production, marketing and distribution activities will lead to considerable costs.	Sales will be normal for a niche strategy. Adapting production, marketing and distribution activities is relatively affordable.	On the basis of the results of competitors, knowledge about the market reaction can be monitored. Introduction can wait until viable market applications emerge. A risk is that other competitors may establish a strong position in the market more quickly.
Scenario 3: Short phases before large-scale production and diffusion		
Sales increase quickly, and because of the available facilities demand can be met and the company can gain a strong position quickly.	Sales will be higher than expected. An inability to meet demands can have considerable consequences (loss of market share) unless production distribution and marketing can be scaled up very quickly.	Diffusion takes off earlier than expected. When competitors adopt a niche or mass market strategy it may be difficult for a company that has opted for his strategy to establish a position in the market.

large investment, a niche strategy requires a relatively large investment and a wait-and-see strategy requires a relatively small investment.

It may be clear that each strategy is optimal for only one of the scenarios. A mass market strategy is the best strategy in scenario 3 where diffusion takes off shortly after the invention. A niche market strategy appears to be the best strategy in scenario 2, where it takes a long period of probing before the diffusion takes off, and a wait-and-see strategy seems to be the best strategy in scenario 1, where it takes a long period before the technology is ready to be marketed. Potential gains or losses also vary for each of the strategies. Mass marketing is a strategy with very high gains in scenario 3, and very high losses in the other two scenarios. Niche marketing has fairly high gains in scenario 2 and limited losses in the other two scenarios. A wait-and-see approach minimises the risk of large losses, although there is a risk that a company that decides to take this approach is too late when the diffusion takes off. The strategies applied by some of the main actors in the three cases are described in the following section.

4. Case Studies

We focus on three cases: the photocopier, video recorder and microwave oven. The main actors who developed and supplied these products (Table 3), the customer segments and applications (Table 4), and the business models (Table 5) are described below.

4.1. *Actors*

By actors we mean the individuals or companies responsible for developing a technology or creating products based on that technology. They play an important role in the pattern of development and diffusion of breakthrough technologies. Whether or not a product is successful depends, at least partly, on the strategies adopted by these actors. The actors in the different phases of the pattern are in Table 3. The approximate year in which they became active is shown whenever possible.

Table 3: Actors during the different phases of diffusion of breakthrough technologies.

Before invention	First phase (invention to initial market introduction)	Second phase (initial market introduction to large-scale production/ diffusion)	Third phase (large-scale diffusion to complete substitution)
Photo-copier 1934 – 1938 Chestor Carlson works on photocopying process	1938 – 1949 Chestor Carlson, Battelle Memorial Institute (1944), Haloid (1946)	1949 ~ 1976 Pioneer: Haloid –Xerox (1949); *Follower:* IBM (1970), Kodak (1975); Smart follower: Canon (1968)	~ 1976 - - *Late entrants:* Minolta, Ricoh, Sharp, GE, etc.
Video-recorder ~ 1951 Ampex forms an in-house team of engineers to develop video recorder	1951 – 1956 Ampex (1954), Toshiba (1959)	1956 ~ 1971 *Pioneer:* Ampex (1956); *Follower:* Toshiba (1959), RCA (1969), CBS (1968), CTI (1971); *Smart follower:* Sony (1962), Matsushita & JVC (1963)	~ 1971 - - *Late entrants:* Philips, Panasonic, Fisher, Sharp, GE, etc.
Microwave oven 1940 ~ 1946 Raytheon develops radar appliances and systems mainly for military use	1946-1947 Raytheon Company (1945)	1947 ~ 1972 *Pioneer:* Raytheon Company (1947); *Smart Follower:* Tappan (~1955), Sharp (1962), Litton Industries (1965), Panasonic (1966), Amana - Raytheon's acquired subsidiary (1967)	~ 1972 - - Late entrants: GE, Magic Chef, Whirlpool, Norelco, Sunbeam Samsung, etc.

Sources: (Chesbrough, 2003; Constantinos & Geroski, 2005; Lurie & Yoffie, 1990; Magaziner & Patinkin, 1989; Rosenbloom & Cusumano, 1987; Wong & Buzzel, 1983) and the website of individual companies.

The actors enter and leave the diffusion process on a varying timeline. They create a new strategy or adopt a strategy based on the one used by the previous actors. Based on this, we can distinguish four different groups of actors, all of which try to supply products in the market. First, the pioneers who are the first to commercialise a new product based on a breakthrough technology. Secondly, the followers, who copy the pioneers' business model and also target niche markets.

Thirdly, the smart followers, who open up a mass market by adopting a different business model than the one(s) used by the pioneers. Fourthly, the late entrants, who copy the mass market strategy of smart followers.

Several observations can be derived from Table 3. *First*, different actors dominate different phases. For instance, while Ampex was most active in phase 1 and 2, other actors like Sony and JVC took the lead later on in phase 3. *Secondly*, there is a real increase in the number of actors just before large-scale production and diffusion. For instance, many firms, including Tappan, Sharp and Litton, entered the market in phase 2 before the large-scale diffusion of the microwave oven, and numerous firms, such as GE, Magic Chef and Whirlpool, entered in phase 3 after the diffusion took off. *Thirdly*, the actors dominating a particular phase were already active in an earlier phase. In other words, the pioneers were already active before the first market introduction, while the smart followers were already active long before the diffusion took off. For example, Haloid (now Xerox) partly sponsored Carlson's experiments in return for proprietary manufacturing rights and thus was already active before products were introduced. Canon, which opened up the mass market for photocopying machines, was actively experimenting with its new product in phase 2. *Fourthly*, none of the pioneers in our cases opened up a mass market. Ampex, for example, introduced the video recorder for a niche market, while companies like Sony, Matsushita & JVC opened up the mass market.

4.2. *Applications and customer segments*

One of the key features distinguishing *the smart followers* from *the pioneers* is the customer segment they target. The customer segments targeted by these two types of actors are presented in Table 4. It is important to study these two types of actors in particular, because they are the ones creating entirely new markets. Although these actors take high risks, the returns of a successful market creation can also be high, as we explained in the introduction.

Based on Table 4, a few observations are worth making. First of all, pioneers target their products at a niche market, which usually means they choose a high end market with specific needs. For example,

Table 4: Customer segment – pioneers and smart followers.

Customer segment—Pioneer	Customer segment—Smart followers
Photocopier: Corporate and government offices, catalogue publishers	Consumer mass market—small and medium-sized businesses (SMEs) and individuals
Video recorder: Broadcasting industry, schools, police department	Consumer mass market—household use
Microwave oven: Institutional market—hospitals, schools, restaurants, railroad and ocean liners	Consumer mass market—household use

Raytheon targeted its microwave ovens at institutional markets such as restaurants, hospitals and railroad liners, where large quantities of food have to be cooked rapidly. *Secondly*, pioneers often stick to their niche market. They keep developing their product to improve its performance in an attempt to serve the needs of their specific customers rather than making a simpler version of their product that can be supplied to a large market. *Thirdly*, the smart followers that create the mass market focus on an entirely different customer group. Rather than a small number of customers, they target at a larger customer group that needs a simple product with basic functionality and that is unwilling to pay a high price.

For example, instead of competing for Xerox's major government or corporate customers, Canon focused on individual people and small businesses unable to afford expensive photocopiers and needing no more than a few copies each day.

4.3. *Business models*

An overview of the different business models adopted by pioneers and smart followers in the cases is provided in Table 5. The pioneers often base their products on past experience, and target them at a niche of high-end customers with which they are familiar. They attempt to use their breakthrough technology to solve a challenge that cannot be solved through traditional technologies. Often, required complementary products and services are either entirely absent or underdeveloped. Pioneers often have to invest in the creation of complementary products and services. Followers, by contrast, gain some lead as they built on the experiences of the pioneers. They have the opportunity to learn from the successes and failures of their predecessors. In addition, they have the advantage of being able to use existing complementary products, services, proven technology, consumer awareness and market research.

Table 5: The business models adopted by pioneers and smart followers.

Pioneer	Smart followers
Photocopier » Xerox: high end niche market; high speed, high volume, quality prints; *lease* model – low lease rate for machine, nominal charge per copy; developed complementary products and services – papers, service parts, dealers, service personnel, financing, etc.	» Canon: mass consumer market; medium speed, medium quality prints, self repairable machine; affordable price; low margin on machine, high margin on cartridges; outsourced servicing, dealership etc.
Video recorder » Ampex: high end niche market; leveraged on convenience of use and cost effectiveness; offered complete package.	» Sony, Matsushita and JVC: started high-end in Japan, moved to consumer market using their expertise; reduced the product size; made product cheap and cassettes widely available.
Microwave oven » Raytheon: focused on institutional market; robust construction; improved performance step-by-step; offered complementary products and services; high margin; used cross-licensing to capitalise on patented technology; acquired Amana to introduce simpler product for consumers.	» Sharp, Panasonic, Sanyo: focused on small size, sufficient heating capacity product for household use; affordable price; dramatically reduced prices based upon the 'Japan tube' costing $1/9^{th}$ the price of Raytheon's magnetron tube.

5. Conclusions

In this article we have focused on whether the actors and their strategies change during the different phases of the pattern.

Actors during the pattern

Depending on the timing of the activity of the actors and on the strategy that they adopt we distinguished four groups: (1) the pioneers; (2) the followers copying the strategy of the pioneers; (3) the smart followers adopting a different strategy aimed at creating a mass market; (4) the late entrants copying the smart followers. We found that actors enter and leave in the course of the entire process and that different actors dominate in each phase. The actors who dominated in a particular phase were also active before that phase. In all cases the number of actors increased considerably immediately before the technology-based products started to diffuse in a mass market. This is consistent with Agarwal and Bayus (2002) who investigated this phenomenon in thirty cases.

Strategies of pioneers and followers

The pioneers in the three cases first targeted their products at niche markets. There are several reasons why pioneers adopt this approach. Firstly, the high-end of the market is willing to pay a premium price. If a pioneer can protect its technology, for example using patents or unique knowledge that is hard to copy, it makes sense to skim the market, *i.e.* first focus on the niches that are willing to pay most and gradually reduce the price to increase the number of customers. Secondly, when a technology is first introduced, complementary products and services still need to be developed. In the case of photocopiers, for instance, special types of ink and paper are required. In the market adaptation phase the technology and the products are still evolving, as are the complementary products and services, because no standards have been set. This means that at first both the products and their complementary products and services are made to measure. This can only be profitable when targeting high-end customers.

In none of the cases did pioneers create a mass market with their breakthrough technology. This statement applies to Xerox (photocopiers), Raytheon (microwave ovens) and Ampex (video recorders). It would seem pioneers have a tendency to stick to the high-end niche markets. They keep developing to improve their products and serve the needs of their demanding customers. An explanation for this phenomenon is provided by Christensen (1997). Companies tend to upgrade their product and focus on the demands of their current customers rather than make a simpler version of their products and supply it to a broader group of customers.

In our cases we found that followers who merely copy the pioneer's strategy do not create a mass market. They have to find other niches with less potential because the pioneer will have a strong position in the 'best' niches. Moreover, they often have to pay for licenses to be allowed to produce and supply their products. Examples of this type of followers are IBM and Kodak in the case of photocopying. Both these powerful corporations decided to leave the market after they incurred considerable losses. Copying the niche strategy of a pioneer only makes sense in very specific circumstances, for example when patents can be circumvented or when the follower has access to a very profitable niche market that can not be served by the pioneer.

Followers who create a mass market adopt a different approach. The customers in a mass market need a simpler product and are usually more price-sensitive than their high-end counterparts. Supplying these markets in a profitable way requires a different business model: a different product design, a different scale of production, a different form of marketing and servicing, and so on. Canon, for example, targeted photocopiers at SMEs. SMEs want simple photocopiers at a low price and are willing to compromise on quality and speed. They also want standard parts and the ability to do most of the servicing of the machine themselves. This strategy entails a more large-scale approach, which is a viable option when more customers want products on the basis of the technology, and when standards for the products have emerged and complementary products and services can be supplied by actors other than the pioneer. Therefore, smart followers wait and see. In practice, it is difficult to indicate clearly when a market is ready for a mass market approach. This means that the followers who try to create a mass market often have to test and see whether the market is ready for their business model. Canon, for example, tried to introduce its photocopiers a number of times before its strategy became a success.

6. Discussion

The strategies used by companies are shown to be different in the stages of the product life cycle (PLC) (Swan & Rink, 1982). We have added to existing knowledge by focussing on the technology incorporated in technology-based products and on the phases preceding the PLC. In some cases it is the followers rather than the pioneers who benefit from the commercialisation of breakthrough technologies (Teece, 1997; Markides & Geroski, 2005). Our contribution has been to distinguish different kinds of followers and to indicate in what ways their strategies and chances for success are different.

We understand that there is a need to look at more cases than the three included in this article. More research is also needed to investigate the causes of the second phase and different types of niche-strategies that can be adopted to cope with this phase.

References

Arthur, W. B. (1996). Increasing Returns and the New World of Business. *Harvard Business Review*, July-August, pp. 100–109.

Abernathy, W. J, and Clark, K. B. (1985). Innovation: Mapping the winds of creative destruction. *Research Policy*, 14(1), pp. 3–22.

Agarwal, R. and Bayus, B. L. (2002). The Market Evolution and Sales Takeoff of Product Innovations. *Management Science* 48(8), pp. 1024-1041.

Burgelman, R. and Sayles, L. R. (1986). Inside corporate innovation. The Free Press, New York.

Carey, J. and Moss, M. L. (1985). The Diffusion of Telecommunication Technologies. *Telecommunications Policy*, 6, pp. 145–158.

Chesbrough, H. (2003). Open Innovation: The New Imperative For Creating And Profiting From Technology. Harvard Business School Press, Boston.

Christensen, C. M. (1997). The Innovator's Dilemma. Harvard Business School Press, Boston.

Clark, K. B. (1985). The Interaction of Design Hierarchies and Market Concepts in Technological Evolution. *Research Policy*, 14, pp. 235–251.

Constantinos, M. and Geroski, P. A .(2005). Fast second. Jossey-Bass, San Fransico.

Debruyne, M., Moenaert, R., Griffin, A., Hart, S., Hultink, E. J. and Robben, H. S. J. (2002). The impact of new product launch strategies on competitive reaction in industrial markets. *Journal of Product Innovation Management*, 19(2), pp. 159–170.

Hultink, E. J., Griffin, A., Hart, S. and Robben H. S. J. (1997). Industrial New Product Launch Strategies and Product Development Performance. *Journal of Product Innovation Management,* 14(4), pp. 243–257.

Kleinschmidt, E. J. and Cooper, R. G. (1991). The Impact of Product Innovativeness on Performance. *Journal of Product Innovation Management*, 8, pp. 240–251.

Leifer, R., McDermott, C. M., O'Connor, G. C., Peters L. S., Rice. M. P. and Veryzer, R. W. (2000). Radical Innovation: How Mature Companies Can Outsmart Upstarts. Harvard Business School Press, Boston.

Lurie, R. Y. and Yoffie, D. B. (1990). The world VCR industry. Harvard Business School Press, Boston.

Magaziner, I. C. and Patinkin, M. (1989). Fast heat: how Korea won the microwave war. *Harvard Business Review*, Jan-Feb, pp. 83–91.

Markides, C. C. and Geroski, P. A. (2005). Fast Second; Being First to Market Doesn't Always Cut it. When it Comes to New Technologies, Rapid Responders Might Just Have the Competitive Edge. Strategy & Innovation, Harvard Business School Case, January-February, pp. 1–5.

Olleros, F. (1986). Emerging Industries and the Burnout of Pioneers. *Journal of Product Innovation Management,* 1, pp. 5–18.

Ortt, J. R. and Schoormans, J. P. L. (2004). The Pattern of Development and Diffusion of Breakthrough Communication Technologies. *European Journal of Innovation Management*, 7(4), pp. 292–302.

Rogers, E. M. (1986). Communication Technology. The New Media in Society. The Free Press, New York.

Rosenbloom, R. S. and Cusumano, M. A. (1987). Technological pioneering and competitive advantage: The birth of the VCR industry. *California Management Review*, 29(4), pp. 51–76.

Swan, J. E. and Rink, D. R. (1982). Fitting Market Strategy to Varying Product Life Cycles. *Business Horizons*, January-February, pp. 72–76.

Teece, D. J. (1997). Managing Strategic Innovation and Change. In: Capturing Value from Technological Innovation: Integration, Strategic Partnering, and Licensing Decisions (Tushman, M. L. and Anderson, P., eds.), pp. 287–306, Oxford University Press, Oxford.

Tellis, G. J. and Golder P. N. (1996). First to Market, First to Fail? Real Causes of Enduring Market Leadership. *Sloan Management Review,* Winter, pp. 65–75.

Tushman, M. L. and Anderson, P. (1986). Technological Discontinuities and Organizational Environments. *Administrative Science Quarterly*, 31, pp. 439–465.

Utterback, J. M. and Brown, J. W, (1972), Monitoring for Technological Opportunities, *Business Horizons*, 15(October), pp. 5–15.

Veryzer, R. W. (1998). Key Factors Affecting Customer Evaluation of Discontinuous New Products. *Journal of Product Innovation Management,* 15, pp. 136–150.

Williams, F., Rice, R. E. and Rogers, E. M. (1988). Research Methods and the New Media. The Free Press, New York.

Wong, K and RD Buzzel (1983). Note on the Microwave oven industry. Harvard Business School Case.

Chapter 14

Industrialization Guidelines for South Africa's Pebble Bed Modular Nuclear Reactor Programme

André Buys

Graduate School of Technology Management
University of Pretoria, Pretoria, 0002, South Africa
andre.buys@up.ac.za

The industrialization of any new technical system requires the establishment of the required industrial infrastructure and supplier base for the system. This chapter presents a new methodology that was used for determining industrialization guidelines for South Africa's Pebble Bed Modular Nuclear Reactor (PBMR) programme. The methodology consists of a number of steps. These were a survey of the required industrial capabilities and an audit of current domestic industrial capabilities. In the next step, the strategic and operational requirements of the programme, current and future market conditions and domestic industrial capabilities were evaluated and the scores were combined to calculate the recommended industrialization guidelines. There were four possible outcomes or guidelines that this methodology could produce: local procurement, overseas procurement, partnership procurement, or competitive procurement. A panel of experts using a modified Delphi technique did the evaluation. It was found that the PBMR systems, sub-systems, units and components should be procured from different sources. However, the option of open international tenders where price is the primary criterion was not the favored option for any of the items considered.

1. Introduction

Most of South Africa's older power stations will reach the end of their life shortly after 2025 and will have to be replaced within the next 20 years. However, even moderate economic growth will demand that South Africa increase it's electricity supply by an additional 20 000 MW by 2025 over and above the currently installed 39 000 MW. The Pebble Bed Modular Reactor (PBMR) technology provides a clean, safe and cost-effective option for meeting this need.

The PBMR is a High Temperature Reactor (HTR), with a closed-cycle gas turbine power conversion system. Although it is currently not the only HTR that is being developed in the world, the South African project is internationally regarded as the leader in this technology. The PBMR combines high efficiency and attractive economics with high levels of passive safety. Not only does it have the ability to generate electricity economically, but also to produce hydrogen for the fuel of the future, desalinated water and industrial or residential process heat.

A PBMR plant will comprise of a module building housing the Reactor Pressure Vessel and the Power Conversion Unit (Figure 1). The RPV is 6,2 meter in diameter and about 27 meter high and holds about 450 000 fuel spheres. It is lined with a 1 meter thick layer of graphite bricks, which serves as an outer reflector and a passive heat transfer medium. The molded graphite fuel spheres containing low enriched uranium particles are about the size of tennis balls. The particles consist of kernels of uranium dioxide coated with silicon carbide and pyrolitic carbon.

Helium is used as the coolant and energy transfer medium, to drive a closed cycle gas turbine and generator system. The helium coolant enters the reactor vessel at a temperature of about 500 °C and a pressure of 9 MPa. The gas moves down between the hot fuel spheres, after which it leaves the bottom of the vessel with a temperature of about 900 °C. The hot gas then enters the turbine that drives the generator through a speed-reduction gearbox on one side and the gas compressors on the other side. The coolant leaves the turbine at about 500 °C and 2,6 MPa, after which it is cooled, recompressed, reheated and returned to the reactor vessel. See Figure 2.

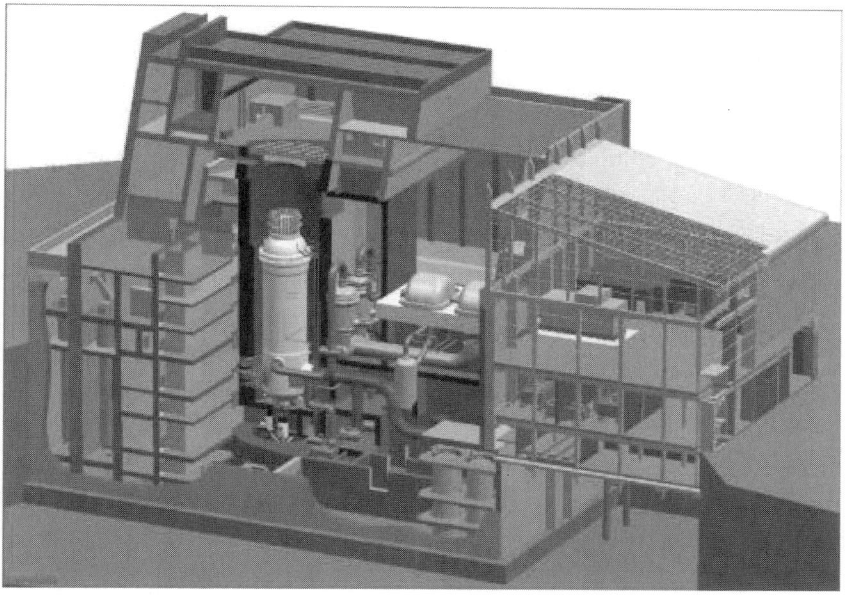

Figure 1: The PBMR power generation plant (Source: PBMR (Pty) Ltd., 2007).

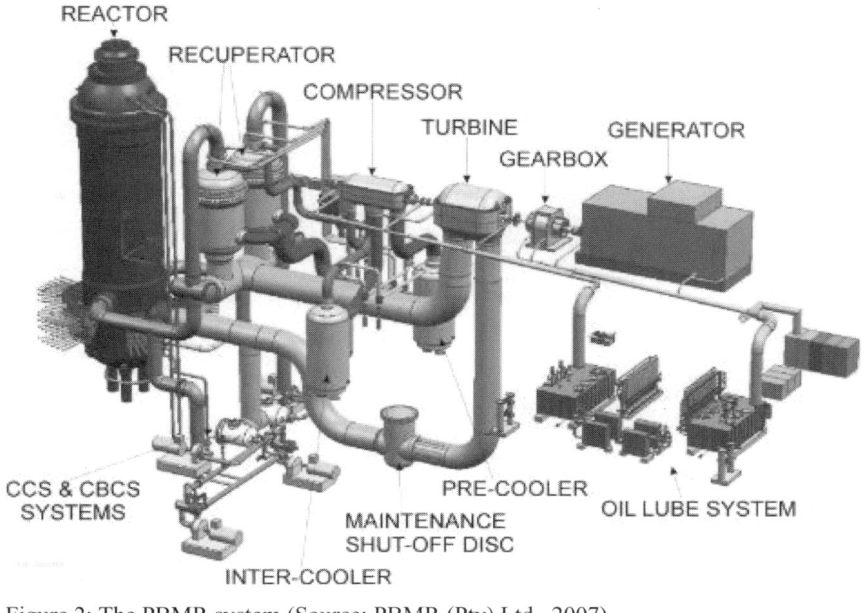

Figure 2: The PBMR system (Source: PBMR (Pty) Ltd., 2007).

The key features of the PBMR system are safety, size, environmental hygiene, short construction periods and cost-competitiveness. Because PBMR plants are much smaller than conventional power stations, the PBMR plants can be erected closer to points of demand, facilitating distribution and lowering costs. Furthermore, the modular construction allows for extension according to demand. PBMR technology has passive safety systems embedded that does not rely on human intervention. Even if there is a failure of the active systems that are designed to shut down the nuclear reaction and remove core decay heat, the reactor itself will stop any nuclear fission and eventually cool down naturally. PBMR technology emits no pollutants into the atmosphere. Radiation is contained and the environmental impact is much less disruptive than with traditional fossil fuel power generation plant.

2. PBMR Industrialization Plan

The PBMR programme is a major technological innovation programme currently conducted by the SA industry. It has the potential to stimulate technology and industrial development in a variety of industrial sectors in South Africa. This could not only lead to economic growth but has the potential to create many jobs in the manufacturing sector of the economy.

The industrialization of a new technical system such as the PBMR consists of the establishment of the required industrial infrastructure and supplier base for the construction, manufacture, operation, support, maintenance and upgrading of the system during its life cycle. To be able to meet the expectations of its stakeholders and customers the PBMR Company has to ensure that its infrastructure and supplier base are adequate and sustainable.

2.1. *PBMR system life-cycle model*

Any industrial system will progress through its life cycle, starting from the inception of the concept to final decommissioning to greenfield site after mothballing. The life-cycle of a nuclear power reactor system is very long and the PBMR is being designed for an operational lifetime of 40 years, followed by a passive period of another 40 years to allow for the decay of the radioactive isotopes in the spent fuel stored in the shut-down module building. With such a long operational lifetime, it could be expected that technological progress would necessitate the upgrading and modernization of the system during its operational lifetime. It is further foreseen that there will be ongoing construction of new modules.

The technological capabilities required during the life-cycle of the system will change, starting with the initial emphasis on design and development capabilities during the early phase, progressing to emphasis on manufacture and construction capabilities during the construction phase, followed by the operational phase when the requirement for operational capabilities, including maintenance and upgrade capabilities, becomes important. It is important to keep in mind that this is a process of accumulation of capabilities as those that are required in the early phases, such as design and development capabilities, do not become superfluous during the later phases, as they are required for ongoing maintenance, upgrade and modernization of the system. The life-cycle model proposed for the PBMR system consists of ten phases as shown in Table 1.

2.2. *PBMR system breakdown structure*

The purpose of the System Breakdown Structure was to obtain a complete overview of the technologies incorporated in the PBMR system. The System Breakdown Structure consisted of a multilayer System Hierarchy down to the level where technical capabilities could be identified, in most cases to that of components, materials and processes as shown in Figure 3.

The System Hierarchy also defined the industry structure in terms of supplier relationships as shown in Table 2.

Table 1: PBMR Life-cycle phases.

Life-cycle phase	Description
Phase 1: Concept definition	The first phase in the initiation of the PBMR programme consisted of the concept definition and is primary a theoretical phase. Various studies are undertaken such as technical feasibility studies and financial and market analyses.
Phase 2: Basic design	Once the concept has been established, the basic system design commences. This is primarily a technical design phase, consisting of theoretical analysis of system operational features and engineering design.
Phase 3: Demonstration module detail design and construction	This phase consists of the detail design, specification, contracting, manufacture and construction of the first operational demonstration PBMR module.
Phase 4: Demonstration module operation and evaluation	After commissioning of the first module, a period of operation, test and evaluation is required to verify the integrity of the system and the operational characteristics.
Phase 5: Standardized module design and construction	This phase consists of the detail design, specification, contracting, manufacture and construction of standardized operational PBMR modules.
Phase 6: Operation	This phase consists of the routine operation and maintenance of operational PBMR modules by the utility company(-ies) operating them.
Phase 7: Upgrade and modernization	This phase consists of the upgrade and modernization of PBMR modules during their operational lifetime, including mid-life and life extension upgrades.
Phase 8: Shut-down and mothballing	After completion of its operational lifetime of 40 years, the PBMR module is shut down, all spent fuel is removed to the storage area and the unit mothballed.
Phase 9: Interim spent-fuel storage	During this 40-year period the spent fuel is stored in the shutdown module/s to allow for the decay of the highly radioactive isotopes in the spent fuel.
Phase 10: Final decommissioning to Greenfield site	This phase consists of the removal of the spent fuel from the module/s, decontamination and demolition of the module building and site rehabilitation.

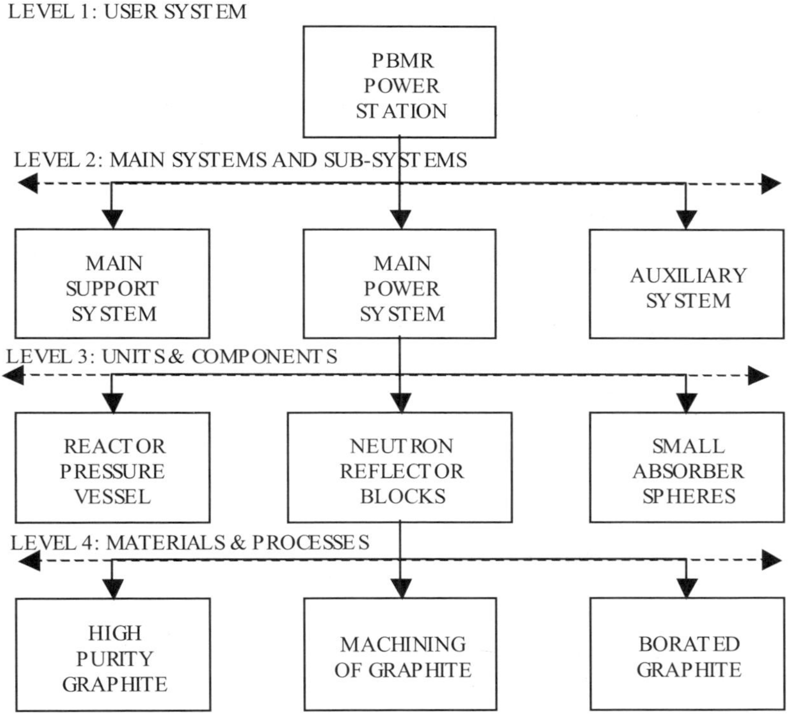

Figure 3: Hierarchy of the PBMR system breakdown structure.

Table 2: Industrial hierarchy.

System Hierarchy	Industrial Hierarchy
Level 1 (a): User System (Power Station)	User organization (Utility company)
Level 1 (b): PBMR Module	PBMR Module supplier (Design authority)
Level 2 (a): Main Systems	Main Systems suppliers
Level 2 (b): Systems	Systems suppliers
Level 2 (c): Sub-systems	Sub-systems suppliers
Level 3 (a): Units	Units suppliers
Level 3 (b): Components	Components suppliers
Level 4 (a): Materials	Materials suppliers
Level 4 (b): Processes	Processes suppliers

2.3. *Survey of required industrial capabilities*

The purpose of the Required Industrial Capabilities Survey was to obtain a complete picture of all the industrial capabilities required for the PBMR programme. It consisted of an assessment of the levels of knowledge and skills, and other aspects such as facilities (for development, production, quality assurance and testing), equipment and processes that will be required for the PBMR programme during the complete life-cycle of the PBMR system. The Required Industrial Capabilities Survey used the PBMR System Breakdown Structure to identify all technological capabilities required for the PBMR programme. The Required Industrial Capabilities Survey therefore used a two-dimensional domain, defined by a life-cycle breakdown and a system hierarchy breakdown.

2.4. *Audit of industrial capabilities*

The purpose of the Industrial Capabilities Audit was to obtain a complete overview of the existing industrial capabilities in South Africa that is relevant for the PBMR programme. It consisted of an assessment of the levels of knowledge and skills, and other aspects such as facilities (for development, production, quality assurance and testing), equipment and processes that could be utilized for the PBMR programme. The Industrial Capabilities Audit also used the System Breakdown Structure to identify all technological capabilities required for the PBMR programme.

The audit was done by a panel of experts. Workshops were conducted at Centurion and Cape Town and the panel consisted of thirty representatives from industry. Some of the firm represented were Bechtel, CBB-Power Steinmüller Africa, CBI, Consani Engineering, Dorbyl Engineering, DSE Structural Engineers & Contractors, EMS., Fluor, GEA, Genrec, Group Five Engineering, Heunes Engineering, John Thompson, KHA Contracting, LTA Autecon EEIP, Mitsui, PBMR, PB-Thermal, Petrel Engineering and the SA Institute of Steel Construction. A database of companies with relevant capabilities was compiled.

The following three areas of industrial capabilities of the local industry were determined:

Technology: Technology can be defined as scientific and engineering knowledge, including both tacit and codified knowledge, and techniques, processes and operating systems to analyze, design, develop, manufacture and maintain technical systems, subsystems, units, components or materials. It therefore resides primarily in the human resources. The degree to which the local industry has relevant technological capabilities was determined.

Infrastructure: Industrial infrastructure refers to physical infrastructure such as buildings, facilities, plant, equipment, machines, and other systems required for the development, manufacture and maintenance of technical systems, subsystems, units, components or materials. The degree to which the local industry has the relevant infrastructure was determined.

Suppliers: An important industrial capability is the existence of a local supply chain or supporting industrial cluster. The degree to which such a capability exists was determined.

The Industrial Capabilities Audit showed that the domestic manufacturing industry has capabilities that could be utilized by the PBMR programme to varying degrees. In some areas the current capabilities are adequate, but in most areas it would require substantial development and technology transfer to enable it to participate in the programme. This creates a dilemma for the PBMR programme. On the one hand the cost-performance requirements for the major sub-system are such that only the most cost-effective and low-risk suppliers and contractors can be utilized. These would mostly be established foreign firms. But foreign procurement of most of the major sub-systems will not contribute to long-term local industrial development and job creation. Imaginative policies that would meet both the short-term cost goals and long-term industrial development goals are required. Some trade-offs will have to be made. Guidelines that indicate the preferred procurement options for the sub-systems of the PBMR would therefore be valuable to assist decision-makers.

3. PBMR Industrialization Guidelines

3.1. *Methodology*

The industrial guidelines methodology used for the PBMR programme was based on a similar technique that was developed for the defense industry by the author (Buys, 1991). This methodology was found particularly useful for decision support in terms of defense R&D spending, technology development or technology transfer projects during peacetime (Buys, 1992). However, the industrialization of any major new technical system requires the establishment of the industrial infrastructure and supplier base for the system. The defense industrial guidelines methodology was therefore adapted for the determination of industrialization guidelines for such programmes (Buys, 2003).

The industrial guidelines methodology was tailored for the PBMR system by the author. The point of departure taken was that the PBMR Company is primarily a systems supplier. It has the overall system responsibility for the PBMR system but subcontracts the subsystems to suppliers in industry. It should strive to provide best value-for-money to its customers, which would be power utilities such as Escom. The system supplier has to support the system over its entire operational life, which in the case of nuclear power stations could be 40 years or more. Considerations, other than price, therefore becomes important in the setting up of a supplier base, such as quality, technical support and long-term viability. Add to this the national imperatives of job creation and economic growth, and the decision-making becomes complex.

The methodology is based on the following premises:

Building an internationally competitive industry. Industrial development should aim at building viable industries that are able to compete in the global marketplace. Subsidizing unproductive and non-viable industries will not result in an internationally competitive manufacturing industry. Michael Porter (1990) argued that international competitiveness is built on first becoming competitive in a demanding domestic market and then stepping out into the international marketplace. Both quantitative (market size in terms of sales volume) and qualitative (competition, demanding customers) domestic market conditions should be present. Increasing globalization has not changed this. Although there is no longer a strictly "domestic" market, the development of competitive industries is still based on favorable local market conditions.

Building on comparative advantages. Successful industries are based on leveraging comparative advantages (Porter, Sachs, and McArthur, 2001). These are conditions that are exploited to obtain marketplace advantages. There are many such factors such as access to scarce resources (raw materials, cheap labor, technology, skilled manpower, managerial skills, etc.) and a supportive environment (financial support, taxation, infrastructure, services, industry clusters, etc.).

Strategic considerations. Economic growth is driven by innovation (Fedderke, 2000; Freeman, 1986). Innovations in turn are driven by three driving forces: technological developments, entrepreneurship and strategy (Sundbo, 1998). It has become clear that strategic considerations have become the dominant force in the current wave of innovation-driven global economic growth. Building sustainable globally competitive industries require a long-term commitment to development and maintenance of capabilities. Local industries have to operate within the domestic environment and simultaneously be globally competitive.

The methodology considers three types of considerations: 1) Strategic Requirements, 2) Market Considerations, and 3) Current Industrial Capabilities. The current industrial capabilities were obtained from the Industrial Capabilities Audit as described in paragraph 2.4. A total of eight evaluations (figures of merit) were required for each item in the System Breakdown Structure. The questions that had to be answered are shown in Table 3.

3.2. *Guideline logic*

The eight factors described above were combined to calculate the recommended industrialization guidelines. There are four possible outcomes or guidelines that this methodology produces. These are: local procurement, overseas procurement, partnership procurement or competitive procurement.

Table 3: Industrialization Guidelines Worksheet.

Do you agree with the following statements? Use 0–9 point scale
Strategic Requirements 1. Strategic Competitive Advantage "Our international competitive advantage with the PBMR system will be based on our unique local capabilities for the provision of this item." 2. Construction and Operational Support Requirements "Cost-effective local construction, operation or maintenance of the PBMR system will require products or services for this item to be supplied by the local industry."
Market Considerations 3. Local Market "The local PBMR programme will result in market demand (in terms of sales volume) that will sustain a viable domestic industry for this item." 4. Export Market "The foreseen export market (in terms of sales volume and competition) for the PBMR system could sustain a viable domestic industry for this item." 5. Related Markets "Related markets (for other applications) both locally and overseas, in conjunction with the PBMR market, could sustain a viable domestic industry for this item."
Current Industrial Capabilities 6. Current Technology "The local industry currently has sufficient technological capabilities (technical knowledge, know-how and techniques) to provide this item to the PBMR programme." 7. Current Infrastructure "The local industry currently has sufficient physical infrastructure (buildings, facilities, plant, equipment, machines, etc.) to provide this item to the PBMR programme." 8. Current Supplier Base "The local industry currently has sufficient suppliers and/or a local supporting industrial cluster to support it's provision of this item to the PBMR programme."

Local procurement: This recommendation is for the utilization of the local industry for the supply of products or services as this is both desirable and possible. Proposals and tenders will only be requested from the local industry. Support in the form of technology development contracts could form part of the contracting as the maintenance and development of the local industrial base is of strategic importance or is in the national interest due to its high commercial potential.

Overseas procurement: This recommendation is for outright overseas procurement as it is neither possible (not sufficient local capabilities) nor strategically required or commercially viable due to an insufficient market to consider a local supplier.

Partnership procurement: This recommendation is for the procurement from the local industry as this is of strategic importance or in the national interest due to its high commercial potential. The current capabilities of the local industry is however insufficient and cooperation with a foreign partner(s) is therefore recommended. As the goal is local industrial development, technology transfer must be a key component of the partnership or joint venture.

Competitive procurement: This recommendation is for open tenders to be issued in cases where there is no clear need for local procurement but where sufficient local capabilities exist. This would ensure best value-for-money through open competition between local and foreign suppliers.

The simple logic used in arriving at the recommendations is shown in Table 4.

Table 4: Guideline logic.

Strategic Requirements*		Market Considerations*		Current Capabilities*	Acquisition guidelines
High	or	Good	and	Good	Local procurement
Low	and	Poor	and	Poor	Overseas procurement
High	or	Good	and	Poor	Partnership procurement
Low	and	Poor	and	Good	Competitive procurement

* See Table 3.

3.3. *Evaluation of input variables (Strategic Requirements, Market Considerations, Current Capabilities)*

Experience with using a similar methodology for the defense industry has shown that the evaluation of the input variables is best done by a panel of experts using a modified Delphi technique (Buys, 1991). The same panel of experts that conducted the Industrial Capabilities Audit did the evaluation of the input variables. The panel members were asked to evaluate each item individually. Individual scores were recorded on worksheets and the calculations were done at a later stage. The panel members were then given the opportunity to review their input after the average scores were calculated and distributed to the panel members.

4. PBMR Industrialization Guidelines

The analysis found that the PBMR systems, sub-systems, units and components should be procured from different sources. The high-level results obtained from this exercise are shown in Table 5.

Table 5: High-level PBMR industrialization guidelines.

Local Procurement	Core barrel, Intercooler & pre-cooler, Valves, Tanks, casks & bins, Pressure vessels, Heat exchangers, heaters, coolers & filters, De-fuelling machines, Pumps & motors, Heavy lifting device (800ton), Hoists & lifting devices, Electrical systems, Control & instrumentation, HVAC, fire detection/protection, Piping, Building, Site and services.
Overseas Procurement	Turbo machinery, Power turbine, Power generator, Electromagnetic bearings, Electromagnetic seals.
Partnership Procurement	Fuel, Reactor & PCU pressure vessels, Auxiliary bearings, Fuel transfer pipes, Graphite blocks, Reactor control & shutdown system, Blowers & compressors, Hot pipes, Recuperator.

The procurement guidelines can be grouped into three major groupings: 1) local overseas procurement, 2) overseas procurement and 3) partnership procurement. The option of open international tenders where price is the primary criterion was not the favored option for any of the items considered.

Local procurement. Items such as the core barrel, tanks, heat exchangers, valves, electrical systems, piping, hoists & lifting devices, building, site and services (see Table 1), are those where the local industry should be contracted for the supply of these products or services as this is both desirable and possible. Support in the form of technology development contracts could form part of the contracting as the maintenance and development of the local industrial base is of strategic importance or is in the national interest due to its high commercial potential.

Overseas procurement. Items such as the turbo machinery, power generator and electromagnetic bearings and seals (see Table 1), are those where outright overseas procurement is recommended as it in neither possible (not sufficient local capabilities) nor strategically required or commercially viable, due to an insufficient market, to consider a local supplier.

Partnership procurement. Items such as the reactor & PCU pressure vessels, fuel transfer pipes, graphite blocks, and recuperator (see Table 1), are those where it is recommended to procure from the local industry as this is of strategic importance or in the national interest due to its high commercial potential. The current capabilities of the local industry is however insufficient and cooperation with a foreign partner(s) is therefore recommended. Investment, training and technology transfer must be key components of the partnership or joint venture as the goal is local industrial development

5. Summary and Conclusions

This chapter presents a new methodology that was used for determining industrialization guidelines for South Africa's Pebble Bed Modular Nuclear Reactor (PBMR) programme. The methodology takes into account the strategic and operational requirements of the programme, current and future market conditions, and current domestic industrial capabilities. There are four possible outcomes or guidelines that this methodology produces. These are: local procurement, overseas procurement, partnership procurement or competitive procurement. In the case of the PBMR system a panel of experts using a modified Delphi technique did the evaluation. It was found that the different sub-systems, units and components should be procured from different sources. However, the option of open international tenders where price is the primary criterion was not the favored option for any of the items considered.

In this chapter the methodology for the determination of industrialization guidelines for the PBMR programme was presented. The merits of the methodology are that it is systematic, consistent and explicit (open to analysis and criticism). Subjective views and judgments are expressed as numerical values and clear unambiguous answers are obtained. This methodology was found to be a very useful decision support tool that provided guidelines indicating the preferred procurement options for the sub-systems of the PBMR system. This methodology can be utilized or adapted for the determination of industrialization guidelines for any major new technical system that requires the establishment of industrial infrastructure and a supplier base for the system.

References

Buys, A. J., (Ed.) (1991). Strategic Evaluation of the South African Defense Industry, Armscor, Pretoria.

Buys, A. J. (1992). An AHP Application: Military Priorities in Peacetime. Proceedings of the Operational Research & Management Science Joint International Conference, Helsinki, Finland.

Buys, A. J. (2003). Industrialization Guidelines Methodology, *Transactions of the South African Institute of Electrical Engineers*, 94, (3), p 22-27.

Fedderke, J. W. (2000). Growth and Innovation Report. National Science and Technology Forum, Pretoria.

Freeman, C. (1986). The Economics of Industrial Innovation, 2nd ed.. MIT Press, Cambridge, MA.

PBMR (Pty) Ltd. (2007). http://www.pbmr.com/. Website visited on 17 January 2007.

Porter, M. E. (1990). The Competitive Advantage of Nations. The Free Press, New York.

Porter, M. E., Sachs J. D. and McArthur J. W. (2001). Executive Summary: Competitiveness and Stages of Economic Development. Global Competitiveness Report 2001-2002, World Economic Forum, Oxford University Press, New York.

Sundbo, J. (1998). The Theory of Innovation: Entrepreneurs, Technology and Strategy, E. Elgar, Massachusetts.

Chapter 15

A Longitudinal Analysis of Inventors' Movements in Technology Clusters

Jiang He and M. Hosein Fallah, Ph.D.

Wesley J. Howe School of Technology Management, Stevens Institute of Technology

E-mail: jhe@stevens.edu

hfallah@stevens.edu

Clustering is one of the key drivers for regional development. The existing literature indicates that the knowledge spillovers, facilitated by the mobility of inventors, play a positive role in the development of clusters. The questions remain as to how the patterns of innovator mobility change while clusters go through different stages of their lifecycles. Using patent authorship data, we construct networks of inventors' movements within two different telecom clusters, in New Jersey and Texas. We observe the evolution of inventor mobility networks overtime, and discuss how the changes in properties of inventor networks can be used to predict the changes in economic or social conditions of the clusters over time.

1. Introduction

Clusters are geographic concentrations of interconnected companies and organizations in particular fields that co-operate but may also compete with each other (Porter, 1998). Clustering is a key driver of regional economic growth and its impact on business competitiveness and regional prosperity has been well documented (Feldman, 1999; Koo, 2005; Porter, 1998). In this study, we contribute to this body of knowledge by investigating how the inventor networks in telecom clusters evolve longitudinally while the cluster go through different stages of its lifecycle and how the network dynamics reflect the changes in social and economic conditions of the clusters.

2. Evolution of Telecom Industry in New Jersey and Texas

Historically, the telecom markets in the United States were tightly regulated monopolies. Before 1984, the US telecom market was the monopoly of the Bell System. Due to the presence of Bell Labs, New Jersey was a global center for telecom innovations. The monopoly of AT&T was broken up in 1984, and the deregulation of US telecommunications market, which was further promoted by the passage of 1996 Telecommunications Act, opened up significant opportunities for start-up businesses and venture investment. Given this favorable environment, New Jersey's telecom sector went through a short period of speedy growth in late 1990s, a phenomenon observed in some other regions of US as well. Texas was one of the fastest-growing states during that period. Figure 1 displays the historical trends for the two states in telecom innovation output measured with annual telecom patents assigned to each state. Following the so-called dot-com bubble burst in 2001, the telecom industry as a whole went into a downturn phase. Almost 5 years after the "bubble burst", the NJ telecom industry still appears to be struggling to recover from the downturn, while the sector in TX has been able to move ahead and replace the NJ leading position in telecom R&D. Figure 1 shows that NJ has been falling rapidly behind TX in the telecom patent output since 2003.

Thus, New Jersey (NJ) and Texas (TX) make for a rather appropriate pair for comparison of cluster development. The former state represents a historical hotbed of telecom innovation but is currently stuck in a stagnant to declining stage, whereas the latter is on a significant growth path and has replaced NJ leadership position in telecom innovation. The shift in advancement of cluster development provides us with an opportunity to observe the dynamics of inventor networks under different stages of cluster development.

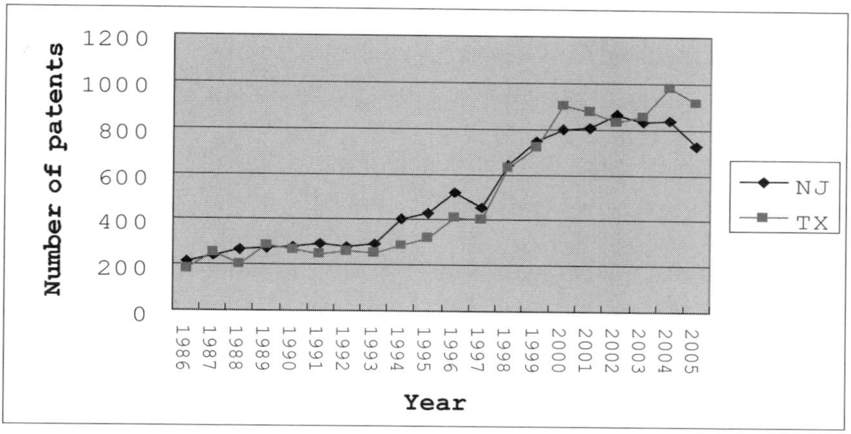

Figure 1: Annual patent output by state.

3. Theoretical Framework

The studies on clusters date back to Marshall (1919), who developed the agglomeration concept based on the cost-saving scale effects brought by industrial localization. Based on Markusen's (1996) work, there are four basic types of industrial clusters: the Marshallian form in which the business structure is comprised of small, locally owned firms; the "hub-and-spoke" form, which is supported by one or more dominant firms; the "satellite platform" which forms an assemblage of unconnected branch plants embedded in external organization links; and the "state-anchored" form, in which the local business structure is dominated by a public or nonprofit organizations such as a military base or a national lab. According to Markusen's model (1996), geographical clusters may exhibit significantly different traits from district to district due to their differences in fundamental typology. From the very beginning of the cluster research, knowledge spillover has been considered a key contributing factor explaining the geographically concentrated innovation activities (Marshall, 1919).

The social network approach provides a good way to examine the patterns of localized knowledge-spillovers and their contribution to the innovation activities. Traditionally, the research on social networks and innovation has emphasized how the network actors benefit from the formal or informal network connections they maintain, these studies are largely focused on the question of how the performance of individual network nodes, which could be measured by a variety of indicators, are associated with the roles or importance of the nodes within the network. Since the discovery of "small-world" network phenomenon, which is characterized by short average path length and high degree of clustering properties (Watts and Strogatz, 1998), researchers started to investigate the "small-world" properties and its implications to knowledge transfer. However, empirical studies have not shown that "small-world" networks significantly boost regional innovation activities (Fleming et al., 2004; Strumsky et al., 2005).

4. Methodology

4.1. *Patent data and knowledge spillover network*

In this study, we regard patents as the carriers and indicators for technological knowledge development and spillovers, as patent data reflect the technological inventions, the mobility of inventors and collaborative activities between individuals and organizations.

The original network data was collected from the Patents BIB dataset issued by the United States Patent and Trademark Office in 2006 (USPTO, 2006). We selected the telecom patents granted to inventors in New Jersey and Texas between 1986 and 2005 for analysis. A patent is considered to belong to either New Jersey or Texas, as long as one or more of the inventors were located within that state. The patents belonging to both states (accounts for about 1%) were excluded from the analysis. In this study only the patents assigned to organizations rather than self-owned were taken into consideration (account for 92% of the total number of patents). Our definition of "telecom" industry followed the approach of categorization introduced by Jaffe (2002), in which 12 main patent classes were grouped into a category labeled "communications".

The patent dataset enables us to develop a bipartite network which consists of two sets of vertices – patent assignees and patent inventors.

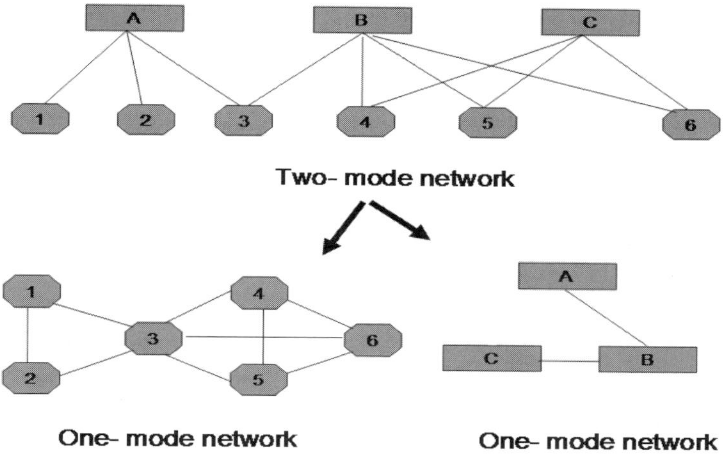

Figure 2: Transformation of a two-mode network to one-mode networks.

Such bipartite networks cannot be interpreted easily though, as the network parameters such as degree distribution have different meanings for different sets of vertices. Therefore we transform the bipartite network into two one-mode networks. Figure 2 illustrates an example of transformation from bipartite to one-mode networks. The nodes represented by numbers are inventors and those represented by letters are assignees. In the one mode network of inventors, the assignees are represented by the links among inventors. Similarly, the nodes in the assignee network are connected by the inventors that are linked among the assignees. In this study, we will be focusing on the one-mode network of assignees.

4.2. *Methodology of Network Analysis*

4.2.1. *Overall network structures: NJ vs. TX*

The first part of our analysis is concerned with the similarities and differences in overall inventor network structures between the two clusters over a long-term period. We create a one-mode assignee network for each cluster by using the telecom patents granted between 1986 and 2005. In such networks, an undirected network tie connecting two nodes (assignees) indicates that there are at least one common inventors delivering patents for both assignees over the time period between 1986 and 2005. In practice these undirected network links can occur in two ways: through co-patenting in which a patent is assigned to multiple assignees simultaneously, or sequential patenting in which inventor(s) deliver patent(s) for both organizations in a time sequence. In either case, we assume there would be a knowledge spillover occurring.

In this step, we assign an undirected and unweighted network tie between two assignees as long as they share at least one inventor over the period between 1986 and 2005. The one-mode network of assignees constructed in this way reflects the dynamics of inventors' movements shaped by inventors' voluntary changes and firm-level adjustments.

Based on Markusen's model (1996), we expect that the inventor networks across the two clusters would be different in multiple aspects due to their difference in fundamental cluster typology. In New Jersey, much of the prosperity of telecom industry has been attributed to the presence of the Bell System. According to Markusen's (1996), the NJ telecom cluster is a typical Hub-and-Spoke cluster in which the Bell System performed as the central hub; whereas the TX telecom cluster features a mixed cluster typology, which exhibit elements of multiple types of clusters simultaneously — Hub-and-Spoke, Satellite platforms, and State-anchored. The reasons are as follows. First, Texas hosts few important IT firms headquartered within the state (e.g. Texas Instruments, Dell), which may act as anchors to the regional cluster; second, the IT innovation in Texas also owes much of its performance to the presence of some large, externally headquartered IT firms (e.g. IBM, Motorola, AMD, Nokia, Nortel); third, the prominence of Austin as a hotbed of IT innovation is to large extent attributed to the presence of University of Texas.

According to Markusen (1996), the labor market serving a Hub-and-Spoke district is relatively inflexible. The business structure would not encourage horizontal R&D cooperation between firms. Usually the hub firms are more attractive to workers than the surrounding smaller and less powerful firms, *"if jobs open up in hub firms, workers will often abandon smaller employers to get onto the hub firms' payroll"* (Markusen, 1996, p. 303). On the other hand, major downturns in the hub industry or declining of the principal firms can dramatically change the patterns of job movements in the cluster (He and Fallah, 2006). In either scenario, we would observe mobility-induced network ties linked to the hub(s). Nevertheless, over the long run, the job mobility among the peripheral companies would stay at a low level.

Therefore, we expect the inventor network of NJ would feature a more centralized structure than the TX one. In terms of overall network connectivity, we expect the NJ network would be less connected than the TX one over the long run.

Also, we expect that the two networks would differ significantly in network efficiency as well. For the NJ network, the presence of the central hub(s) would better facilitate the flow of communication by reducing the average "distance" between nodes[1].

Despite of its advantage in efficiency, we contend that the centralized inventor network of NJ would not be in favor of the cluster development over the long run, because the diversity of the "spilled knowledge" is limited as most of the mobile inventors are somewhat affiliated with the hub firm(s).

4.2.2. *Evolution of inventor networks over time*

The following analysis will investigate how the inventor networks evolve over time while the clusters go through different stages of their lifecycle.

Based on the same approach of constructing bipartite network and the transformation from two-mode to one mode, we construct a set of consecutive assignee networks for TX and NJ, respectively. Each sub-network corresponds to a 3-year period. As the time-window is being moved forward from 1986 to 2005, we examine how the properties of the inventor networks change overtime and how the processes differ between the two clusters. For each cluster, the whole patent dataset covering

---

[1] In practice, the "shortest-path distance" between two nodes is measurable only if the nodes are connected, directly or indirectly, within a certain component. Nodes located in disconnected graphs should not be taken into consideration because they are not reachable anyway. For this reason, we will extract the largest component from each network (NJ, TX) for efficiency measurement. This approach is more valid for networks in which the main components are significant.

period 1986-2005 is divided into multiple segments to enable the "window-based" analysis to be performed. Then in each block of the dataset covering 3 years, we update the list of assignees according to the companies' history records collected via internet search or interviews. By doing so, we remove the network ties corresponding to merger and acquisitions activities or common ownership from the original network, and focus on voluntary job mobility and collaborating R&D activities between firms/organizations.

With the assumption that both voluntary job movements and collaborating activities are positive indicators for cluster development, we expect that the inventor network would become better connected while the corresponding cluster is experiencing a growing stage, and the network would become less connected when the cluster is declining. Considering unique business structure for Hub-and-Spoke clusters, we do not expect that such a simple relationship would hold true for this type of clusters, as the principal firm(s) has overwhelming power to influence the connectivity of the entire network, either by its success or failure.

5. Results and Interpretations

5.1. *Overall network structures – NJ vs. TX*

Figure 3 and 4 illustrate the network visualization results for the complete one-mode network of assignees. One observable difference in network structure between the two networks is the centralization of network. The network ties for NJ are heavily clustered around two central hubs, whereas the TX network shows a more random distribution of network ties. The descriptive statistics on the degree centralization[2] of network confirmed the difference between the two networks (Table 1).

---

[2] Degree is number of connections associated with a node. Degree centralization, ranging from 0 to 1, is defined as the "variation in the degree of vertices divided by the maximum degree variation which is possible in a network of the same size" (De Nooy et al., 2005, P. 126). Put it differently, a network is more centralized when the vertices vary more with respect to their centrality. A star-shape network typology is an extreme with degree centralization equaling one.

Table 1: Descriptive statistics for one-mode network of assignees: NJ vs. TX.

	NJ	TX
Total number of vertices (assignees)	841	1210
Density of network ties	0.002	0.001
Average degree of the network	1.75	1.48
Percentage of nodes which are non-isolated	0.48	0.46
Degree centralization	0.12	0.03

Table 2: Descriptive statistics for the main components extracted from Figures 3 & 4.

	NJ	TX
Relative size of the main component	0.34	0.31
Total number of vertices connected in the main component	288	376
Density of network ties	0.02	0.01
Average degree of the network	4.56	4.05
Average path length	3.22	4.79
Degree centralization	0.34	0.09

In terms of network's overall connectivity, contrary to what we expected, the descriptive network statistics show that the NJ network is slightly better connected than the TX one, based on the measurements of network density, average degree of nodes, and proportion of non-isolated nodes in the whole network. Given the significant role of the hubs in maintaining the NJ network, we interpret the relatively high level of connectivity of the NJ network to be largely due to the break-up of the monopoly of the Bell System, which led to a great deal of redistribution of employees and patents.

In terms of network efficiency, Table 2 reports the descriptive statistics for the largest components (extracted from Figure 3 and 4). For each cluster, the nodes connected within the largest component account for about one-third of the total nodes. Consistent with our expectation, compared with the counterpart of TX, the main component of the NJ network features a shorter average path length. We believe that this network feature is attributed to the presence of central hubs in the network of NJ. The centrality measure suggests that NJ one-mode network of assignees is more centralized than in TX. For the NJ network, although such a hub-supported network structure would allow for

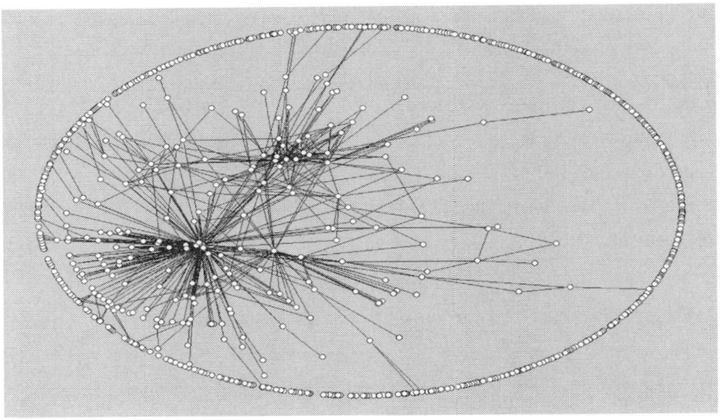

Figure 3: Inventor network for NJ over the entire period (one-mode of assignees).

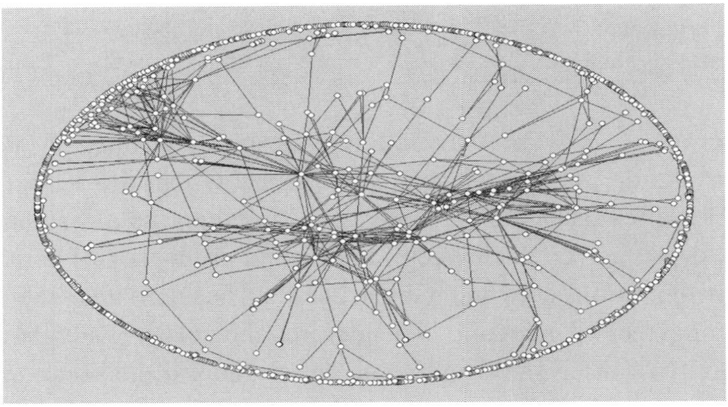

Figure 4: Inventor network for TX over the entire period (one-mode of assignees).

efficient information flow, it might constrain the potential benefits from knowledge spillovers, when the majority of knowledge transferred over the network originally comes from a common source (e.g., the Bell System), which is limited in diversity of knowledge.

5.2. *The evolution of patent inventor networks*

Based on the consecutive sub-networks, we calculate the key measures of network structure including the density and centrality on each window period for each cluster. Projecting the network parameters on their corresponding time-window periods allows us to observe the changes in network properties over time.

To measure the overall network density on each window period, we compute the average network degree and proportion of non-isolated nodes over complete members of the corresponding sub-network. Figure 5 and 6 present the historical trends in these two measurements. The trends observed from the two charts are consistent with each other. The charts indicate that the historical trends in network connectivity considerably vary across the two geographical clusters. As the time went by, the resulting curve for NJ exhibits a U-shape profile which suggests higher instability in network density; whereas the TX curve shows a gradual transition in the early period and a growing trend catching up with NJ in the later period.

The historical trends in network connectivity for the Texas cluster demonstrates a very similar pattern of growth to the corresponding curve shown in Figure 1, which suggests that the network connectivity is positively associated with innovation output of clusters. For the NJ inventor network, it appears that the network connectivity has a negative association with the cluster's innovation output – in the mid 1990s, the NJ network was poorly connected when compared to TX, even though the telecom cluster in NJ was considered to be strong during that period (see Figure 1); the NJ network became better connected than the TX counterpart in the recent years during which NJ was constantly declining and losing its leadership in telecom R&D to TX.

From Figure 7, we observed that the inventor networks across the two clusters show very different historical trends in network centrality. Over the entire period of observation, the centrality data for the TX network yields, roughly, a straight line, suggesting the TX network always maintained a decentralized structure regardless of the developing stage of the cluster. Compared with the TX network, the NJ network is more centralized during most of the observation periods, and it shows a tremendous growing trend in degree centrality in the later years. Furthermore, the hub was growing quite fast in the later years and reached its peak on the window period 2000-2002.

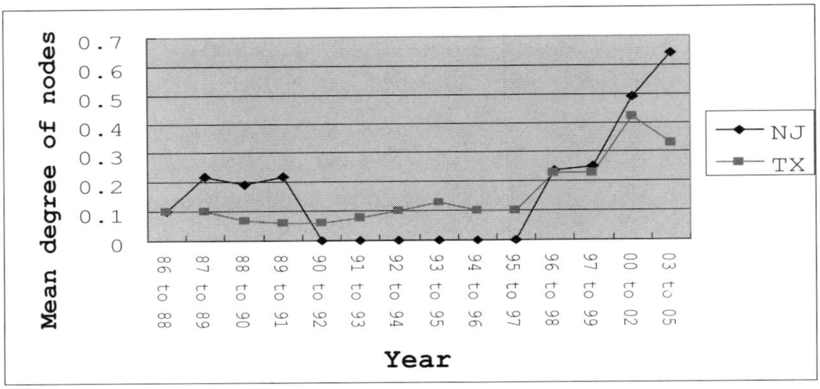

Figure 5: Average degree of nodes in inventor network (one-mode of assignees).

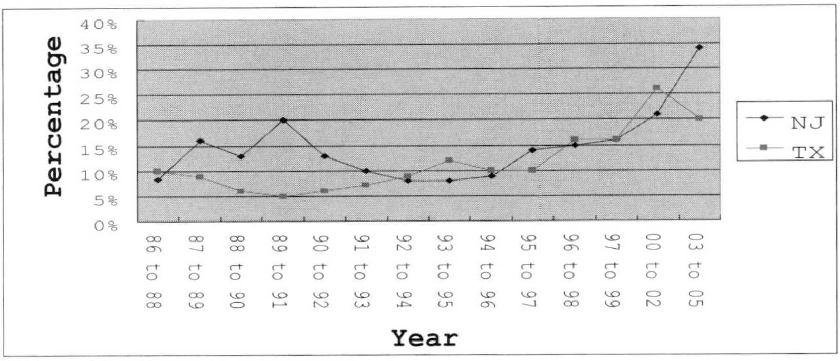

Figure 6: Percentage of non-isolated nodes in inventor network (one-mode of assignees).

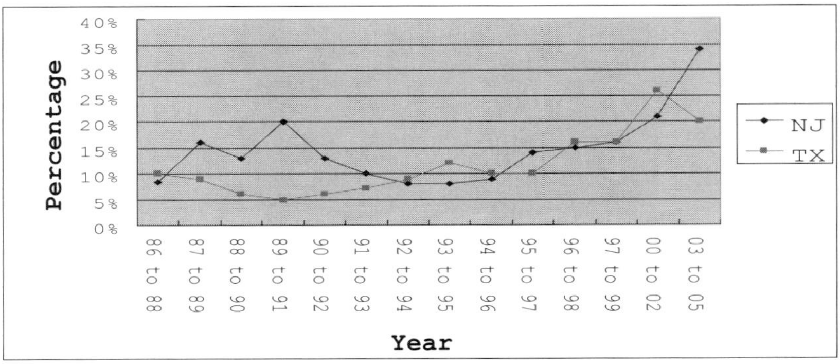

Figure 7: Network centrality for inventor network (one-mode of assignees).

When the connectivity data and centrality data are put together for interpretation, we conclude that in the recent periods of observation during which the NJ telecom cluster was declining, the hub of the NJ network became rather significant and the hub-connected component accounted for a great proportion of the total connectivity of whole network. We suspect that the majority of network ties emerged in the NJ network in recent years may be related to the declining of the incumbent telecom player(s) in the state. Following the implementation of the 1996 Telecom Act, professionals started to move out of the hub companies (e.g. Lucent Technology, AT&T) for exploring new opportunities in the market, creating, many of the NJ-based telecom start-ups. As the "bubble burst" which came shortly after, it forced the hub firms to push out more professionals. As a result, the inventor network for NJ became highly centralized within a relatively short period.

6. Conclusions and Future Research

This paper examines the dynamics of inventor networks and its implications for cluster evolution. The analysis results highlight why the telecom cluster in New Jersey has been falling behind that of Texas in innovation in recent years. In terms of overall network structure, it turns out that the NJ network is much more centralized than TX. It suggests that most of the mobile inventors in NJ have been with the incumbent sometime in their careers. Indeed, the role of AT&T in supporting the NJ inventor network is much more significant than expected, which makes the NJ network, on the average, better connected than that of TX. As for network efficiency, the inventor networks of the two clusters perform differently as well. Theoretically, hub-based network structure would enable information to be transferred easily from one to another within the NJ main network component. In practice, this situation is suggestive of the limitation of the NJ innovation network over the long run– the diversity of the knowledge and vulnerability of the mobility network.

Longitudinally, our analysis results suggest that the dynamics of job mobility networks within geographical clusters do not follow a simple linear relationship with the clusters' innovation performance; the fundamental "typology" of cluster may be a key factor determining the relationship.

Our results are suggestive of the associations between cluster evolution and inventor network dynamics, yet the generalizability of the finding is limited as only two clusters, one with hub-based typology and the other with mixed typology, in one industry were examined in this study. Indeed, real-world clusters may be of any one of the three types of typology defined by Marsuken (1996). Our future research will focus on an empirical study covering a greater variety of cluster typologies and industries.

References

De Nooy, W., Mrvar, A. and Batagelj, V. (2005). Exploratory social network analysis with Pajek. Cambridge University Press.

Feldman, M. P. (1999). The new economics of innovation, spillovers and agglomeration: a review of empirical studies. *Economics of Innovation and New Technology*, 8(1–2), pp. 5–25.

Fleming, L., King, C. and Juda, A. (2004). Small worlds and innovation. Harvard Business School, mimeo.

He, J. and Fallah, M. H. (2006). Mobility of innovators and prosperity of geographical technology clusters: a longitudinal examination of innovator networks in telecommunications industry. Proceedings of International Conference on Complex System, Boston, MA, June 24-30

Jaffe, A. B. and Trajtenberg, M. (2002). Patents, citations, and innovations: a window on the knowledge economy. MIT Press, 2002.

Koo, J. (2005). Technology spillovers, agglomeration, and regional economic development. *Journal of Planning Literature*, 20(2), pp. 99–115.

Markusen, A. (1996). Sticky places in slippery space: a typology of industrial districts. *Economic Geography*, 72(3), pp. 293–313.

Marshall, A. (1919). Industry and Trade. Macmillan, London.

Porter, M. E. (1998). Clusters and the new economics of competition. *Harvard Business Review*. Nov./Dec.

Watts, D. J. and Strogatz, S. H. (1998). Collective dynamics of "small-world network", *Nature*, 393, pp. 440-442.

Chapter 16

Technology Mining of Gulf Coast Intellectual Assets: Discovering Regional Assets for Economic Development

Cherie Courseault Trumbach, Sandra Hartman and Olof Lundberg

Department of Management, University of New Orleans,
2000 Lakeshore Dr,
New Orleans, La, 70148, USA
ctrumbac@uno.edu, shartman@uno.edu, olie.lundberg@uno.edu

The Gulf Coast is facing significant challenges in rebuilding after Hurricane Katrina. Post disaster perceptions of blight and crime have severely harmed the bread-and-butter industry of the area: tourism. As a result, the region must take inventory of its intellectual assets in order to determine new areas for economic development. This chapter first discusses the importance of absorptive capacity in economic development. It then presents results from a technology mining study conducted on the intellectual assets (publications and patents) along what is known as the I-10 Corridor in Louisiana, Mississippi, and Alabama. These results reveal indicators of the economic development struggle of the region. More importantly, they reveal the technology areas, largely economically untapped, where the region exhibits strong research capabilities and educational focus, indicating high levels of absorptive capacity and thus, are areas prime for economic development. In addition, the paper demonstrates how technology mining can be used as a tool to aid in economic development decision-making.

1. Introduction

The Gulf Coast is facing significant challenges in rebuilding after Hurricane Katrina. Prior to the storm, the region was sluggish in its attempt at economic development. After the storm, the challenges are even greater. Post disaster perceptions, especially of blight and crime, have severely harmed the bread-and-butter industry of the area: tourism. Where should the region look for rebuilding a shattered economy and infrastructure?

This research provides a profile and analysis of the regional knowledge assets along what is known as the I-10 Corridor, including southeast Louisiana, southern Mississippi, and southern Alabama. The study provides a profile of the research conducted along the Gulf Coast, including an analysis of publication and patent records, in order to provide a tool to assist in economic development activities. Bibliometric analysis and text mining of publications and patents have been used primarily for national policy decisions and corporate decision-making (Kats & Hicks, 1997; Porter & Cunningham, 2005). However, we contend that such analysis is beneficial at other levels as well. These methods are utilized to provide a profile that can support economic development activities.

Regions vary in the ability to convert that research into commercial innovation. The differences may rest in the availability of large R&D intensive firms (Agrawal and Cogburn, 2003). Jaffe (1989, 1993) finds that companies benefit, particularly in the realm of patenting, from a close geographic location to research facilities. Technology mining can be used to build a profile of a region's innovation potential to support funneling company investment into a region. Shapira et al. (2003) found that, not only are publication and patent counts useful for characterizing a region's innovation potential, but that they also provide leading indicators of technology employment, a key goal in economic development. In this study, we utilize bibliometric analysis to provide information that can be used to influence organizations that are considering business opportunities along the Gulf Coast and economic development investment decision-makers. In considering how this information can be used, we also incorporate ideas from developing countries using technology innovation as the basis for economic development.

2. Lessons from Less-Developed Nations

The I-10 Corridor can be thought of as a region which, post-Katrina, is encountering economic conditions that in many ways make it resemble a less-developed country more closely than a thriving industrialized locale. We consider insights offered by research on the relationships among economic, scientific and technological development in less-developed nationsto see what may be applicable. In general, issues in this literature center upon whether developing nations should simply try to "catch up" by imitating the technology and methodologies of developed countries or whether they should engage in technology development in specific niches where they have the capability to lead. Are there lessons that the I-10 Corridor can learn from the successes of developing countries?

Nelson (2004) notes a number of factors that have been crucial to developing economies attempting to catch-up. He notes the importance of indigenous research efforts, along with a focus on higher education and engineering training as factors for rapid catch-up. With highly scientific technologies, advanced education is required for absorbing that technology. He also states that, in the past, successful growth in developing economies has been characterized by a considerable "...cross-border flow of people," where citizens go elsewhere to learn new technologies and then return to implement them or where experts come into the area to serve as mentors/advisors. Something similar to this phenomenon could occur along the I-10 Corridor, if residents displaced by Katrina can be induced to return and if they return with new ideas. The cross-border flow is likely since we are addressing regional innovation within an already innovative nation, where the parties with the potential to lead development can move freely to the region. Even in absence of the cross-border flow of people, Nelson hails the research producers as the critical element in the catch-up process.

Lazonick (2004) cited the development of "indigenous innovation," in an analysis of China's success in "leaping" into the information age.

He, like Nelson, points to the need of the nation to absorb outside technologies into their own economy. Lazonick noted that a key factor permitting China to move rapidly into the information age was that foreign computer companies had not yet mastered the problem of Chinese-language word processing. When Chinese companies were able to achieve mastery in this area, they controlled a dimension of computing which enabled them to become world leaders. The lesson here rests in China's ability to combine mastery of outside technology and its own uniqueness into a basis for competitive advantage. In the same vein, the I-10 Corridor must identify its own unique strengths that can result in a regional advantage. China used its uniqueness to provide an opportunity to engage in more far reaching innovative research. Are there potential niches where research along the I-10 Corridor can result in leadership? Is the area already a research leader in technologies that are not being harnessed for economic advantage? In this research, we use technology mining methods to consider these possibilities.

3. Assessing Regional Absorptive Capacity

Absorptive capacity is the ability to acquire external information, assimilate it and exploit it for commercial purposes (Koo, 2005). It can be developed by both education and experience. In this research, absorptive capacity refers to the capability of the region defined as the I-10 Corridor to transfer research from research-producing institutions into the community where it can be acted upon and used to build business ventures. In the knowledge economy, knowledge-based industry clusters reflect the absorptive capacity of a region. The knowledge-based industry clusters reflect areas in which the area already possesses rich knowledge. Therefore, the region will have a greater capacity to assess the value of new information, assimilate that information, reduce its uncertainty, give it meaning, and apply it in its knowledge-based clusters (Julien et. al, 2004) Innovation requires sufficient knowledge to understand, interpret, and realize the benefits of external sources. Indigenous R&D is necessary to increase absorptive capacity as a means of economic development and increasing innovation capacity, particularly since knowledge and its spillover into the community are the considered the most important factors in long-term economic growth (Kearns and Görg, 2002, Barkeley, 2006).

Seeking outside R&D that correlates with indigenous R&D optimizes the potential for successful knowledge spillover. Therefore, an understanding of the indigenous knowledge infrastructure, including the organizations, researchers, collaboration networks, and knowledge-based clusters provides insight into the absorptive capacity of the region. For example, research in universities provides insight into the areas in which students are exposed either in the classroom or with assistantships. Those students then take that learning into the workforce. Collaboration networks demonstrate areas that expand across institutions. Therefore, the development and use of the knowledge within an area is not held by an isolated individual. It may also indicate areas that are strategically important for the region as a whole. As tools, bibliometric analysis and text mining of publications and patents have recently been shown in research to provide the needed macro-level perspectives of the R&D infrastructure (Katz & Hicks, 1997; Porter & Cunningham, 2005; Becker & Sanders, 2006, Jaffe, 1989, 1993). In this research, we use bilbiometric analysis and text mining to identify knowledge-based clusters and collaboration networks in publication and patent databases for the Gulf Coast region.

4. Methodology

This research examined publications and patents through technology mining methods. In general, publications are identified as an indicator of research activity and patents are identified as an indicator of development activity. Patent indicators are considered precursors to commercialization. We look at patent numbers as indicators of how well entities are doing in turning research activity into economic development activity. In this study, the Gulf Coast I-10 Corridor is defined as four Metropolitan Statistical Areas (MSAs): New Orleans, Baton Rouge, Biloxi-Gulfport-Pascagoula, and Mobile. This area includes 573 zip codes and 310 cities. The assets analyzed include technical publications collected from the Compendex databases and patents from January, 1988 to the December, 2005. Compendex covers the whole spectrum of Engineering/Technical disciplines. It contains over 5 million references with 500,000 added annually from 5,000 journals, conferences and technical reports (University of New Orleans Library Database List, 2007). Compendex was chosen because it is more generic in its scope than more specialized databases which may be better suited to research in a particular field. Our analysis includes over 10,000 technical publications and over 2,500 patents. In conducting this research, VantagePoint software was utilized. VantagePoint is a text mining software package designed specifically for analyzing technology resources.

Bibliographic information in patents includes both the location of the inventor and the location of the assignee. The assignee is the holder of the patent. We analyzed the patents where the assignee was listed as being from a regional organization because it is the assignee location (i.e., along the I-10 Corridor) that provides the opportunity for the region to economically benefit from the patent. However, given the later discovered imbalance between inventors and assignees along the I-10 Corridor, inventors perhaps should have also been analyzed to determine absorptive capacity. Had inventors been included, our definition of regional absorptive capacity would have also included the knowledge spillover into the community that takes place from inventors with outside assignees.

Our findings from this analysis are divided into two primary areas:

Knowledge Assets Profile: These sections contain the basic information on "who is doing what." It lists the top publishing organizations and patent assignees with their main topic areas. There is also a collaboration map showing what organizations are publishing or inventing together.

Technology Profile: These sections include a patent economics indicator, profiles the knowledge-based clusters discovered in the research, and compares patent activity to publication activity. Each sub-cluster is individually profiled with a map of the sub-clusters in each area, the main affiliations associated with each sub-cluster, a life cycle chart, and a comparison of the region's main subtopics versus those of the general technology.

In addition to the two primary areas above, we also provide a summary of the impact of Katrina on the most affected asset areas.

5. Results and Analysis

This analysis is based on 10,139 publications attributed to approximately 400 entities and approximately 12,000 authors along the I-10 Corridor. There are 2,542 patents assigned to approximately 450 organizations that identified their location as in the Corridor. There were 2,521 inventors named as being on an inventing team. Our full results are available on the web page: fs.uno.edu/ctrumbac/GulfCoast.

5.1. *Knowledge assets profile: Organizations profiles*

Below are two tables that show the Top 10 publishing organizations (Table 1) and the Top 10 patenting organizations (Table 2). The website "Organizations" page includes a profile of the top 59 most frequently publishing organizations in the Gulf Coast Region and the top 54 patent holders. The profile includes the top authors or inventors within those organizations and the main topics for publications or patents. The website also includes a profile of the top 46 most frequently published authors in the Gulf Coast Region and the top 46 inventors. The profile includes the organization with which the individual is associated and the main topics for publications or patents.

Table 1: Top 10 publishing organizations.

#	Organization	Top 5 Keywords
1	Louisiana State University [4472]	Heat transfer [35]; Parallel processing systems [27]; Environmental engineering [26]; Soils [24]; Thin films [23]
2	Tulane University [1313]	Biomedical engineering [15];Engineering education [14]; Titanium dioxide [12]; Nanostructured materials [11]; Biomechanics [11]
3	University of New Orleans [958]	Nanostructured materials [41]; Induction motors [11]; Thin films [10]; Shipbuilding [8]; Perovskite [7]
4	University of South Alabama [442]	Engineering education [14]; Pattern recognition [12]; Resin transfer molding [10]; Automatic target recognition [9]; Optical correlation [9]
5	Stennis Space Center [287]	Oceanography [80]; Remote sensing [15]; Ocean engineering [14]; Sonar [6]; Naval warfare [6]
6	LSU Agricultural Center [274]	Forestry [13]; Wood products [12]; Spreaders [10]; Wood [10]; Plants (botany [9]
7	USDA [267]	Cottonseed oil [12]; Cotton fabrics [11]; Cotton fibers [10]; Cotton [9]; Cotton Fabrics -- Dyeing [9]
8	Southern University [193]	High temperature superconductors [7]; Superconductivity [6]; Oxide superconductors [6]; Engineering education [6]; Composite materials [4]
9	Naval Research Laboratory [164]	Oceanography [33]; Remote sensing [12]; Ocean engineering [10]; Flow of fluids [4]; Sonar [4]
10	USDA ARS [159]	Cotton [9]; Cotton fibers [7]; Cotton fabrics [4]; Cellulose [4]; Plants (botany) [3]

Table 2: Top 10 patent assignee organizations.

	Assignee	Top 5 Abstract Phrases
1	Babcock & Wilcox and Comp [646]	flue gas [56];one end [23]; steam generator [22]; inner surface [17]; fiber optic [17]
2	Louisiana State University [295]	insulin resistance [22]; similar species [19]; day dependent [18]; Type II diabetes [18]; neural centers [18]
3	Tulane Educational Fund [84]	present invention [15]; pharmaceutically acceptable salt [10]; amino acid [9]; 3 alkyl [7]; peptide bond [7]
4	Laitram Corporation [59]	conveyor belt [9]; link ends [9]; pivot rod [8]; preferred embodiment [8]; alphabetic characters [7]
5	Plant Development Services INC [30]	distinct variety [25]; new variety [21]; unique blooming time [13]; Azalea plant [10]; globose shaped plant [8]
6	S. Alabma Med. Sci. Foundation [25]	present invention [8]; tetrahydrofolic acid [7]; natural isomer [5]; polyglutamyl derivatives [5]; reduced folate [5]
7	Lockheed Martin Corporation [23]	solid state repair process [3]; one embodiment [3]; one inch [3]; flight hardware [3]; composite tank [3]
8	DSM Polymer, Inc [21]	high ethylene content ethylene [4]; sheared blend [4]; simultaneous blending [4]; improved lower temperature properties [4]; lower ethylene content ethylene [4]
9	Lew Childre Sons, Inc. [21]	fishing rod blank [2]; rod blank [2]; resilient material [1]; different density molding resins [1]; dimensioned molds [1]
10	University of New Orleans [19]	heterocyclic carbene [6]; least one N [6]; aryl halide [6]; halogen atom [5]; hydrocarbyl group [5]

5.2. *Knowledge assets profile: Collaboration networks*

The Collaboration Networks are two maps that identify which organizations work together. One map identifies organizations that publish together and other map shows organizations that invent together.

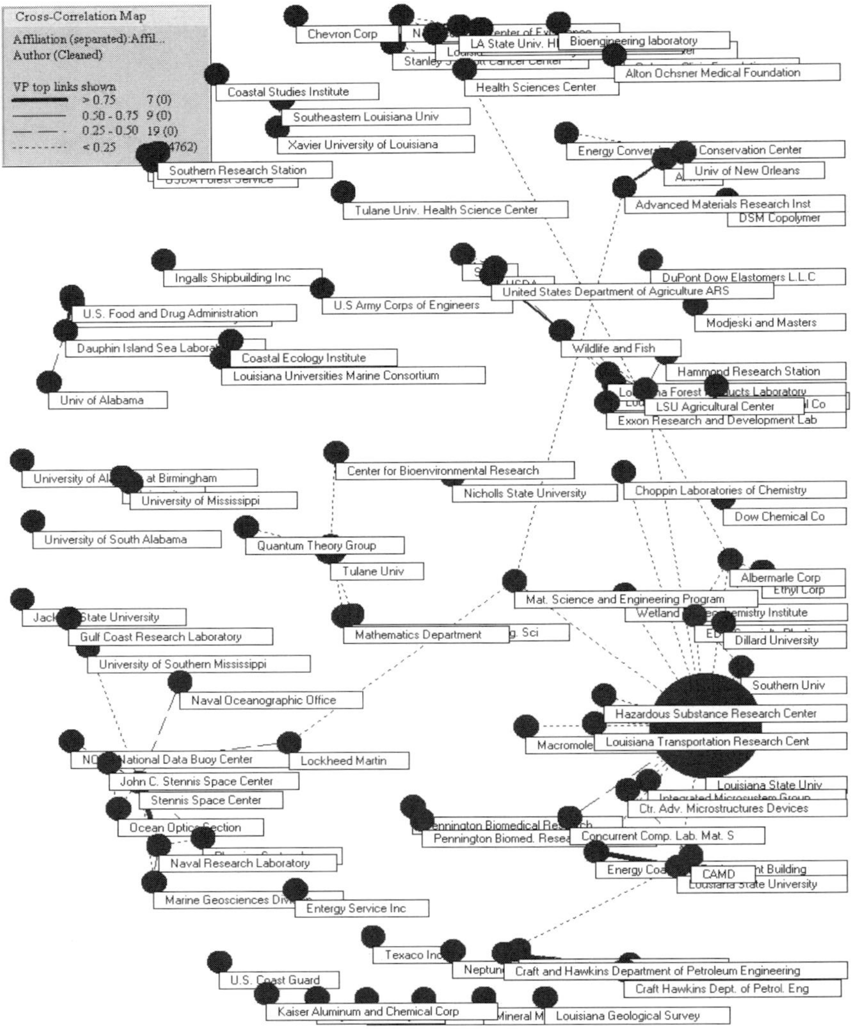

Figure 1: Publication collaboration map.

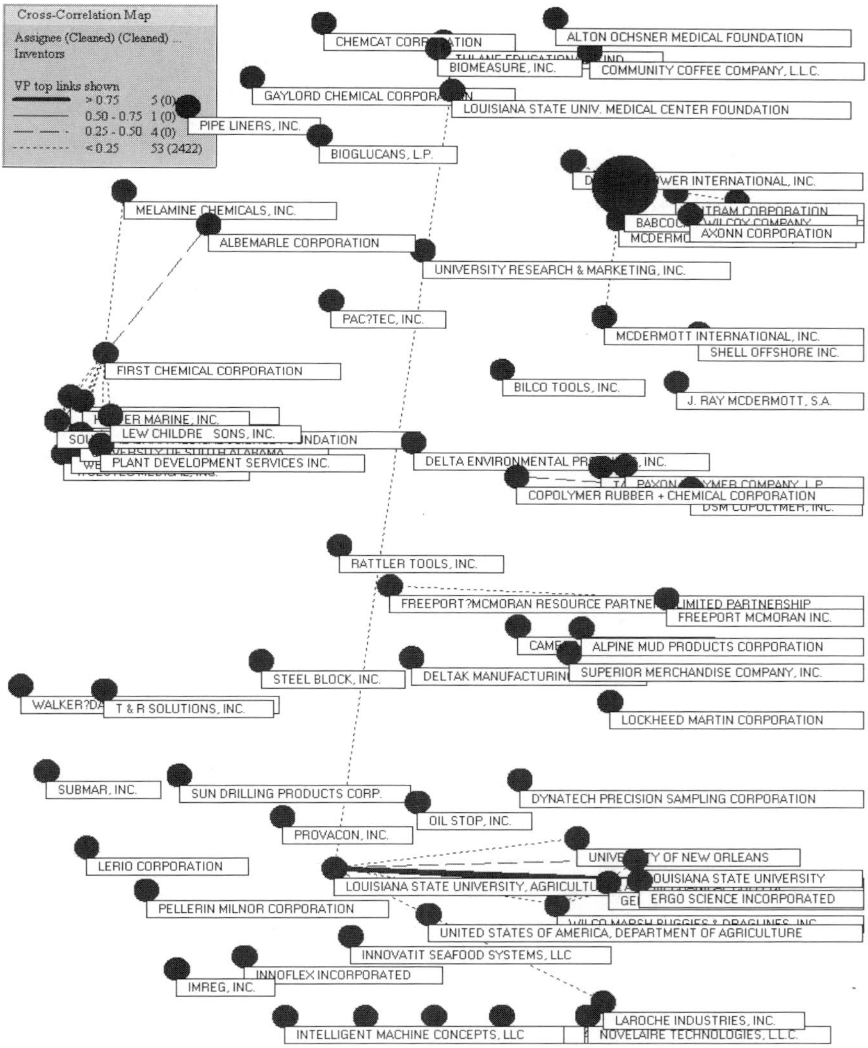

Figure 2: Patent collaboration map.

Collaboration between research institutes and industry is important to increase knowledge spillover, increase absorptive capacity, and increase the potential for innovative developments (Cooke, 2005). For the most part, the publication collaboration map demonstrates a *lack* of collaboration among academic centers, government, and industry, with some notable exceptions. Most of the links noted in the publication map are various components of the same entity. For example, the top of the map shows various representations for Louisiana State University (LSU) Health Science Center collaborating or the large LSU circle with numerous LSU centers radiating from it. Some notable collaboration activities include a link between Xavier, a private university, and Southeastern, a public university and links among USDA, Gulf Coast Seafood Laboratory and Dauphin Island Sea Laboratory. The University of Southern Mississippi has significant collaboration with Stennis Space Center. Lockheed Martin works both with Stennis Space Center and the University of New Orleans (UNO). LSU works with Albermarle Corp, IMG. LSU also works in a trio with EDO Specialty Plastics and Southern University. Once the internal collaboration links have been considered, it is clear that there is more of an indication of collaboration in patenting than in publishing. In patenting, there is a web of companies who (which?) all work with each other. These companies include Ingalls Shipbuilding, Halter Marine, Lew Childre Sons Incorporated, South Alabama Medical Science Foundation, University of South Alabama, Plant Development Services, Wellcutter Incorporated, Wuestec Medical, and First Chemical Corporation. There is also significant co-patenting between LSU and ERGO Science Incorporated. LSU has also co-patented with UNO, Wilco Marsh Buggies and Draglines, Incorporated, General Hospital Corporation, the LSU Medical Center Foundation, and Biomeasure Inc. There also exist strong inter-company relationships, such as between Novelaire Technologies and Laroche Industries and also between Babcock Wilcox and McDermott.

5.3. *Technology profile: The patent story*

An indicator frequently used by economic development practitioners is the ratio: the number of patent assignees (the organization to which the patent is assigned) in a state to the number of patent inventors in a state. The number of inventors indicates the level of intellectual productivity occurring in the area. The assignee indicates the organization that is benefiting from the patent. The graph below (Figure 3) was constructed using the number of patents for assignees and inventors for the states listed. The yellow line is the actual value for the indicator. This indicator was calculated for 23 states, based on the entire state, not just the I-10 Corridor regions. Alabama, Louisiana, and Mississippi are at the bottom in the AS/IN ratio. While Alabama and Louisiana, in particular, have a number of inventions comparable to other states, there are significantly fewer patents assigned to organizations in these states. **This result indicates that the region encompassing Alabama, Louisiana, and Mississippi is disproportionately producing patents benefiting other states.** This indicator is the first insight into the problem that the region has in bringing its research into commercialization.

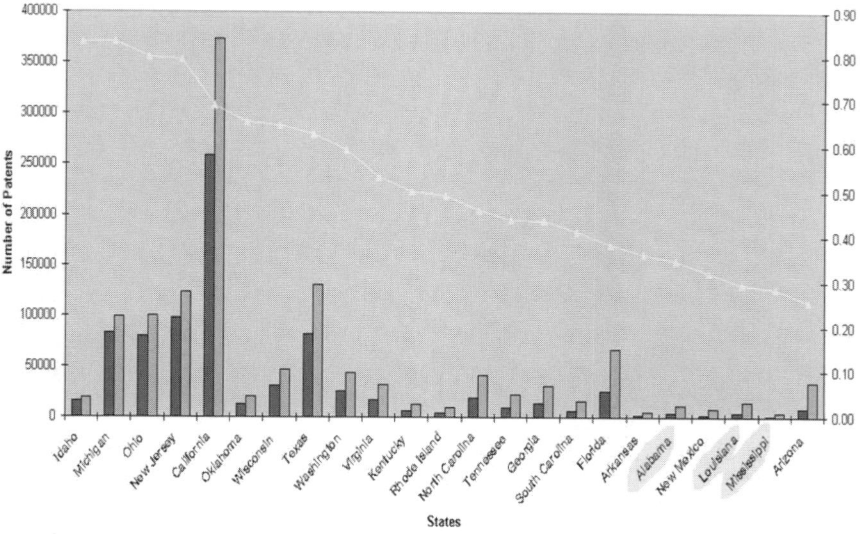

Figure 3: Assignee-to-Inventors ratio by state.

5.4. *Technology profile: Knowledge-based clusters*

The next step in the process entails creating conceptual maps based on the co-occurrence of terms in the abstracts for both publications and patents. It is interesting to note the lack of correlation between the

Publication Map and the Patent Map. The lack of patent activity in the main publication areas indicates that the development of many "indigenous" research areas is not occurring, at least not within the region. There are 19 knowledge-based publication clusters that reveal regional research strengths. Since academic institutions are conducting research in many of the clusters, they represent areas where knowledge spillover from researcher to student is likely to occur. Yet, the region is not capitalizing commercially on many of these capabilities. This lack of capitalization is further confirmed in a closer look at the patenting activity in the specific cluster areas. Table 3 shows the search string derived from the defining cluster terms utilized to identify the specific regional strengths and make comparisons to patenting behavior.

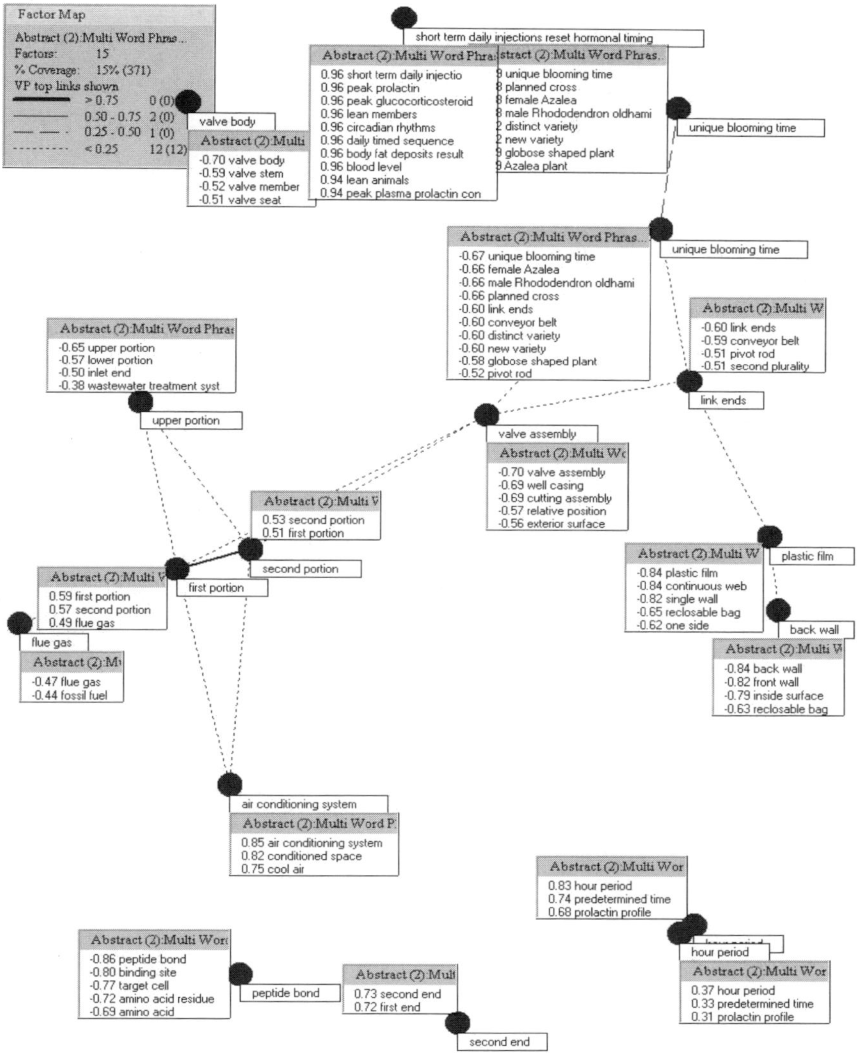

Figure 4: Knowledge-based patent term clusters.

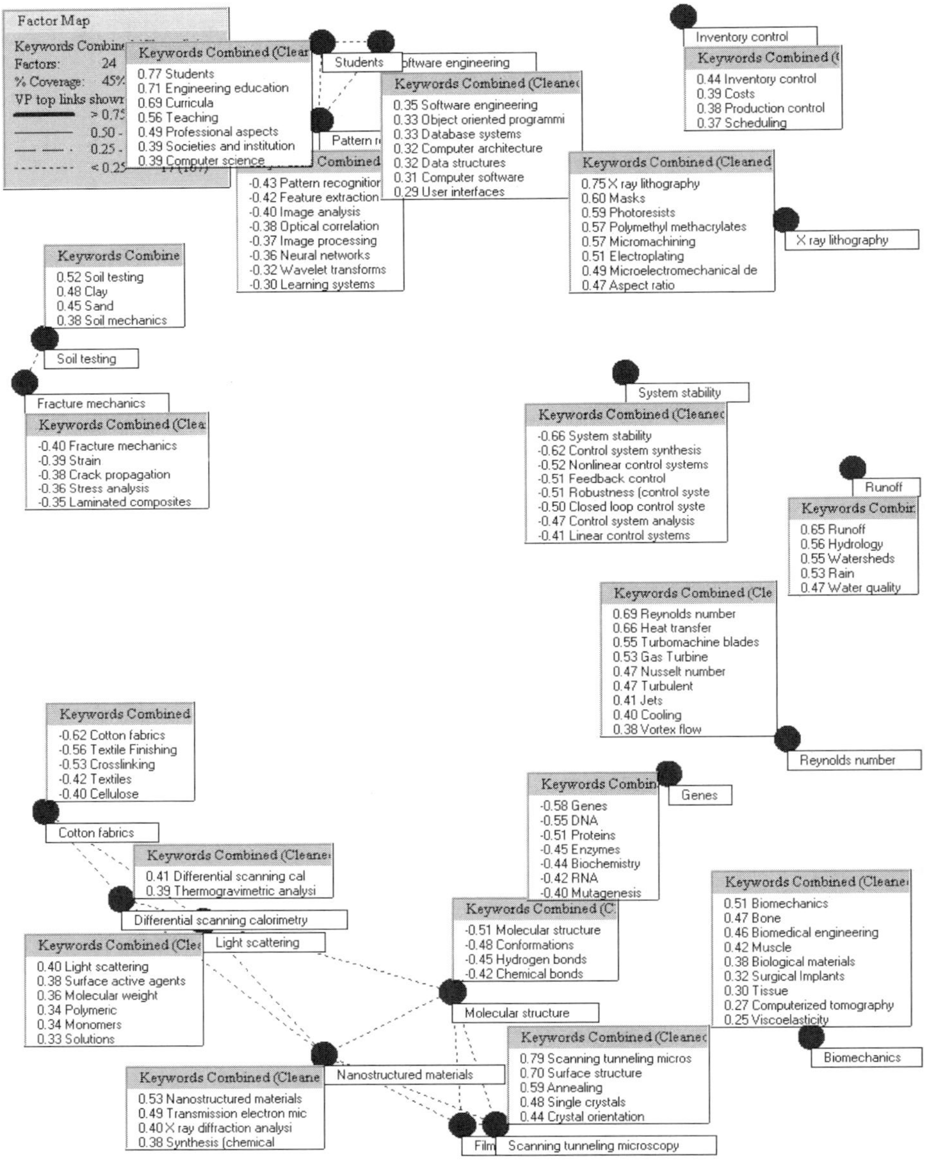

Figure 5: Knowledge-based publication term clusters.

Table 3: Cluster-defining search strings.

Cluster Label	Search String
Biomechanics	(biomechanics or "biomedical engineering") OR (tissue or muscle or bone or "biological material")
NanoStructured Materials	("nanostructured materials" and "transmission electron microscopy") OR ("synthesis chemical" and catalysts) OR ("nanostructured materials" and "X ray diffraction analysis")
Light Scattering	("surface active agents" or solutions or "light scattering" or "molecular weight" or moromers or polymeric) and ("nanostructured materials" or fluorescence or "molecular structure" or "composition effects" or "sodium compounds" or "nuclear magnetic" or "organic polymers" or "reaction kinetics" or viscosity)
Reynolds Number	(("heat transfer" or "reynolds number") and (turbulent or jets or cooling or "vortex flow" or "gas turbine" or turbomachine or "nusselts number")) OR ("heat transfer" and "reynolds number")
X ray Lithography	(("microelectromechanical devices" or "x ray lithography") and (electroplating or microstructure or methylcrylates or micromachining or microstructure or "aspect ratio".)) or (micromechanical devices" and "x-ray lithography") or (micromachining and (microstructures or methylcrylates or masks or photoreists))
System Stability	("control systems" and (robustness or synthesis or nonlinear or analysis or linear or "closed loop")) or ("system stability" and ("control systems" or mathematical models")) or ("feedback control" and "mathematical models")
Inventory Control	(Costs AND ("inventory control" OR "production control" OR scheduling)) OR ("inventory control" AND "production control)
Fracture Mechanics	((strain or "stress analysis") and ("fracture mechanics" or "crack propagation")) or ("laminated composites" and "mathematical models")
Soil testing	("soil mechanics" or clay or sand)
Runoff	(("water quality" or hydrology or runoff) and (rivers or watersheds or wetlands or "mathematical models")) or (runoff and ("water quality" or hydrology)) or (runoff and rain)
Software Engineering	((computer software" or "database systems" or "computer architecture" or "data structures" or "user interfaces") and (environmental or coastal or map or imaging or traffic))
Differential Scanning Calorimetry	("differential scanning calorimetry" or "thermographic analysis") and (chemical or "x ray" or "nanostructured materials" or spectroscopy or microscopy or thermal or epoxy)
Scanning Tunneling Microscopy	((crystal or surface) and ("scanning tunneling microscopy" or annealing))

Genes	(proteins and (biochemistry or DNA or RNA)) or (enzymes and (biochemistry or DNA or genes)) or (mutagenesis and DNA)
Pattern Recognition	("neural networks" or "learning systems" or "pattern recognition" or "wavelet transforms") AND ("image analysis" or "image processing" or "feature extraction" or "optical correlation")
Textiles	(textiles and ("cotton fabrics" or "crosslinking" or cellulose)) or (("cotton fabrics" or cellulose) and crosslinking)
Thin Films	("thin films" or monolayers) and (deposition or "film growth" or nucleation or "atomic force microscopy")

5.5. *Technology profile: Publications and patent activity comparisons*

The results in Table 4 demonstrate the patenting discrepancy in the technology cluster areas for the region. Note that the "Specific Area" corresponds to the search strings that can be found in Table 3. "General Term" corresponds to the broader area that is identified as the title. Where there is only one number, it is understood to be the Specific Area as described in the Technology Profiles. The "Specific Areas" below are all areas where at least one Gulf Coast organization is a top publisher in the US or the World. **These results show that even in areas where Gulf Coast entities are top publishers, the research is not advancing to the development stage or those developments are not being protected through patents.**

Table 4: Comparison of patenting activity.

Technology Concept: Publication Cluster Title	Total US Patents in Technology Concept	Total (LA-MS-AL) Patents in Technology Concept
Thin Films	Specific area: 452 General Term: 1882	Specific Area: 0 General Term: 3
Nano-structured Materials	Specific area: 16 General Term: 277	Specific area: 0 General Term: 1
Control Systems	Specific Area: 317	Specific area: 0 General Term: 7
X ray Lithography	Specific Area: 17 General Term: 2263	Specific Area: 0 General Term: 4
Soil Mechanics	9576	76
Genes	2840	14
Heat Transfer	2088	19

5.6. *Summary of Katrina impact*

In analyzing the I-10 Corridor, it is important to review the impact of Katrina on these knowledge assets. The Biotech Sector has suffered significant losses with the damages to LSU and Tulane facilities. The loss of facilities has also resulted in a loss of individual researchers, but it is difficult to assess the impact of individual losses. Pennington Biomedical Research Center, a significant publishing hospital, is in intact. In general, the other knowledge-based clusters have endured losses, primarily the loss of individuals, but remain fairly strong.

6. Conclusions

This research provides a basic profile of the work being completed in the Gulf Coast region. The results show that the Corridor is not lacking in intellectual productivity. However, we have shown that the Corridor lacks ownership of intellectual assets and the ability to capitalize on intellectual efforts, and this lack may represent a long-standing problem. Consistent with our observations about developing countries, what is suggested is that the region *does* possess unique competencies, but that it is not fully exploiting them.

When we consider the impact of the storms, the greatest Katrina impact has come in the Biotech area. This area is severely threatened. Others technology areas such as Materials and Surfaces, Soil and Water Runoff, and the underlying data analysis emphasis which runs through a number of clusters, appear to have been affected, but have fared better than initially expected. These results show that the I-10 Corridor has significant expertise in a number of advanced technology areas, indicating significant potential absorptive capacity. It is in these areas where the region is most likely to excel. They provide the strongest possibilities for long-term economic growth and economic development.

Additionally, the information in the study can be used to identify organizations that might benefit from being in close geographic proximity to the research conducted along the I-10 Corridor. Further research providing an analysis of the region in comparison to the whole of the technology area and other regional areas could be used to further bolster claims of expertise along the Corridor. Thus, this research provides a tool for economic development officials to use in planning and positioning the area.

References

Agrawal, A. (2003). The anchor tenant hypothesis: exploring the role of large, local, R&D-intensive firms in regional innovation systems. *International Journal of Industrial Organization*, 21(9), pp. 1227–1253.

Barkeley, DL, MS Henry, S Nair (2006). Regional innovation systems: implications for nonmetropolitan areas and workers in the south. *Growth and Change*, 37(2), pp. 278–306.

Becker, H. A. and Sanders, K. (2006). Innovations in meta-analysis and social impact analysis relevant for tech mining. *Technological Forecasting and Social Change*. 73. pp. 966–980.

Cooke, P. (2005). Regionally asymmetric knowledge capabilities and open innovation: exploring 'globalization 2' — A new model of industry organization. *Research Policy*, 34,. pp. 1128–1149

Jaffe, A. B. (1989). Real effects of academic research. *American Economic Review*, 79(5), pp. 957–970.

Jaffe, A. B, Trajtenberg, M. and Henderson, R. (1993). Geographic location of knowledge spillovers as evidenced by patient citations. *Quarterly Journal of Economics*, 108(3), pp. 577–98.

Julien, P.-A., Andriambeloson, E., Ramangalahy, C. (2004) Networks, weak signals and technological innovations among SMEs in the land-based transportation equipment sector. *Entrepreneurial Regional Development*, 16(4), pp. 251-269

Katz, J. S. and Hicks, D. (1997). Desktop scientometrics. *Scientometrics* 38(1), pp. 141–153.

Kearns, A. and Görg, H. (2002). Linkages, agglomerations and knowledge spillovers in the Irish electronics industry: The regional dimension. *International Journal of Technology Management*. 24(7), pp. 743–763.

Koo, J. (2005). Knowledge-based industry clusters: Evidenced by geographical patterns of patents in manufacturing. *Urban Studies*, 42(9), pp. 1487–1505.

Lazonick, W, (2004). Indigenous innovation and economic development: Lessons from China's leap into the information age. *Industry and Innovation*, 11(4), pp. 273–297.

Nelson, R. R. (2004). The challenge of building an effective innovation system for catch-up. *Oxford Development Studies*, 32(3), pp. 365–374.

Porter, A. L. and Cunningham, S. W. (2005). Tech Mining: Exploring New Technologies for Competitive Advantage. John Wiley & Sons, Hoboken, NJ.

Shapira, P., Youtie, J. and Mohapatra, S. (2003). Measures for knowledge-based economic development: Introducing data mining techniques to economic developers in the state of Georgia and the US south. *Technological Forecasting & Social Change*, 73, pp. 950–965.

University of New Orleans Library Database List retrieved from http://library.uno.edu/database/c.html on February 10, 2007.

Chapter 17

South Korean System of Innovation: From Imitation to Frontiers of Technology, Successes and Limitations

Aouatif El Fakir[1]

Institut de Recherche Internationale, Dauphine University, Paris, France

In this chapter, we use an analytical framework based on the "national system of innovation" approach to understand how South Korea acquired technological capabilities, increased the effectiveness of its national system of innovation and became a major player in many high-tech industries.

1. Introduction

Some developing countries have emerged as serious competitors in new areas of technology. How did they acquire the necessary capabilities to use technology and even to develop it? And how did they make their systems of innovation more sophisticated?

This chapter will examine the case of South Korea to understand how it has acquired technological capabilities and increased the effectiveness of its national system of innovation to become a major player in many high-tech industries.

We use an analytical framework based on the "national system of innovation" approach to clarify this process. We assume that countries

---

[1] The author is grateful to Mostafa Hashem Sherif for the support received in the preparation of the manuscript.

acquire capabilities to produce new goods, services and technologies through interactive learning spaces (ILS). The effectiveness of these spaces depends on how technologies, institutions and organisations evolve simultaneously.

Section 1 contains a critical review of the literature on the acquisition of technological capabilities. In section 2, we present the analytical framework based on the national system of innovation approach. In section 3, we use this framework to examine the case of South Korea. We conclude with a discussion on developing countries in terms of the relationship between technology and development.

2. The Acquisition of Technological Capabilities in Developing Countries: Theories and Approaches

The study of technology in developing countries started in the 1960's. Until the mid seventies, the main corpus of literature was called the *Theory of Technological Dependency*. This theory considers that technologies that are already developed elsewhere should not be reinvented but must be exploited under the following two conditions: the technology should be appropriate to a country's needs and it should be available at a reasonable cost. The theory also highlights the role of foreign investments and multinationals companies in the transfer of technology (Gerschenkron, 1962; Helleiner 1975).

From the mid seventies to the mid eighties, many scholars assumed that indigenous science can be more responsive to national needs than technology imports. The effectiveness of the National System of Science, Technology and Innovation (S&T&I) depends on some conditions like long term investments, efforts to accumulate knowledge, and skills and know-how, selection of the right R&D fields, effective institutional support and international (Jones, 1971; Cooper, 1973; Herrera, 1973; Sagasti, 2004).

The 1980's witnessed the emergence of the *Theory of Technological Capabilities*. It stipulates that imported technology cannot be used if it does not match the scale, skills and material needs of the new environment. Therefore, developing countries can not take advantage of imported technologies without a threshold of technological capabilities (Bell, 1984; Dahlman, Ross-Larsonn and Westphal, 1987; Fransman and King, 1984; Katz, 1987; Lall 1987).

At the end of the 1980's, the *Windows of Opportunity Theory* assumed that co-evolution of technology, institutions and organisations is essential for catching up and growth. The time needed to learn and the perspectives for growth depend on the technology. Developing countries must identify clearly the current stage of the technology and acquire the suitable combination of productive and institutional infrastructure, scientific/technical knowledge and experience. The cost of these components changes in each stage of the technology life cycle. It is lower in the emerging and mature stages of technology and higher in the growth phase (Perez and Soete, 1988; Perez, 2001).

While all these approaches and theories have illuminated many fundamental aspects on the technology transfer to developing countries, they lack an overall vision of how developing countries can acquire capabilities to use, adapt, improve and even develop technologies.

As a consequence, we propose to use the *National System of Innovation* (NSI) approach, because it could help to identify the actors and the mechanisms that determine the success of innovation and technological development processes. This approach emerged in the 1980's and the 1990's as an analytical framework in developed countries. It assumes that innovations result from interactive learning processes and the co-evolution of technologies, institutions and organisations. Here, innovations mean using new technologies, adapting, improving and even developing them (Lundvall, 1992; Nelson, 1993).

3. The National System of Innovation (NSI) Approach

According to NSI approach, innovations (not inventions) occur within a specific system in which stakeholders interact and learn from each other to take advantage of technological and market opportunities.

Interactive learning is the main process leading to innovation because it enables individuals and organisations to acquire new knowledge and new skills to improve their performances, to seize opportunities and to face pressures (Lundvall, 1992; Mckelvey, 1997) Moreover, this approach emphasizes the uncertainty of the innovation process and the need for an appropriate institutional framework (Nelson and Winter, 1982; Saviotti, 1997).

Although the NSI approach highlights systemic and dynamic aspects of the innovation, it has some limitations. It is based on the observation of sophisticated and complex innovation systems in developed countries. As a consequence, we had to adapt it to the context of developing countries because of the following reasons. First, the behaviour and objectives of the corresponding organisations and institutions differ. Second, the NSI indicators for developing countries deal with systems under construction. Finally, developing countries are operating in the context of a globalized R&D where technological alliances can play a significant role in technology acquisition.

Empirical studies of NSIs in developing countries, such as the comparative study of Dahlman and Nelson (1995), have confirmed that the macroeconomic environment and incentives regime determine technological and economic performance. In addition, resources like appropriate human capital, foreign technologies, industrial infrastructure, public support and funding are crucial for the acquisition of technological capabilities.

Arocena and Sutz's theoretical works about innovation in developing countries highlighted that developing countries lack "*interactive learning spaces*" which are "*interactive activities and processes where individuals and organisations generate, transfer and use systemically knowledge to enhance their ability to learn and to resolve problems.*" (Arocena and Sutz, 2002). These interactive learning spaces (ILS) facilitate the resolution of pressing problems of production by the encounter of actors owning the problems with "knowledge" actors

So, we assume here that these ILS cannot emerge if there are no incentives and pressures on firms to be more competitive and if they do not have the resources to participate in processes to generate, transfer and use knowledge. We also assume that in these learning spaces, firms acquire new technological capabilities and the national system of innovation emerges. As these spaces become systemic, firms strengthen their technological capabilities and the NSI becomes more sophisticated and effective.

4. The Acquisition of Technological Capabilities in South Korea

South Korea achieved a very fast development process in the last forty years and emerged as a competitive supplier in many medium and high tech industries. First, we note the incentives/pressures that pushed Korean firms to enter in ILS. We then examine the resources they used to resolve production and innovation problems. We will observe how spaces, resources and capabilities have evolved from 1960 until now. Our study is based on the works of Kim (1980, 1993, 2000), Westphal, Rhee and Pursell (1985), Westphal, Kim and Dahlman (1985), Kim and Yi (1997), Kim and Lee (1987) and Lim (1999). Table 1 provides figures on the growth the GDP (gross domestic product) and GDP per capita in South Korea from 1960 to 2000. Table 2 shows the evolution of exports from 1966 to 2000. It seen that, in addition to their dramatic growth over the last four decades, their technological content has continuously progressed to higher values.

The South Korean technological, industrial and economic progress has gone through 4 stages. In the 1960's, South Korea was oriented to import substitution before switching to export promotion. This phase witnessed the set up and strengthening of production capabilities[2]. The 1970's was the period of increasing exports, the establishment of heavy industries and the development of investment capabilities[3]. High-tech industries and local R&D activities emerged during the 1980's. In the third phase, industries and activities were strengthened during the 1990's despite the financial crisis and some industrial and institutional failures. In the current phase, which started in 1997, South Korea has undertaken

---

[2] Production capabilities include the supervision of production facilities, raw materials control, production planning, quality control, production problems resolutions, process and product adaptation, facilities repairing and maintenance and marketing (Westphal Rhee and Pursell, 1985).

[3] Investment capabilities include labour force training, feasibility studies, projects implementation, management and engineering, specific studies, basic and specific engineering and purchasing. Ibid.

many reforms which put it on a new growth virtuous spiral, but some limitations prevent it from becoming a knowledge and innovation economy.

Table 1: GDP and GDP per capita of South Korea 1960-2000.

Year	GDP (current billion. US$)	GDP per capita, Purchasing Power Parity (current international ($)
1960	3.9	155
1965	3.0	108
1970	8.9	275
1975	21.5	599
1980	63.8	1632
1985	96.6	2290
1990	263.8	5893
1995	517.1	10850
2000	511.7	9763

Source: World Bank.World Development Indicators Database.

Table 2: Structure of exports of South Korea 1966-2000.

Type of Export (%)	1966	1976	1987	1995	2000
Primary products	37.9	10.1	6.4	3	2
Resources based manufactures	12.5	6.2	3.4	7	11
Low tech manufactures	41.8	55.5	52.3	20	17
Medium tech manufactures	0.4	4.4	13.7	35	33
High tech manufactures	1.3	6.7	21.8	30	36

Source: Calculated from international trade statistics, SITC Rev.2. OECD. 2005.

4.1. *Keys of success*

This success cannot be objectively explained without speaking about the U.S. assistance during the cold war and the major role of Korean government nationalism and culture. In fact, during 1950's, American aid was vital for the reconstruction and industrialization of South Korea after Korean War. Until 1961, 70% of public investments were funded by U.S. aid (military aid from 1953 to 1961 amounted to 1561 million U$) (Toussaint, 2006). In the 1960's, South Korea firms supplied the

U.S. military in the region with construction services, machinery and some finished products (tire, wood). This helped them improve their management, organizational and technical skills (Westphal Rhee and Pursell, 1985).

The U.S. has remained an important source of technology. Moreover, thanks to Korean Diaspora in the U.S, joint ventures with U.S. firms and research structures allowed South Korean firms to assimilate new technologies, especially in electronics, particularly during the 1980's.

During the totalitarian regime (1961-1987), the South Korean government pursued an active industrial development policy which was continued after the democratic transition in 1987. Reforms of the agricultural regime, investments in infrastructure and education, public funding, subsidies, tax incentives and strong protectionism were the main tools that the South Korean government used to achieve technological and industrial development.

South Korean nationalism and culture played an important role, particularly in the early stages (Koo, 1995). According to Song (1997), new Confusionism in South Korea encouraged personal achievement (education and discipline), family values, patriotism, harmony and community spirit.

We will now highlight the ILS that improved technological capabilities of Korean firms and lead to the production of new goods and the use of new technology in each period. Our objective here is to clarify the link between ILS and the technological and industrial development of South Korea.

4.2. *The 1960's: The acquisition of production technological capabilities*

In this decade, South Korea had three major handicaps: a small local market, lack of natural resources and low technological capabilities to compete locally or abroad. The Korean government switched rapidly from import substitution to export promotion to encourage the acquisition of production capabilities in many new industries. Public incentives (subsidies, domestic market protection, fiscal incentives and so on) and attractive market opportunities encouraged Korean firms to acquire technological capabilities even in industries in which they had no competitive advantage. In general, the technology in these industries was mature except for electronics.

Korean firms took also advantage of qualified workers who were able to assimilate new knowledge and to acquire new skills rapidly. They benefited from available foreign technologies in the form of capital goods, turnkey plants, technical assistance and licences. Large firms (chaebols or Chaebols, please be consistent throughout the paper) benefited from attractive financial terms from Korean banks. Lastly, the government set up many measures to increase the bargaining power of Korean firms.

To acquire technological capabilities, Korean firms entered deliberately in many ILS that involved different actors according to the production process and scale as well as the size of the firms.

In small batch industries (shipbuilding, machinery), large firms favoured capital goods, licences imports and technical assistance. They were involved in interactive learning spaces with foreign suppliers and experts through training and *in situ* technical assistance. Korean engineers and technicians learned how to manufacture effectively large quality of quality products with varying technical specifications. Small firms applied imitative reverse engineering. In addition, they interacted with local users to improve their copies of foreign products. Lastly, in new industries, technical staff moved from the pioneer firms to latecomers and interacted with their inexperienced personnel. Thus, foreign technologies were rapidly disseminated.

In large batch industries (electronics, automobile), large and small firms set up ILS with foreign suppliers of sophisticated products to acquire knowledge through training and technical assistance. Small firms preferred imitative reverse engineering of simple products. Foreign technologies have been disseminated in the same way as in small batch industries.

In industries using process production (chemicals, cement, pharmaceutical), firms purchased turnkey plants to acquire initial production capabilities. Korean technical personnel entered in interactive learning spaces with foreign technical experts to assimilate technology and to perform production, maintenance and repair operations.

In summary, in this period, the learning spaces were dominated by foreign sources of technology. Large firms imported technologies to acquire production capabilities and small firms preferred imitative reverse engineering. However, all tried to improve their capabilities to be more autonomous and to enhance quality and sophistication to go to next stages.

4.3. *The 1970's: The emergence of innovation capabilities*

During the 1970s, the Korean companies improved rapidly their technological capabilities in production and in product design. Interactions with foreign customers enhanced the know-how of local companies and new technologies, some mature and some growing, were acquired.

The output of large production facilities set-up in many industries to achieve economies of scale economies was larger than the local market could absorb, so Korean firms had to open new markets. To have competitive exports, they had to reduce costs, differentiate products and improve their quality. A large percentage of Korean firms worked under Original Equipment Manufacturing (OEM) agreements and had to meet tight specifications. To do so, they had to acquire more sophisticated engineering capabilities. Also, OEM relationships constituted precious sources of information about international markets.

Finally, the South Korean government provided attractive incentives in heavy industries like chemicals, the automobile industry, steel or shipbuilding.

The combination of all these factors resulted in a more technically qualified labour force. Korean firms could then import or assimilate more complex and newer technologies. Even while they continued to use some of the old spaces, Korean firms could also engage in new interactive learning spaces in the following manner.

First, in many established industries, new firms hired the technical personnel of pioneering firms to benefit from their experience and skills. Moreover, the new entrants avoided the inherent risks of technology selection by taking advantage of adaptations achieved by the pioneers.

Secondly, large Korean companies involved in OEM agreements entered in a specific interactive learning space with their costumers.

Lastly, large Korean firms continued to import sophisticated and complex technologies through specially licences. Hence, interactive learning space between licence suppliers (especially Japanese) and Korean firms aimed at acquiring production capabilities in new technologies and/or design capabilities in established areas. In short, the acquisition and mastery of production technologies progressed during this phase through "learning by doing" and "learning by using" as well as by adaptations of the foreign technologies. The mobility of qualified workers and other formal and informal mechanisms made the dissemination of technologies easier. Simultaneously, the acquisition of design capabilities continued through the transfer of foreign knowledge, NSI was more and more embedded in economic system, interactions between actors became more systemic and Korean firms reinforced their international competitiveness and enhanced the range of their products.

4.4. *The 1980's: The emergence of high-tech industries and the expansion of R&D*

At the beginning of the 1980's, South Korea faced a dramatic change. The economic crisis pressed the U.S. and Europe to set up protectionist measures against new industrialised countries. South Korea also lost its competitive advantage in labour intensive industries as real wages increased. Moreover, foreign suppliers of technology became reluctant to supply it to their Korean competitors. Lastly, imitative reverse engineering became more difficult as South Korea tightened regulations regarding the protection of intellectual property. Many emerging high-tech industries, however, provided new opportunities. Among them, South Korea chose semiconductors, optical fibre, robotics, computer and aircraft.

During this decade, large Korean firms had to increase their reliance on indigenous resources in R&D with government assistance through exportations subsidies, technology monitoring, financial assistance and so on. Yet, the supply of highly qualified engineers and scientists at this time was not adequate for the needs. To fill this gap, Korean firms attracted Koreans expatriates (from the United States, Europe and Japan). Korean-American engineers and scientists provided the necessary bridges with foreign experts on specific R&D projects. Moreover, large firms like Samsung, Hyundai and LG invested in foreign technological outposts to monitor technological changes and work with foreign firms and R&D institutes within technology clusters. Korean foreign subsidiaries and outposts contributed to the dissemination of foreign knowledge and know-how particularly in electricity, electronics and chemicals. All these interactive learning spaces, particularly those that involved Korean and U.S companies, were extremely beneficial in leading-edge technologies such as semiconductors, computer and biotechnology.

In many industries, such as robotics and computers, companies (notably Chaebols) continued to use their usual ILS. New interactive learning spaces involved Chaebols and multinationals firms (MNFs) through joint ventures. The latter were interested in technological cooperation with Korean firms with large production capabilities. While Korean firms aimed to strengthen their innovative capabilities.

In this decade, Korean companies started to work with government research institutes (GRIs). National R&D Projects (NRP) and Industrial Base Technology Development Projects (IBTDP) set up by the Korean government involved the GRIs and firms in selected technologies.

In short, during the 1980s, Korean companies (particularly Chaebols) moved from reverse engineering to international cooperation and local R&D to acquire the knowledge associated with the emerging industries. Reverse engineering and internal R&D was for relatively simple technological tasks in robotics, computers at the beginning. Interactions with foreign companies and indigenous R&D structures were reserved for more complex cases. The ILS helped to establish new industries to replace those that lost their competitive edge and to take advantage of new opportunities. The NSI became more autonomous, innovations more endogenous. As a consequence, the Korean economy was radically transformed in this critical decade.

4.5. *The 1990's: The expansion of high-tech industries and the strengthening of innovation capabilities*

The Korean slowdown during the 1990's had many reasons, both internal and external. Nevertheless, after the financial crisis, favourable conditions accelerated Korean recovery. Foreign investors' confidence was re-established, many large companies were restructured and currency was re-stabilized. Moreover, favourable global demand, and information and communication technology expansion encouraged Korean firms to increase their efforts towards more creativity and inventiveness in order to take advantage of new innovation-intensive industries.

In addition, large investments in basic research in universities, venture capital availability, attractive public incentives and the needs of more flexible firms in many industries encouraged the setting-up of high-tech and innovative SMEs.

By and large, Korean firms achieved internal R&D and kept their relationships with Korean diaspora to improve their innovative capabilities. They also maintained cooperation with MNFs and GRIs. In addition, the government continued to support firms through subsidies, financial assistance and so on. Lastly, in the case of more complex capabilities, Korean companies increased their patent portfolio to enhance their negotiation position with international parties through cross-licensing and similar intellectual protection schemes.

Korean firms, notably Chaebols, reorganized themselves to encourage new "creative spaces." For example, to stimulate innovation, employees with new ideas could set up experimental spaces involving persons of various domains to find profitable projects in the short and medium terms.

Furthermore, outsourcing needs of Chaebols strengthened semi-permanent relations between them and their SME suppliers of components and machinery. This cooperation resulted in cost reduction, and improved the design and production capabilities for complex technologies.

Also, Korean SMEs were involved together in many interactive learning spaces through R&D cooperation, common equipment and inputs purchasing, common marketing in order to improve their efficiency and quality.

Korean companies continued to be involved with GRIs in new high-tech industries, such as artificial intelligence, advanced robotics and so on. Lastly, Chaebols kept interactive learning spaces through joint ventures with MNFs to strengthen their innovative capabilities.

In this stage, ILS led to the acquisition of more complex engineering and innovative capabilities in order to keep pace with technology frontiers and to generate new technologies in specific industries.

Figure 1 summarizes the technological catch-up process of South Korea.

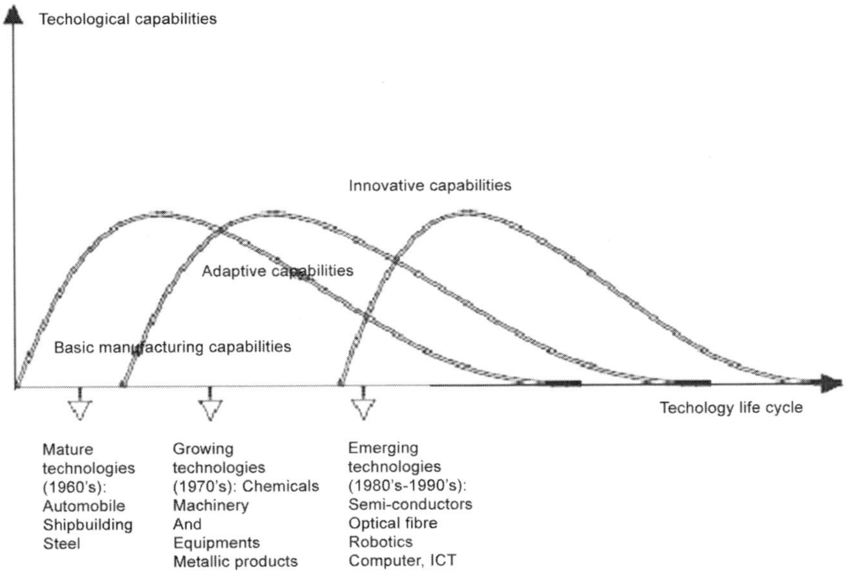

Figure 1: Technological development process of South Korea.

5. The 2000's: South Korean System of Innovation Limitations

Many studies and reports in the 2000's have indicated that South Korea is not a producer of radically new knowledge. Chaebols do not have enough capabilities to innovate radically and there are only few technological SMEs. Basic and applied researches are not promoted enough and scientists and researchers are not equally distributed between GRIs and universities (Kim, 2000; OECD, 2005).

To fill the gaps of its system of innovation, OECD recommended that South Korea focus on three major areas of improvements. First, basic research, especially within universities, must be strengthened to increase the chances of new discoveries and to develop new technologies. More interactions between different innovation actors (firms, GRIs, universities) are also required. Moreover, scientific and technological cooperation with others countries must be intensified to take advantage of their previous experience. Second, education policy must be reformed to diversify higher education and develop creative and high quality expertise in strategic areas. Third, industrial policy must lead to a more balanced structure between SMEs and Chaebols and to more intensive learning and innovation through more business services, venture capital development, support networks, enhancement of creativity culture and so on.

In recent years, Korea has followed many of these recommendations. In fact, R&D expenditures reached 2.63% of GDP in 2003 (24321.3 million U.S.$), higher than the OECD average. New R&D public programs and higher universities budgets (to intensify basic research) lead to the rise of public expenditures. Moreover, the number of researchers reached 6.2 per 1000 workers, which is close to the OECD average in 2002 (OECD, 2006). Furthermore, in the early the 2000's, South Korean government restructured the national system of innovation to encourage creativity an innovation such as upgrading academic and industrial laboratories.

Consequently, South Korean scientific and technological outputs increased significantly in recent years. As shown in Table 3, patent applications and patents granted to South Korean inventors increased dramatically between 1995 and 2005. The percentage of high- and medium tech exports rose respectively 36.1% and 32.2% of the total South Korean exports. In contrast, low and medium-low tech exports declined to respectively 11.4% and 20.3% (OECD, 2006; STAN database).

Table 3: Patent applications and patents granted of South Korea 1995-2005.

	Patent applications				Patents granted			
Year	1995	2000	2002	2005*	1995	2000	2002	2005*
Number	78,499	116,886	106,136	160,921	12,512	34,956	45,046	73,512

Source: World International Property Organisation WIPO. * Last data available.

6. Discussion and Conclusions

The common pattern of Asian success in manufacturing (Japan, South Korea, China…) shows two main characteristics. First, firms are able to integrate and evolve from an interactive learning space to another to get the technological capabilities required for each stage of development. Second, institutional framework and organisational arrangements supported the technological effort.

Like Japan earlier, South Korea imitated and integrated the existing technological systems. Similarly, Chinese firms in the field of mobile telephone technology, entered in several ILS and had joint ventures with Multinational companies to acquire basic manufacturing skills and hone their adaptive capabilities (Von Zedtwitz and Jin, 2007). Unlike Korean firms, however, Chinese firms moved quickly to establish joint ventures and R&D to keep pace with rapid change of 3G technology.

South Korean strategies were particularly effective in the assimilation, adaptation and improvement of technologies developed outside Korea. Organisation arrangements and the institutional framework were adapted to fit the technology needs. This experience illustrates, however, that a successful pattern of technological assimilation does not necessarily help in producing new technologies and that contributing to radically new knowledge and technologies requires another set of strategies.

The failure of other developing countries can be attribute to the lack of interactive learning spaces and to the absence of a supportive institutional framework. Firms in these countries rarely have the internal and external resources to integrate and evolve from one interactive learning space to another. Thus, they cannot solve their pressing production problems; stagnate at a level of low technological capabilities and insufficient competitiveness in local and international markets. This inability to integrate interactive learning spaces and to evolve among them results from the disjunction of the institutional framework and organisational arrangements from technology evolution.

Accordingly, we propose that technological and economic development is possible only if developing countries make their institutional framework and organisational arrangements compatible with the technology life cycle. More studies of catch up processes are needed to clarify whether it is the political regime, institutional arrangements and culture aspects that determine the accumulation of capabilities and the production of new knowledge or whether the latter depends on each national experience and the history of the various national innovation systems.

References

Arocena, E. and J. Sutz, (2002). Innovation Systems and Developing Countries, Danish Research Unit for Industrial Dynamics, DRUID Working Paper No 02-05.

Bell, M. (1984). Learning and the Accumulation of Industrial Technological Capacity in Developing Countries. In: Technological Capability in the Third World (Fransman, M. and King, K., eds.), Macmillan, London, pp. 187-210.

Cooper, C. (1973). Science, Technology and Development: The Political Economy of Technical Advance in Underdeveloped Countries. F. Cass, London.

Dahlman, C.J., Ross-Larsonn, B. and Westphal, L. (1987). Managing Technological Development: Lessons from the Newly Industrializing Countries. World Bank, Staff Working papers. No. 717.

Dahlman, C. and. Nelson, R.R (1995). Social absorption capability, national innovation systems and economic development. In: Social Capability and Long-Term Economic Growth (Koo, B. and Perkins, D., eds.), Macmillan, London, pp. 82–122.

Fransman, M. and King, K. (eds.) (1984). Technological Capability in the Third World. Macmillan, London.

Gerschenkron, A. (1962). Economic Backwardness in Historical Perspective: A Book of Essays. Belknap Press of Harvard University Press, Cambridge, MA.

Helleiner, G.K. (1975). The Role of Multinational Corporations in the Less Developed Countries' Trade technology. In: The Economics of Technology Transfer (Lall, S., ed.), pp. 161–171, Edward Elgar, Cheltenham.

Herrera, A. (1973). Social Determinants of Science in Latin America: Explicit Science Policy and Implicit Science Policy. In: Science, Technology and Development:

The Political Economy of Technical Advance in Underdeveloped Countries (Cooper, C. ed.), F. Cass, Londonm pp. 19–38.

Katz, J. M. (ed.) (1987). Technology Generation in Latin American Manufacturing Industries. Macmillan, London.

Kim, L. (1980). Stages of development of industrial technology in a developing country: a model. In: Learning and Innovation in Economic Development (Kim, L., ed), pp. 4-26, Edward Elgar, Cheltenham.

Kim, L. (1993). National System of Industrial Innovation: Dynamics of Capability Building in Korea. In: National Innovation Systems: A Comparative Analysis (Nelson, R. R., ed.), pp. 357–383, Oxford University Press, New York.

Kim, L. (2000). Korea's National Innovation System in Transition. In: Technology, learning and Innovation (Nelson, R. R. and Kim, L., eds.), pp. 334–360, Cambridge University Press, Cambridge (UK).

Kim, L. and Lee, H. (1987). Patterns of Technological Change in a Rapidly Developing Country: A Synthesis. In: Learning and Innovation in Economic Development (Kim, L., ed.), pp. 109–124, Edward Elgar, Cheltenham.

Kim, L. and Yi, G. (1997). The dynamics of R&D in Industrial Development: Lessons from Korean Experience. In: Learning and Innovation in Economic Development (Kim, L., ed.), pp. 167–82, Edward Elgar, Cheltenham.

Koo, B.H. (1995). Socio-cultural Factors in the Industrialization of Korea. In: Social Capability and Long-Term Economic Growth (Koo, B. and Perkins, D., eds.), pp. 181–202, Macmillan, London.

Korea Ministry of science and technology, "Science and Technology Policy Directions for the 21st century". http://www.most.go.kr/.

Korea Ministry of Commerce, industry and energy. http://www.mocie.go.kr/main.html

Jones, G. (1971). The role of science and technology in developing countries, London, New York, Published for the International Council of Scientific Unions by Oxford University Press.

Lall, S. (1987). Learning to Industrialize: The Acquisition of Technological Capabilities by India, Macmillan, London.

Lim, Y. (1999). Technology and Productivity, the Korean Way of Learning and Catching Up. MIT Press, Cambridge, MA.

Lundvall, B.-Å (1992). National systems of Innovation: Towards a theory of innovation and Interactive learning, Pinter, London.

Mckelvey, M. (1997), Using Evolutionary theory to Define Systems of Innovation. In: Systems of innovation: growth, competitiveness and Employment (Edquist, C. and Mckelvey, M., eds.), Edward Elgar, Cheltenham, Vol II, pp. 140–162.

Nelson, R.R. (ed.), (1993). National Innovation Systems: A Comparative Analysis, Oxford University Press, New York.

Nelson, R. R. and Winter, S. (1982). An evolutionary Theory of Economic Change. The Belknap Press of Harvard, Cambridge, MA, pp. 3-22.

OECD (2005). OECD economic reviews, Korea, October 2005.

OECD (2005). International Trade statistics. www.oecd.org.

OECD (2006). Main Science and Technology Indicators. www.oecd.org.

OECD (2006). STAN database. www.oecd.org.

Perez, C. (2001). Technological change and opportunities for development as a moving target, CEPALR Review 75.
Perez, C. and L. Soete (1988). Catching up in technology: entry barriers and windows of opportunity. In: Technical Change and Economic Theory (Dosi, G. et al., eds.), pp. 458–479, Pinter, London.
Sagasti, F. (2004). Knowledge and Innovation for development, the Sisyphus challenge of the 21st Century, Edward Elgar, Cheltenham.
Saviotti, P. P. (1997). Innovation systems and Evolutionary Theories. In: Systems of innovation: Technologies, Institutions and Organizations (Edquist, C., ed.), Chapt. 8, pp. 180–199, Pinter, London.
Song, B.-N. (1997). The Rise of The Korean Economy, Oxford University Press, Hong Kong.
Toussaint, E. (2006). Banque Mondiale, le coup d'Etat permanent, CADTM, Liège.
Von Zedtwitz, M. and Jin, J. (2007). Process of Technological Capability Development: Cases from China's Mobile Phone Industry. In: Management of Technology: New Directions in technology Management (Sherif, M. H. and Khalil, T. M., eds.), Elsevier, Amsterdam, pp. 311–326.
Westphal, L.E, L. Kim, and C. Dahlman, (1985). Reflections on the Republic of Korea's Acquisition of Technological Capability. In: International Technology Transfer: Concepts, Measures, and Comparisons (Rosenberg, N. and Frischtak, C., eds.), Praeger, New York, pp. 167–221.
Westphal, L.E, Y.W Rhee, and G. Pursell, (1985). Sources of Technological Capability in South Korea. In: Technological capability in the third world (King, K. and Fransman, M., eds.), Macmillan, London, pp. 279–300.
World Bank (2007). World Development Indicators Database. web.worldbank.org
World International Property Organisation. (2007). Patents Data. www.wipo.int

Chapter 18

On Creating Value in Various Positions in the Value Chain – The Pulp and Paper Industry in China

Ou Tang, Jaana Sandström**, Hanna Kuittinen**, and Kalevi Kyläheiko***

**Linköping University, Sweden*
***Lappeenranta University of Technology, Finland*

The global pulp and paper industry is undergoing a major structural change. Due to the varying comparative advantages of different countries, such as raw material costs and market potential, firms are revisiting their global manufacturing and sourcing strategies. China offers a huge market potential, and companies in the pulp and paper industry (PPI) are actively investing in manufacturing in China. This raises interesting questions concerning the orchestration of the value chain in this emerging market, as the comparative advantage of China does not lie in raw materials of the PPI. In this paper, we use data collected in the Chinese PPI industry to describe the current determinants of creating competitive advantage in various parts of the PPI value chain in China. The most interesting factors in the Chinese PPI are the increasing demand and production volumes and, in particular, the lack of wood-based pulp. This means that sourcing can be central to creating competitive advantage in this emerging market. The results of this study can help PPI firms to make strategic decisions.

1. Introduction

The global players of the pulp and paper industry (PPI) are seriously revisiting their strategies (Autio et al., 1997; Christensen and Caves, 1997; Sande, 2002; Pesendof, 2003; Cenatempo and Nutt, 2005; Perks and Jeffery, 2006). Due to the cyclical nature of the demand and the fragmentation of the industry combined with the overemphasis on heavy investments and inability to sense the rise of keen global competition, the industry has for some grades of papers been driven to global overcapacity problems (Diesen, 1998; Siitonen, 2003). Also, the industry is struggling with declining demand for major grades in the traditional markets (North America and Europe), while the emerging and fast growing markets of China, India, Russia and South America offer lucrative business opportunities (e.g., Li, Luo and McCarthy, 2006).

The overcapacity and stagnated consumption are not the only challenges the traditional areas are facing; their strategic importance as raw fiber suppliers is also lessening. South America has rapidly become one of the most important pulp exporters because of their relative advantage in forest growth, especially the Eucalyptus fiber (Hurmelinna-Laukkanen et al., 2006; Välikauppi et al., 2006). In the near future, Russia and other East European countries are expected to generate new demand, with Russia playing also an important role as a roundwood and pulp provider.

The structure of the industry seems to shift away from the diversified, vertically integrated, and locally or regionally based enterprises towards the global firms that are more focused horizontally and vertically but more diversified geographically. In this new situation the strategic focus of the firms needs to be revisited. Strategies based on cost minimization and consolidation have not significantly improved the performance of PPI firms. Thus, the sources of competitive advantage must be searched elsewhere, for example in the orchestration of the global value net (Björk, 2006).

Motivated by the above, we investigate the PPI sector and provide an overview of the current status of this industry in China. By examining its market, production, sourcing and relevant government policies, we try to identify the critical factors in value creating in various parts of the value chain of this emerging market.

2. Posing the Issue – PPI Value Creation in China

It has been forecast by RISI that in 2020, the total consumption of paper in China will reach 100 million tons compared to the 2005 level of 59 million tons. Consequently, the Chinese PPI market attracts the majority of global investments in new lines, machines and mills. During recent years about half of all the global paper and board machine investments have gone to China. The development is striking since the "power region of the PPI" in the 1980s and 1990s, namely North America, is dramatically losing its strength as a manufacturing region and shutting down its capacity. The Scandinavian global players are maintaining their Scandinavian capacity for the European markets and simultaneously sharpening their strategic focus and foothold at the global market place. The emerging markets interest global PPI firms, and probably India might be the next interesting region with quite similar features with Chinese PPI markets. India is similar to the Chinese market in that it has a huge potential but lacks a sufficient base of raw fibers. Despite the interesting India we see that China is at the moment the fastest growing and developing region with interesting and challenging strategic options in the PPI. The uncertainties are also remarkable, however, for example raw material and energy supply and the development of government policy regarding foreign actors in China. Figure 1 presents the value chain of the pulp and paper industry (PPI), and in this paper we try to identify the upside potential as well as the downsides of various positions in the value chain.

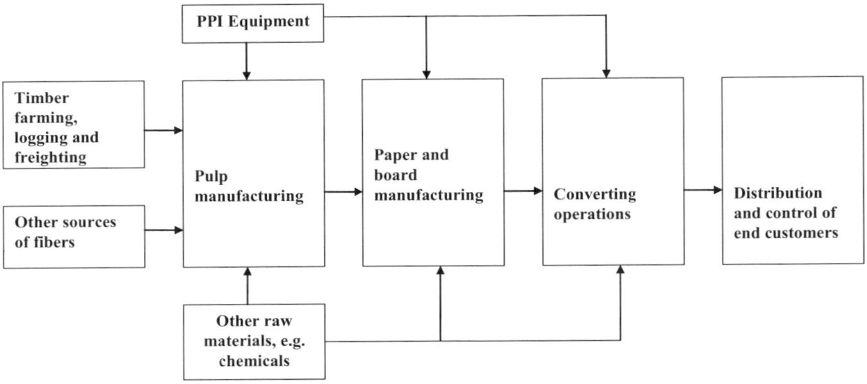

Figure 1: The value chain of the pulp and paper industry (PPI).

2.1. *Huge growth and investments in the Chinese PPI*

Many previously state-owned firms have been privatized. Even though the total output is increasing, the total number of mills is decreasing and economies of scale is a key parameter in production planning. There were approximately 250 pulp, paper and board related investment announcements in China between 2001 and 2005. In 2006, there were about 80 announcements (excluding closures), and this shows the increasing pace of development in China (Data source: Paperloop Mill Projects Database). In the 2001–2006 interval, these PPI investment announcements in China included:

- 203 new paper/paperboard machines (45.2 million tons/year)
- 13 new pulp mills (5 million tons/year)
- 70 new pulp/recycled fiber lines (17.4 million tons/year)
- 25 rebuilds of paper/paperboard machines (0.8 million tons/year)
- 9 rebuilds of pulp/recycled fiber lines (0.5 million tons/year)
- 9 restarts of idled paper/paperboard machines (0.7 million tons/year).

We should note that not every announcement had the planned capacity in tons in their announcements, but still these figures show that the majority of new lines and tonnage announced in 2001–2006 were for paper and board manufacturing. Consequently, the combined pulp/

recycled fiber lines tonnage was only about half of the announced paper/paperboard capacity (22.9/46.7 tons per annum, tpa). This indicates an increased need for fiber imports as paper and board cannot be manufactured without raw materials, i.e. fiber. Consequently, the rising pulp prices can seriously harm the profitability of non-integrated paper and board mills. Production capacity has been expanded very quickly with an annual rate of 15%. Assuming that all the announced investments are realized, paper and board capacity can reach nearly 80 million tons per annum at the end of 2008 compared with 30 million tons and 50 million tons in 2000 and 2004, respectively. To obtain production efficiency, new projects and expansions seem to focus on large, modern, high speed equipments. The technology improvement in the industry is obvious, supported by government loans and subsidies and the foreign investment and technology brought into the country. In most of the new and expanded paper mills, the technology is state-of-the-art; for instance many new established machines have a speed of 2,000 meters/minute, which is a world class efficiency rate.

Shangdong, Zhejiang, Guangdong, Jiangsu and Henan are the top five provinces in terms of production volumes. Output from these five provinces accounts for 67% of the production of the country. Except for Henan, the other four provinces are situated along the coastal area. In addition, these provinces attract most foreign investment (including in the PPI sector), because of it has a well developed infrastructure, transparent local government, and a well-trained labor force. The above-mentioned provinces have different raw material bases. For instance, in Shangdong, the main resource is non-wood pulp due the availability of large reserves of straw. In the Zhejiang province, the primary raw material is recycled paper and the major product is board; whereas the Jiangsu province uses great amounts of imported wood pulp and its major output is printing and writing papers.

Even though total production is growing, the total number of firms is decreasing. On the one hand, paper mills with small capacity are closing down due to strict environmental regulation and, on the other, big firms are improving efficiency through economies of scale (major producers often aim at a capacity of up to 1 million tons/year). The three largest domestic firms are JiuLong with a production capacity of 2 million tons

(Guangdong province), ChengMing with 1.44 million tons, and JingLong with 2 million tons (the Jiangsu province). Smaller mills with a capacity of less than 10,000 ton/year are facing pressures of shut down.

Table 1: Top provinces with the highest production volume in 2004 and new announcements in 2001–2006. Sources: Dictionary of China Paper Industry, 2006 sand Paperloop.

Province	Capacity, 2004 in millions of tons	Announcements 2001–2006	Change of capacity 2001–2006 in millions of tons
Shangdong	9.97	79	11.6
Zhejiang	7.48	32	5.7
Guangdong	5.98	55	14.8
Jiangsu	5.06	46	12.3
Henan	4.95	9	1.4
Hebei	3.13	9	0.2
Hunan	1.68	12	3.5
Fujian	1.66	16	1.7
Anhui	1.05	6	1.0
Sichuan	1.03	10	0.8
Guangxi	< 1	14	7.5
Hainan	< 1	5	2.9

If we examine the number of announcements and the total announced capacity for the various regions, we again notice that Shandong, Guangdong, Jiangsu and Zhejiang attract most of the new projects. In addition, Guangxi and Hainan are rapidly increasing their capacity though they have not been important manufacturing regions before. The aggressive increase of the capacity is due to the re-forestration programs near Guangxi and Hainan.

2.2. *Sourcing – A problem to be solved to create value with investments in the Chinese PPI*

Sources of fiber to be used in pulp can be wood, recycled paper, and non-wood fibers. In 2004, the consumption of the above three types of pulp were 22%, 52% and 26% respectively, with a total volume of 44.55 million tons. Moreover, the first two categories can be either imported or domestic, but the last one is mainly domestic. Thus, future increases in production require remarkable pulp volumes, and this demand is so strong that it probably affects the global balance of pulp trade flows. The effect can already be seen from the hike in late 2006 in the increasing prices of pulp. Figure 2 shows the monthly price development for Jan. 2004–Feb. 2007 and the steady rise for both the imported pulp to China. The increasing demand for recycled paper has not had a similar effect on the prices.

Next we briefly elaborate the various fiber sources in the Chinese PPI.

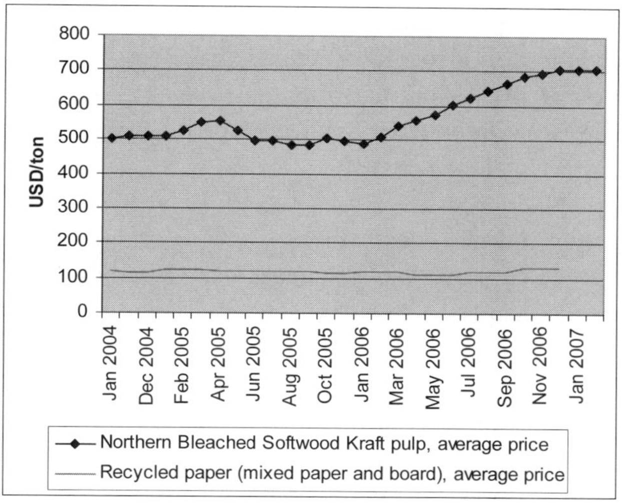

Figure 2. Pulp (NBSK) and recycled paper (mixed paper and board) import prices to China (Source: Paperloop price history database).

Domestic wood pulp

Domestic forestry can only provide about 5% of the needed wood pulp in China. Due to the high and fluctuating market price of pulp, the Chinese government has a policy to encourage the development of domestic pulp production. The government has sanctioned the forestation/re-forestation of 2.667 million hectares for the pulp and paper industry. This is the equivalent of 32 million m^3 of roundwood, which would give 8 million tons of wood-based pulp. But such projects will need about 10 years. Thus, the Chinese PPI will still rely heavily on imported pulp.

Imported wood pulp and roundwood

In 2004 about 16% of the needed pulp was imported. The pulp is made from roundwood but it is not economically wise to transport roundwood for pulping from a long distance. If the pulp is made in China from imported wood, the wood is probably transported in chips. Russia is the most important source of roundwood and according to the estimates by RISI the total need of roundwood will be 150 million m^3 in 2015. At the moment imports from Russia are 18 million m^3. In 2007, Russia has announced a plan to increase the export tax for roundwood, and this may lead to some changes in the Chinese sourcing policy from Russia. The Chinese might have plans for pulp mill investments or chipping activities in Russia. Indonesia is also an interesting fiber source for China because of its tradition of silviculture with foreign firms and short distance from China. If pulp is imported from South America or Europe, transportation will add about 3–5% to the costs.

Recycled paper and non-wood fibers

Imports of recycled paper have increased drastically (Figure 3). The imported recycled paper comes from USA and Japan. The recycled paper from USA has a high quality compared to that of Japan. Recycled paper is cheap and together with low container transportation rates and low labor intensive sorting costs in China the imports are cost-effective. The import tax and value added tax are again 0% and 17% respectively. Firms are required to have import permits to obtain return paper from aboard.

Non-wood pulp is produced domestically and its share is diminishing. Raw materials for non-wood pulp include bamboo, reed, dregs of cane and straw. Currently, high quality fiber bamboo contributes less than 5% of the total pulp consumption. The problem of reed and straw is their low fiber quality. Using non-wood pulp is also problematic due to its pollution. In the waste water treatment of wood mills, black liquor can be extracted to the extent of up to 99%, whereas non-wood mills seldom exceed 90%, due to the high silica content and high viscosity of the raw materials (Ren, 1998).

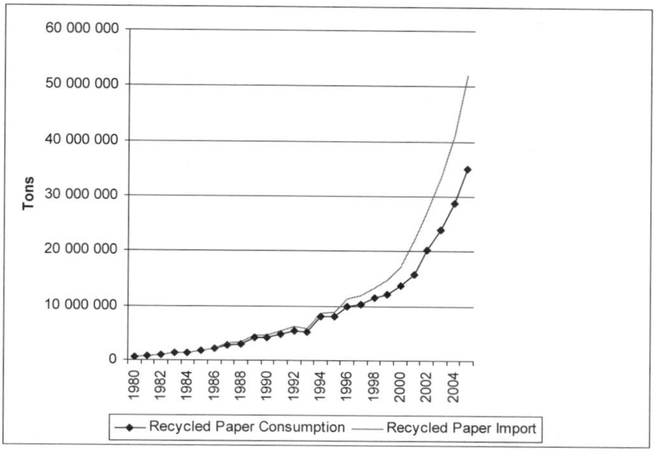

Figure 3: Consumption of recycled paper and recycled paper import, tons (Source: Paperloop).

The large new mills operating with modern technology use wood pulp. In China, almost all chemical additives in the pulp and paper process can be produced domestically. The quality of these chemicals is not very consistent, and thus foreign firms making quality papers use imported substitutes.

3. Value Creation in the Chinese PPI – Some Considerations

3.1. *Technology imports helps technological learning*

Foreign and private firms are playing an active role in the paper industry, and many state-owned paper mills are facing difficulties due to the poor production efficiency. Foreign PPI firms invest in China often through joint ventures and direct investment, and there are more than 10 large foreign firms operating this way. The major competitive advantage of foreign firms is their technological capabilities and product quality that can be transferred to local PPI firms.

Besides firms in joint ventures and direct investments, also technology intensive PPI equipment manufacturers have entered the Chinese market. Voith established a manufacturing plant in Shengyang in 1995, and in 1994 and 2002, opened sales offices in Beijing and Shanghai respectively. Metso Paper entered the Chinese market through the Metso-Chinese joint venture, Valmet-Xian, in which Metso has a holding of just under 50%. In 15 years Metso Paper has penetrated the PPI machinery market. Other international PPI technology providers have also launched operations China.

3.1.2. *From import to export: Technological learning in newsprint*

Newsprint illustrates the passage of the Chinese paper industry from import to export. In the late 1990s China was a large importer of newsprint. By early 2000s, low quality domestic production substituted the imported tonnage efficiently (price RMB1000/ton less than for imports). In 2006, large domestic firms made quality newsprint with a price level of RMB800/ton less than that of imported newsprint.

In the international market, the newsprint price has increased from \$465/ton (2002) to \$675/ton (2006). This provides a new market for Chinese newsprint producers. The net exports of newsprint were 19,800 tons in 2005 and 340,000 tons in 2006, and it is estimated to be 500,000 tons in 2007 (Figure 4).

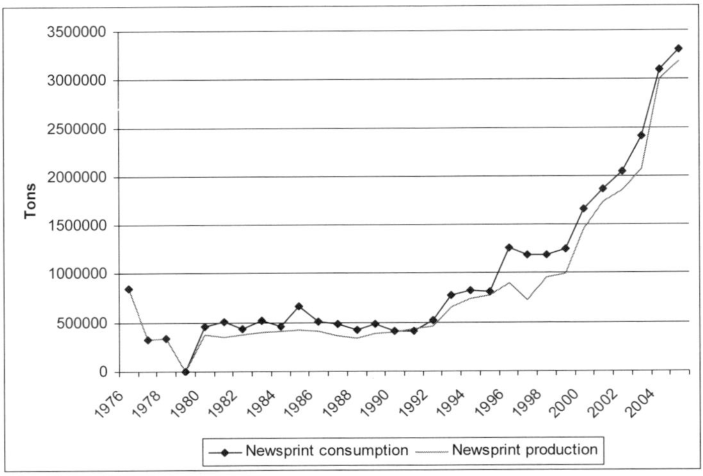

Figure 4: Newsprint consumption and production in China (Source: Paperloop industry statistics).

3.2. *Other factors affecting value creation in the Chinese PPI*

Environment issues and energy

The major challenge facing PPI is that it is an energy-intensive and polluting industry. The energy consumption is about 0.577 ton (standard coal) per ton of paper, and it has been rather constant in the past years. With respect to pollution, water treatment has been improving: by implementing new technology, and by reducing the use of non-wood fiber. Due to the limited water resources of China and stricter environmental regulation, there is a trend of (re)locating pulp/paper mills. In the Jiangsu province for instance, the pulp and paper mills used to be along the Yangtze River; most of the new mills have been established in the costal area of the North Jiangsu province.

Governmental activities

The government policy in the PPI sector is to encourage the usage of wood fiber, paper recycling, and improvements to the technology of non-wood fiber.

In addition, the government is encouraging re-forestation particularly, eucalyptus, poplar and pine. Because ten years are needed to get the planted trees cut, in the near future the Chinese PPI will continue to import wood pulp and recycled paper.

The government is also encouraging centralized production of pulp and a relatively decentralized production of paper and board. This dis-integration of pulp and paper mills will improve the usage of forestry resources. On the other hand, wood chips are considered as one alternative for raw material for pulp.

Tax and tariff

Foreign firms plan their activities at global scene and the PPI firms should remember that China is active in supporting or difficulting selected issues with regulation. In 1997, the anti-dumping tax was added to newsprint and art paper, with South Korea, Canada and USA as targets at rates of 4–71%. Later, an anti-dumping tax was added to almost all grades of imported paper to protect the domestic paper industry. In 2004, the anti-dumping tax was renewed for another 5 years. However during the same period of time, the import tax for pulp and return paper has been zero for about 10 years. Thus many foreign paper producers have started building paper mills in China instead of importing paper to the Chinese market.

Since China's admission in the World Trade Organization, import taxes have consistently been reduced for many products, including PPI products. For certain grades, e.g., newsprint, sliding-scale duties have been used for several years: China imposes high duties on low-price shipments of newsprint and much less on high-price shipments. This tariff protected the Chinese newsprint production at the early stages of the Chinese learning curve in their newsprint story. In 2007, the tariff on paper and paper board has been reduced to 5–7.5%, depending on the grade. It should also be noted that only recently (2007), China has removed/reduced the tax subsidy for re-export products, including PPI products.

4. Discussion and Conclusions

In this paper, we have provided an overview of the Chinese pulp and paper industry along its value chain. Details of the PPI market, production, sourcing, government policy and related issues have been presented as determinants of value creation potential in this industry. We believe that due to the vast growth in consumption and production capacity, the Chinese PPI can be characterized by increasing demand for wood pulp. We believe that controlling the Valuable Rare Inimitable Non-substitutable (VRIN) resources related to the orchestration of the pulp value net gives the best value creation potential. This can also have an effect on the global market price levels of roundwood, wood chips and wood pulp. Also, recycled fiber can complement virgin wood based fibers especially in the coated paper and board grades and in the newsprint and products whose strength properties are modest.

Through this investigation, we have observed that the Chinese have learnt how to be efficient in manufacturing; high volume efficient domestic production with high quality is only a few years away. This means that Chinese PPI manufacturers may start to export aggressively certain paper and board grades. From the global PPI firms' perspective, it is imperative to respond to this competition either by owning some VRIN resources and capabilities or by controlling some critical resources, such as wood-based pulp.

To conclude regarding the future of the Chinese PPI, we expect:

- Paper and board consumption will still increase in China. Technically good quality paper/board grades should be the ones with the highest growth potential. This means investments in large modern PPI mills and efficient paper and board manufacturing lines.
- Chinese domestic production capacity increases drastically, and with certain grades capacity has already exceeded consumption. This increases the Chinese exports of these grades.
- Due to the upgraded technology and strict environment regulation, demand for wood pulp is increasing. Since the domestic fiber resources are very limited, the Chinese PPI relies more heavily on international roundwood, wood chip, pulp and the recycled paper market.
- Long-term projects have been launched to increase forestation for the PPI; however, the outcome cannot be seen in the short term. Chinese firms have also indicated an interest in investing in pulp production in Indonesia and Russia.
- Less taxation for international trading in the PPI.

Figure 5 collects the above ideas along the PPI value chain.

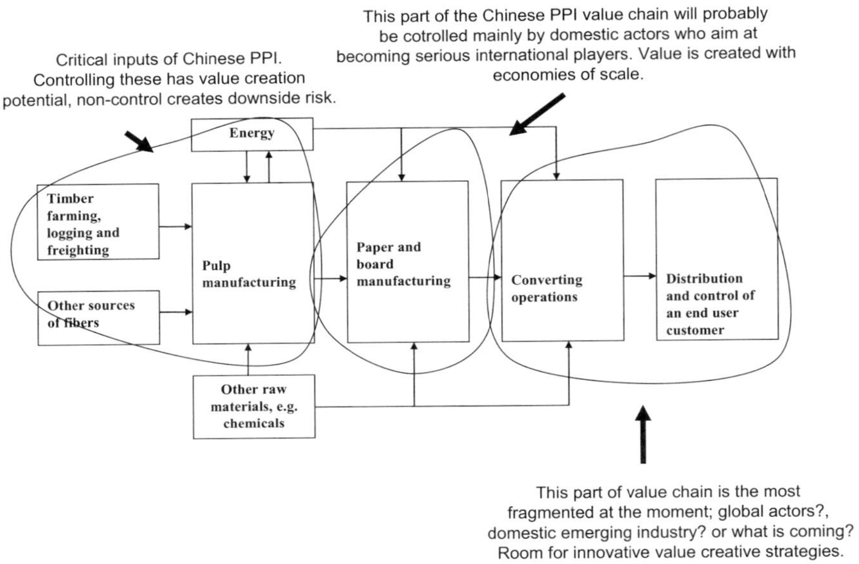

Figure 5. Value creation in various parts of the PPI value chain in China.

References

Autio, E., Dietrichs, E. Führer, K, and, Smith, K. (1997). Innovation Activities in Pulp, Paper and Paper Product in Europe. Studies in Technology, Innovation and Economic Policy (STEP Group). Report 04. ISSN 0804-8185.

Björk, K-M. (2006). Supply Chain Efficiency with Some Forest Industry Improvements. TUCS Dissertations, No. 78, October 2006.

Cenatempo, D. and McNutt, J. (2005). State of the North American Pulp & Paper Industry - An Update & Outlook. Industry competitiveness & the innovation imperative. Presentation at the TAPPI Fall 2005 Technical Conference Session, August 29, 2005. [21st Feb 2006, http://www.cpbis.gatech.edu/news_events/news/040504StateOfTheIndustry_InnovationImperative.pdf]

Christensen, L. R. and Caves, R.E. (1997). Cheap talk and investment rivalry in the pulp and paper industry. *Journal of Industrial Economics*, 45(1), pp. 47-60.

Dictionary of China Paper Industry (2006). China Light Industry Press (in Chinese).

Diesen, M. 1998. Economics of Pulp and Paper Industry. Paper Making Science & Technology. Fapet, Helsinki.

Hurmelinna-Laukkanen, P., Kuittinen, H., Sandström, J. and, Barbosa-Lima-Toivanen, M. (2006). Governing global innovation affecting industry structure–The role of eucalyptus fiber in pulp and paper industry. 13th International Product Development Management Conference (IPDMC), June 11-14, 2006, Milan, Italy

Li, H. Z., Luo, J .F. and McCarthy, P. (2006). Economic transition and demand pattern: Evidence from China's paper and paperboard industry. *China Economic Review*, 17 (3), pp. 321–336.

Liu R.-Q. and Huang X.-J. (2005). Handbook of Chinese Paper, China Light Industry Press (in Chinese),

Long Z. (2005). Newsprint, Chemical Industry Press (in Chinese).

National Bureau of Statistics of China (2006). China Statistical Yearbook, China Statistics Press.

Perks, H. and Jeffery, R. (2006). Global network configuration for innovation: a study of international fibre innovation. *R&D Management*, 36(1), pp. 67–83.

Pesendorf, M. (2003). Horizontal mergers in the paper industry. *RAND Journal of Economics*, 34(3), pp. 495–515.

Ren, X. (1998). Cleaner production in China's pulp and paper industry. *Journal of Cleaner Production*, 6, pp. 349–355

Sande, J. B. (2002). Restructuring and globalization of the forest industry: a review of trends, strategies and theories. Agricultural University of Norway, Ås, Norway.

Siitonen, S. (2003). Impact of Globalisation and Regionalisation Strategies on the Performance of the World's Pulp and Paper Companies. Helsinki School of Economics. Doctoral dissertation. HeSe Print, Helsinki.

Välikauppi S., Kuittinen H. & Puumalainen K. (2006). Global fiber flows in the pulp and paper industry—A gravity model approach. 15th International Conference on Management of Technology (IAMOT), 22-26 May, 2006, Beijing, P. R. China.

Xu Huai, Chao P-H and Chao Z-L (2003). The Development Strategy of Raw Materials in Chinese Paper Industry, research Report by Chinese Economy and Trading Association, China Paper Association, Chinese PPI Research Institute (in Chinese).

Chapter 19

The Internationalization of R&D at Petrobras

Ivete Rodrigues, Eduardo Vasconcellos and Roberto Sbragia

School of Economics, Business Administration and Accounting of the University of São Paulo (FEA/USP)

Av. Prof. Luciano Gualberto, 908 – São Paulo – SP – Brazil – CEP: 05508-900

iveter@fia.com.br,epgdvasc@gmail.com;rsbragia@usp.br

In recent years, several Brazilian firms have conducted their R&D activities globally to be closer to the market and to gain access to know-how. This paper aims at understanding the internationalization process of R&D activities at Petrobras and its technological alliances. The study is qualitative and seeks to throw some light upon the reasons and results of decisions related to the internationalization of R&D. We hope that the findings will be useful to other companies, Brazilian or otherwise, that are undergoing similar R&D internationalization processes.

1. Introduction

Since the 1990s, when the Brazilian economy opened up to foreign investments, Brazilian companies had to become competitive on an international level. These companies discovered that the technological innovation process in this global environment is increasingly complex. New products, processes or services have to meet global requirements as well as the needs of the local consumers in the countries where they do business. In other words, instead of merely adapting their products to new markets, companies need to invent and innovate. To achieve this goal, many companies have been conducting their R&D activities in a global manner to come closer to their markets and gain access to know-how and technologies through partnerships with universities, suppliers, and even competitors.

One of the indicators of the growth of the internationalization of R&D is the increase in the number of cooperative arrangements. Multinational companies seek alliances with other companies and academic institutions, due to a growing awareness that no company is capable, on its own, of generating all the competencies necessary for the development of innovative products and services. In line with this trend, Petrobras has sought to adapt its competitive strategy to the oil industry's new institutional environment, with the objective of ensuring the company's leading position in the Brazilian market and increasing its profitability by diversifying its activities and internationalizing its business.

The current mission of Petrobras is to conduct its business activities safely and profitably, in line with social accountability and environmental responsibility, in the areas related to the oil, gas and energy industries, in national and international markets, by providing products and services that suit the needs of its clients. Thus, the company aims to contribute to the development of Brazil and of the countries where it has business activities. The vision for 2015 presents Petrobras as an integrated energy company, with a strong international presence and leadership in Latin America, as well as a focus on profitability and social and environmental responsibility (Petrobras, 2005). The issue of international expansion strongly underlies the company's corporate strategy. The company has a strong technological background; hence, one of its current strategies focuses on the establishment of international technological alliances. However, no moves have been made yet towards acquisitions or the opening of R&D centers abroad

The objective of this chapter is to understand the internationalization process of R&D activities at Petrobras, under the model proposed by UNCTAD (2005). The paper is based on a case study and it has a qualitative approach. Primary data were gathered by means of interviews with strategic technology management professionals from the R&D Center of Petrobras (CENPES). Secondary data were obtained from previously published academic articles on Petrobras, reports, articles in the press and specialized sites.

2. Conceptual Background

Two main categories of theories seek to explain the internationalization process of companies. Economic models use a macroeconomic analysis to understand companies' investment decisions, while the behaviorism models bring the company to the core of the issue. The School of Uppsala is one of the leaders of the behaviorism school of thought, while the eclectic approach of Dunning (1988) is one of the most significant contributions of the economic school.

The School of Uppsala defines internationalization as a gradual acquisition of knowledge about the external market through a learning process. One of the first concepts defined by this line of thinking is the "psychic distance." According to Johanson and Vahlne (2003), psychic distance is a combination of factors, such as language, education, and cultural differences, legal and business practices, etc., that interfere in the flow of information between markets. Initially, companies tend to do business with markets having a smaller psychic distance, before they enter other markets. Thus, a company increases its commitment to other international markets in a gradual manner, through progressive stages, with a higher allocation of resources for each stage achieved (Alem and Cavalcanti, 2005). For example, manufacturers start by sporadic exports to specific countries. After a certain period of time, they draw up agreements with local sales representatives. The next step is to open up sales subsidiaries and finally, in some cases, to build their own factories in the foreign country. Each new commitment leads to a higher level of learning and skills, which expand the ability to identify new opportunities (Johanson and Vahlne, 2003).

Accordingly to Hofdeste (2003), culture consists of the standards of thoughts, feelings and potential actions that each individual carries as a result of a continuous learning process. Sirmon and Lane (2004) have distinguished between organizational culture and individual culture. The first refers to aspects that the company has absorbed from its native country and is transferring to the host country. The second is more focused on the expatriates and the cultural assets they have absorbed through their experiences. Thus, cultural differences must be considered when organizations want to apply solutions on a world-wide basis, mostly when goods and services need to be adapted to the host country.

The eclectic paradigm of Dunning (1988) points out that companies guide their internationalization process based on three advantages. These advantages are known as OLI, an acronym that stands for Ownership advantages, Locational advantages, and Internalization advantages. The Ownership advantages include those related to tangible and intangible assets — such as brands, technological skills and labor qualifications. The Locational advantages are country-specific — such as natural resources, labor, infrastructure and market size. The Internalization advantages refer to the capacity and the desire of multinational companies to transfer assets across national borders, under their own control.

Both the School of Uppsala and the eclectic paradigm of Dunning approach a company's internationalization from the point of view of the business as a whole. R&D activities have their specific characteristics, and studies that take into account the motivations for their internationalization are still rare. Traditionally, such activities are characterized as having scant mobility, due to the complexity and implied nature of technological knowledge. This makes it difficult and costly to break down R&D activities and/or relocate them. In addition, researchers in general need personal interactions to promote the exchange of information and ideas. The proximity to universities and research centers also generates locational advantages. As a consequence, companies typically base their R&D activities in specific places, generally in the country of origin, where adequate competences are available.

However, recent trends suggest that this scenario is changing, leading to a dispersion of R&D activities. Factors such as competitive pressures, the shorter life span of products and the need to reduce the cost of innovation are pushing companies to reorganize their R&D activities (UNCTAD, 2005). This is reflected by the current trend of investments flowing from developed to developing countries. In addition, companies from developing countries are expanding the horizons of their R&D activities to access foreign markets and centers of excellence. For example, a growing number of American and European companies are outsourcing significant R&D tasks to India, Russia, and China. Texas Instruments has also been granted 225 North American patents through its subsidiary in India (Friedman, 2006). The development of information technologies, associated with the Internet, has contributed favorably to the decentralization of R&D, thus facilitating the coordination among various units. The ERJ 170-190 aircraft family, produced by Embraer, was developed with the collaboration of 400 engineers from 16 companies, located in several countries around the world (Nascimento and Vasconcellos, 2006).

The World Investment Report for 2005 (UNCTAD, 2005) states that there are different ways of internationalizing innovation, as shown in Table 1. In the first category, the national and transnational companies engage in the international trade of a technology, developed in the domestic market. The second category refers to the technical and scientific cooperation among domestic and international companies, public or private institutions, universities and research centers. The third category refers to international innovations generated by transnational companies and involves research activities and the establishment of R&D centers, not only in the domestic market but also in the other countries where the company operates.

3. The Petrobras Case

3.1. *Determinants of the internationalization process at Petrobras*

A prospective study by a British consulting company gave the following conclusions with respect the challenges facing the Oil Industry (UNICAMP, 2006): a) there will be an international race for offshore prospecting in deep waters and for alternative energy sources due to a shortage of oil; b) although offshore oil prospecting in deep waters predominates in Brazil, in international terms, this kind of prospecting is still insignificant, accounting for 3% of the total volume exploited. Deep water prospecting is expected to increase by nearly 300% until 2015 and it will account for approximately 7% of the total volume exploited; c) global investment in deep waters is expected to increase from $14.1 billion US in 2005 to $22.1 billion US in 2010; d) expenses with offshore oil wells will dominate the global oil industry investments; and e) the construction of fixed and floating platforms will be a significant part (32%) of the investment. The installation of pipelines and controls of drain lines will account for 16% of the investment. Offshore drilling equipment will account for 9% of the investment, while the drilling of oil wells on land will account for 2% of the expenses.

Table 1: The taxonomy of the internationalization of innovation.

Category	Players	Forms
International exploration of nationally produced innovations	National and transnational companies and individuals	- Exporting of innovative products - Assignment of licenses and patents - External production of innovative products and services developed internally
International scientific and technological collaborations	- Universities and public research centers - National and transnational companies	- Joint scientific projects - Scientific internships abroad - International flow of students - Joint ventures in specific projects - Production agreements with the exchange of technical information and/or equipment
International generation of innovations	Transnational companies	- Research and Development and other innovative activities both in the domestic market and in countries where the company has business activities - Acquisition of existing R&D centers or the creation of own R&D centers in the countries where the company has business activities

Source: UNCTAD, 2005.

Petrobras is well positioned in the offshore oil exploitation sector as a result of successive investment efforts to develop the necessary technology internally with the help of national and international universities and research centers. Indeed, in 2000 Petrobras launched the

Procap 3000 — *Programa Tecnológico de Águas Profundas* (Deep Water Technology Program) with the objective of discovering oil wells up to three thousand meters deep (Com Ciência, 2005). Presently, Petrobras has a central R&D unit (CENPES) located on the campus of the Federal University of Rio de Janeiro. In order to meet growing R&D demands, especially in view of the internationalization of its business activities, Petrobras is building additional labs, thus, doubling the present size of CENPES (Petrobras, 2005).

Dantas and Bell (2006) identified four phases in the innovation process of Petrobras, as shown in Figure 1. Their study reveals that the central R&D unit of Petrobras coordinated a network of Research Institutes and Universities to complement its capabilities. At the end of these four phases, the company moved from a passive knowledge network to a strategic innovation network. The first phase corresponds to the beginning of the Petrobras offshore operations in 1960 and ends with the discovery of the deepwater oil fields in 1984. The second phase covers the period of the first formal technological program, the Procap. The third phase corresponds to the last years, when Petrobras still held the monopoly over the national oil industry; and the last phase corresponds to the transition period from a monopoly to the opening up of the sector in Brazil.

The distinct technological expertise of Petrobras in the field of oil exploitation and production in deep waters was concurrent with the company's evolution in establishing both national and international cooperation networks. The search for equipment and services adapted to the company's operating needs were the predominant factor in the cooperation network of the first period. The authors define the collaboration network in this phase as a passive network, due the fact that activities were focused on assimilating methods, equipment, services and operational know-how. The main sources of knowledge were the suppliers.

The cooperation networks of the second period, denoted by authors as learning networks, were characterized by the active intention of using the collaborative networks previously established to accumulate scientific and technological knowledge. The suppliers were no longer the only source of knowledge and Petrobras began to include a wider range of players and collaborative arrangements, such as engineering consultants, technical assistance for projects, participation in joint industrial projects, inter-organizational movements of technical personnel and collaborative training programs.

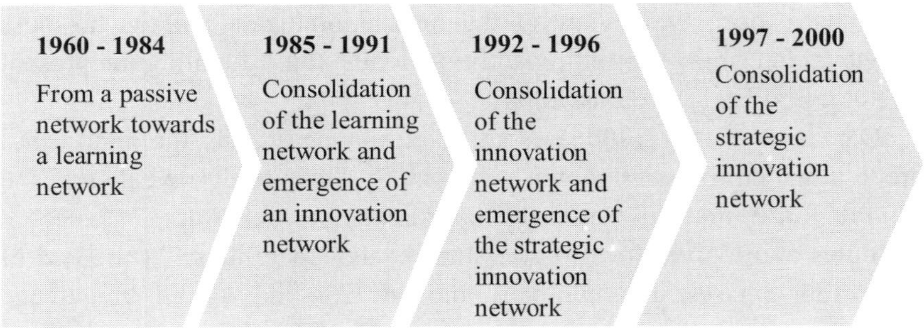

Figure 1: Evolution of cooperation networks at Petrobras (Prepared by the authors, based on Dantas and Bell, 2006).

The third period was characterized by the active intent to use the knowledge networks for innovation-related objectives. The networks are labeled as innovation networks because of the change in the nature of the technology accumulation activities, with the involvement not only of supplier firms, but also of science and technology organizations and other oil companies. The activities lead to the joint development of new technologies such as wet Christmas trees, flowlines and risers, etc. The networks of the last period, named by the authors as strategic innovation networks, reflected the company's awareness of it own knowledge bases that could be attractive to other companies. Simultaneously, there was the recognition that many capabilities and expertise necessary for innovation lie outside its organizational boundaries. Thus, there was an increase in technology transfers with the large, global oil companies. An "open technological innovation system," coordinated by CENPES, was established to include national and international players, such as universities, research institutes, and competitors.

According to Dantas and Bell (2006), the existence of the company's distinctive competency (deepwater exploitation) made it possible for the company to participate in more complex cooperation networks, including those with international competitors. In other words, the R&D investments of the past allowed the company to position itself in a positive way, within the present global scenario. The skills acquired along many years were Petrobras' "ticket" to participate in the latest scenario.

3.2. *The internationalization R&D in Petrobras within the context of strategic innovation networks*

The international business activities of Petrobras have expanded in the last few years. The company's activities in foreign countries already encompass the entire operations chain of the oil and energy industry. The activities range from the exploitation and production of oil and natural gas, refining, gas processing, distribution of by-products, sales, transportation through pipelines to the production of petrochemicals and the generation, distribution and transmission of electric energy. Presently, the international assets, operations and business activities of Petrobras are in 18 countries, on three continents. Organizationally, the International Area comprises six Business Units that act as companies in Argentina, Angola, Bolivia, Colombia, the United States, and Nigeria. In addition, the company does business in twelve other countries: Venezuela, Mexico, Ecuador, Peru, Uruguay, Tanzania, Iran, Libya, Equatorial Guinea, Turkey, China and Paraguay (Petrobras, 2005).

The previously mentioned technological challenges in the oil industry have fueled the constant evolution of the deepwater exploitation and production technology. The collaboration with other organizations, both national and international, has contributed to cost reduction, risk sharing and knowledge development and allowed technological leaps. The company's efforts to master deepwater exploitation and production have generated a growing number of agreements with other major companies in the oil business, as well as with suppliers, universities and research institutes. The development of these alliances has allowed the company to be a leader in a cutting-edge technology, previously available only to multinational companies of developed countries. Table 2 analyzes the Petrobras case, according to the UNCTAD (2005) model for the internationalization of R&D.

Thus, the main vehicle currently adopted by the company for the internationalization of its R&D activities is the establishment of international alliances. Different kinds of agreements have been drawn according to the technical development stages. Today, Petrobras has alliances with Borneman, Westinghouse, Leistritz, Amoco, Chevron, Agip, Conoco, Mobil, Shell, Statoil, among others. The main objective of this kind of partnership is to keep up with the state-of-the-art of the emerging technologies, which are still in the pre-competitive stage, as well as to identify potential partners for future cooperation agreements. Referring to the model of Dantas and Bell (2006), it is possible to infer that Petrobras has adopted a more strategic approach to knowledge, since the purpose of the alliances has changed from simple assimilation to innovation and the accumulation of strategic corporate assets.

In terms of the management of these alliances, the interviews indicated several aspects that can be seen as critical success factors. They are:

Institutional: a) contractual flexibility: the disagreements are solved through discussions and the search for consensus at semester meetings, which are sovereign over the contracts; b) convergence and clarity of the institutional objectives for each of the partners; and c) long-term relationships, which contribute to good relations.

Organizational: a) clear identification of the alliance's interlocutors; and b) defined scope of the project and the responsibility of each party.

Cultural: a) acceptance of cultural differences between the partners; b) the existence of values such as honesty, ethics, and transparency; and c) similar skills and a considerable technological compatibility between the partners.

Operational: a) use of mechanisms such as face-to-face meetings, conference calls, e-mails and sharing tools through the Internet; and b) establishment of an annual interaction calendar.

Table 2: The international activities of Petrobras, according to UNCTAD (2005) classification.

Category	Presence	Activities of Petrobras
International exploration of innovations produced nationally	Yes	The company exports innovative products and provides, externally, innovative services developed internally, at its research center – CENPES
International scientific and technological collaborations	Yes	This has been the preferred way through which Petrobras conducts the internationalization of its R&D activities, either by means of joint scientific projects, scientific internships abroad, and international flow of professionals, joint ventures in specific projects and production agreements, which include the exchange of technical information and/or equipment.
International generation of innovations	No	The company does not develop R&D activities in the countries where it conducts business activities – the R&D needs of the international units are met by CENPES. There are no plans at the moment to acquire or create own R&D centers in the countries where the company conducts its business activities. Nowadays, the company has small decentralized R&D units located in Brazil.

Although the company's international business operations are considerable, it does not intend, at least for the time being, to acquire or implement R&D centers in the countries where it conducts business activities. On the contrary, nowadays, the company invests in the expansion of its R&D center in Brazil (CENPES), which, among other responsibilities, coordinates the cooperative efforts with other institutions. A virtual reality center, including facilities for all the fields related to Biotechnology, will be built at the new unit previously mentioned in this paper. The new Center will be a reference of Petrobras for constructive methods of environmental performance and energy conservation.

During the interviews, the reasons for the company not yet having achieved the third category of Table 2 were also explored. According to the interviewees, international decentralization of R&D, through the establishment of research centers in other countries, depends on a set of factors that relate to the costs and benefits of doing R&D outside the main nucleus. A critical factor to be considered is the extent to which the physical presence elsewhere contributes to the research activities, in terms of cost and adaptation to the regional environment. In terms of costs, Brazilian engineers in the oil sector are very qualified and cost-competitive. As to local adaptation, Petrobras deals with technologies whose development does not demand physical presence at a scale that warrants the creation of other centers. For example, one important constraint for oil production in deep waters is the development of materials resistant to high pressures, beyond 2000 meters in depth. These studies are performed with sophisticated computer simulation technologies, which replace the local, physical presence.

4. Conclusions and Final Remarks

The Petrobras study indicates that the company was able to generate a technological competency that allowed it to react positively to changes in the industry, especially after the state monopoly was lifted. It was also able to insert itself competitively in the international market. Without the technological accomplishments in the field of deep water technology for exploitation and production, the company would certainly not have been able to become a player in this fiercely competitive global game. It is important to state that, at present, the company does not have any plans to physically decentralize its R&D activities. In fact, the company needs to carry out an in-depth study on the advantages and disadvantages of decentralizing its technological activities, since it might eventually lead to a loss of competitiveness in the future.

The absence of R&D centers abroad has led the international technical partnerships to play an important role in the internationalization process of Petrobras. These alliances are becoming increasingly sophisticated in terms of the complexity of the interactions that have been established.

The experience of Petrobras should be considered as a contribution to the complex task of deciding about the level of R&D decentralization worldwide. A challenge for further research is to compare the way Petrobras is conducting the internationalization of its R&D activities with firms doing the same in other contexts (other countries, other industries, etc.), to establish more general conclusions.

References

Alem, A. C. and Cavalcanti, C. E. (2005). O BNDES e o apoio à internacionalização das empresas brasileiras: algumas reflexões. *Revista do BNDES*, 12 (24), pp. 43–76.

COM CIÊNCIA (2006). *Revista Eletrônica de Jornalismo Científico - Número Especial Petróleo*. São Paulo, Unicamp. LABJOR. Available at http://www.comciencia.br/reportagens/petroleo/pet07.shtml. Accessed on 02/Oct/2006.

Dantas, E. and Bell, M. (2006). Latecomer firms and the development of knowledge networks: the case of Petrobras in Brazil. Proceedings of SPRU's 40th Anniversary Conference on "The future of Science, Technology and Innovation Policy", Brighton, Sep. 11–13.

Dunning, J. (1988). The eclectic paradigm of international production: a restatement and some possible extensions. *Journal of International Business Studies*, Spring, pp. 1–31.

Friedman, T. L. (2005). O mundo é plano—Uma breve história do século XXI. Editora Objetiva, Rio de Janeiro.

Hofstede, G. (2003). Culturas e Organizações – Compreender a nossa programação mental. Edições Sílabo, Lisbon.

Johanson, J. and Vahlne, J. E. (2003) Building a model of firm internationalization. In: Learning in the internationalization process of firms (Blomstermo, A. and Sharma, D., eds.), pp. 3-15, Edward Eldar, London.

Nascimento, P. T. S. and Vasconcellos, E. (2006). The Fractal Structure for Integrated Product Development: a new metaphor based on the case of EMBRAER. National Association of Business Administration Graduate Programs Conference, Salvador/Bahia, Brazil, September, pp. 1-14.

PETROBRAS (2005). Annual Report, 2005. Available at www.petrobras.com.br. Accessed on 02/Oct/2006.

Sirmon, D. and Lane, P. (2004). A model of cultural differences and international alliance performance. *Journal of International Business Studies*, 35 (4), pp. 306–319.

UNCTAD—United Nations Conference on Trade and Development. (2005) World Investment Report 2005—Transnational Corporations and the Internationalization of R&D, Geneva.

UNICAMP—Campinas State University (2006). *Boletim Eletrônico da Biblioteca Virtual de Engenharia de Petróleo*, 7(7). Available at http://www.dep.fem.unicamp.br.

Chapter 20

R&D, Entrepreneurship and Innovation in Brazil: Where is the Missing Link?[1]

Paulo A. Zawislak, Cristina Castro-Lucas** and Eda Castro Lucas De Souza***

**School of Management - Federal University of Rio Grande do Sul*
***Graduate Center of Management - University of Brasilia*

The present chapter is devoted to the relationship among R&D expenditures, entrepreneurial cultural traits and the lack of innovative efforts in the Brazilian economic reality. From two different research sources – international data on the impact of R&D on economic development in Brazil and selected countries and an opinion survey of 450 Brazilian businessmen – we will show that the absence of an innovative driver among Brazilian entrepreneurs is a major reason why Brazilian R&D efforts do not lead to significant innovations.

1. Introduction

R&D can be defined as the capability of a company to use, adapt and develop technology to preserve and, whenever it is possible, to enhance its competitive position. Entrepreneurial behavior is the creative search for both new technological and new commercial venture opportunities. Thus, R&D capabilities and entrepreneurship are linked with innovation to form a tripod that is essential for a company's competitiveness.

---

[1] We would like to thank Dr. Hashem Sherif and Dr. Gerhard Jacob for suggestions made in this chapter.

The Brazilian reality, however, indicates that the aforementioned tripod is, to say the least, unbalanced. On the one hand, over the last few decades, the level of R&D expenditures has remained under 1% of GDP. On the other hand, and despite of this low development national trajectory, Brazil, was found to be quite entrepreneurial in a study of 42 countries (Bosma and Harding, 2006).

One possible explanation of this paradox is that most new business ventures in Brazil replicate former success stories. From the hotdog seller to the shopping mall boutique, going through the corner bakery store, most Brazilian entrepreneurs try to build things based on familiar knowledge and techniques, avoiding technological uncertainty and lower R&D expenditures

Recent data collected by the IPEA (*Instituto de Pesquisa Econômica Aplicada* — Applied Economics Research Institute), and summarized in De Negri and Salerno (2005) show that among 70,000 surveyed Brazilian companies, only 1.7% could be considered technologically innovative and having new product development activities. In that same study, the authors attempted to identify what made these few Brazilian companies, world-class in terms of their entrepreneurship.

In these pages, we address the reverse question: Why are there so few R&D- and entrepreneurship-driven companies in Brazil? Our attempt to answer this question is based on the national science and technology statistics over the last decades and on a recent survey of 450 Brazilian businessmen.

The chapter is divided into four sections in addition to the introduction. The next section establishes the overall conceptual basis for the subsequent analysis. Section 3 presents comparative data over thirty years for R&D investments and GDP per capita growth for selected countries including Brazil. The Brazilian entrepreneurial reality will be investigated in Section 4 using the results of the aforementioned survey detailed analysis. Finally, in Section 5, we propose an explanation for the lack of innovation in Brazilian firms.

2. R&D, Entrepreneurship and Innovation

For Schumpeter (1942), the very essence of economic progress is that the new replaces the old. Within this approach, the process of creative destruction promotes innovation. The combination of information, knowledge and creativity will lead to new outcomes (new products, new processes, new techniques, etc.), which if economically successful, will allow extraordinary profits for the innovative company.

According to Nelson and Winter (1982), with specific knowledge and technology, as well as with managerial perception and skills, any firm should be able to establish new technological capabilities to sustain its competitiveness. Filion (1999) shows that the capability to generate and to absorb innovation is, more than ever, the key for transforming any company into an innovative organization. This capability of changing its own internal structures is known as Research and Development (R&D).

From this point of view, the R&D department is the organizational structure responsible for the innovation process. And, like any other investment, the investment in R&D should give returns, not only for the companies themselves, but the whole national development efforts.

The question arises whether these efforts are only a matter of policy, coordination, organization and strategy or if they are the very result of a different kind of thing?

The Schumpeterian view of the entrepreneurship culture and behavior as the agglutination element for innovation should help us to better understand this question. Behind any R&D department, there should be an entrepreneur, "the one that sets new patterns, values and behaviors, just because of his creative attitude" (Cruz, 2005, p. 38), Behind the organizational structure and the various business processes there is a vision fuelled by a risk-taking spirit. In fact, to Carland et al. (1984), what distinguishes entrepreneurial ventures from typical business ventures is precisely innovation. In other words, without entrepreneurship there is no innovation; without entrepreneurial ventures, As Schumpeter (1912) himself points out:

> "The dream and the will of building his own empire... the willing of conquest; the impulse of fighting, of becoming superior, of becoming successful... the joy of creating things or simply using his own energy and ideas... someone that search for hardness, that transforms himself in order to make transformation possible, that enjoys new ventures and for whom the economic return is the very expression of success."

Conversely, the lack of innovation reveals the absence of a real entrepreneurial spirit.

The survival and development of the innovative companies in the face of uncertainties and challenges depend, more on the capabilities of its coordinators-entrepreneurs than on the R&D structure itself. Those capabilities are much more than knowledge and skills; they are creativeness and self-accomplishment, which are the very core of entrepreneurship (Souza, 2005).

To understand the phenomena of entrepreneurship is to understand what motivates the individual itself. McClelland (1972) attributes the entrepreneurial characteristics to four major drivers: self-accomplishment, planning, power and innovation.

"Self-accomplishment" is the driver for taking initiatives, accepting risks, commitment, persistence and the continuous search of new opportunities and efficiency.

"Planning" is related to managerial strategic routines – such as the search for information, the drawing of scenarios and goals, and the systematic appraisal and monitoring –, which are in fact applied by the entrepreneur.

In some sort, "power" is the driver to lead. It denotes how independent, self-confident, persuasive and socially networked is the entrepreneur.

The fourth driver, "innovation", is what will give "life" to some new venture. Creativity, new solutions, new products, new market issues, new technology are the most expected outcomes whenever an entrepreneurial venture is managed. And, in this context, one should expect that the entrepreneur should always aims for those outcomes.

In summary, supplementing R&D capabilities with entrepreneurial behavior is what makes an innovative company. Therefore, by considering the entrepreneurial features of the company and the way it makes its R&D investments gives us a way to evaluate its potential for innovation. At a national level, this will give an indication of the potential for economic development in any country.

The upcoming sections are dedicated to present the results of two different studies, which will help the understanding of the linkages between R&D and entrepreneurship in the Brazilian case. The next section, which is mostly derived from Ruffoni, Zawislak and Lacerda (2004), will contain three sets of data on different R&D investment behavior and GDP per capita outcome for selected countries, with special attention to the particular case of Brazil. Section 4, is based on the extensive 2002 survey by Souza-Depieri (2005) of 450 Brazilian businessmen from different industrial and service sectors on Brazilian entrepreneurial traits.

3. R&D Expenditures in Selected Countries

In the footsteps of Schumpeter (1912) and his "Theory of Economic Development", the so-called neo-Schumpeterians (such as Nelson and Winter, 1982), started to characterize economic development as a process in which revolutionary changes occur in firms' organizations and in technologies.

Archibugi and Coco (2003) maintain that the technological capability has always been a fundamental component of economic growth and the well being of a given society. To reach a certain technological level, specific efforts become necessary and would be in the origin of the GDP differences among countries.

Fagerberg (1988) has presented regression models to demonstrate the positive relation between the level of innovation activities (measured by the percentage of GDP devoted to R&D expenditures) and productivity (for which GDP per capita is a proxy) in different given countries. A more recent work (Fagerberg and Verspagen, 2002) analyzes the existence of a significant positive relation between the level of economic growth and the level of technological development of a country.

The relationship among the aforementioned variables was tested by Fagerberg (1988) for the period ranging from 1973 to 1983. The sample of countries used to verify the relationship between R&D and GDP per capita (GDPpc)[2] was composed of twenty OCDE nations plus India, South Korea, Brazil and Argentina. Figure 1 presents these results.

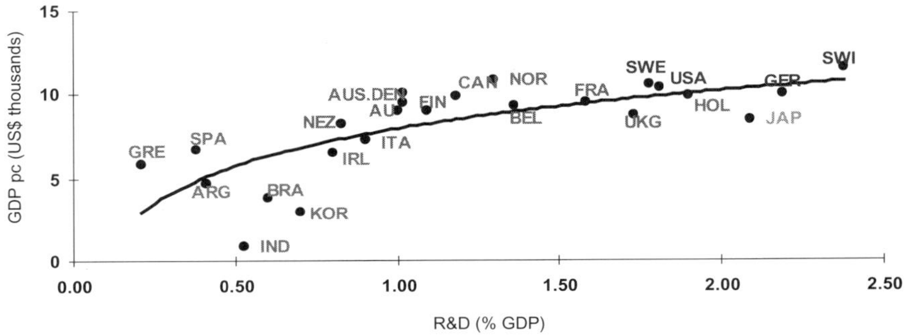

Figure 1: Relationship between the average value of R&D expenditures as a percentage of the GDP and the GDP per capita for selected countries from 1973 to 1983 (Source: Fagerberg, 1998).

Another cluster of countries, such as Belgium, Norway and Canada, comprises those with high economic growth and a lower degree of national technological activity. This can be explained by the peculiar characteristics of these countries, such as their size and the industrial structure.

Recent industrialized countries, such as Brazil and South Korea, form the last group (depicted on the far left side of Figure 1), Their main characteristics are both low levels of R&D expenditures and GDPpc.

[2] Fagerberg (1988) adjusted the GDP *per capita* using the purchasing power parity (PPP) index using the 1980 $US (GDP implicit price index).

Further more, Ruffoni and Zawislak (1998) showed that in the period ranging between 1983 and 1993 the general pattern of a direct relationship between the two variables was kept. The curve estimated for the 1983-1993 period is similar in shape to the one of the previous period. Meanwhile, it is remarkable how South Korea has changed its technological behavior, while Brazil maintained its position, as shown in Figure 2.

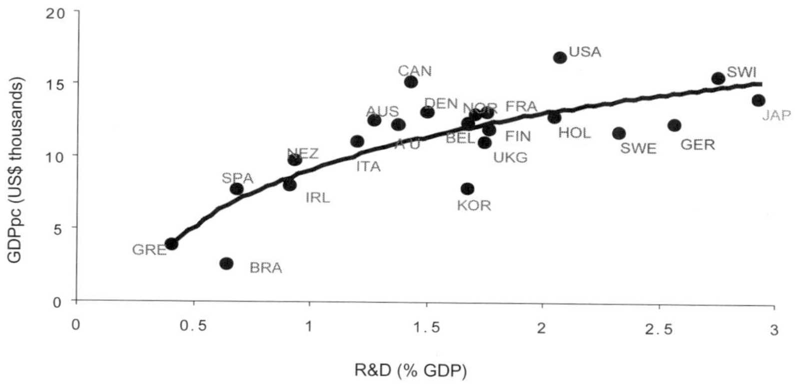

Figure 2: Relationship between the average value of R&D expenditures as a percentage of the GDP and the GDP per capita for selected countries from 1983 to 1993 (Source: Ruffoni and Zawislak, 1998).

Figure 3 show the analysis for the last considered period, from 1991 to 2001. The major difference, when compared to the previous figures, is the upward shift of the whole curve. However, it is possible to identify some sort of development convergence among most of the analyzed countries, even if important technological gaps still persist (from 0.5% to 3.0% of R&D expenditures!). The analytical result shows that, under this period, the statistics provided a falling adjusted correlation coefficient (R2) equal to 0.22.

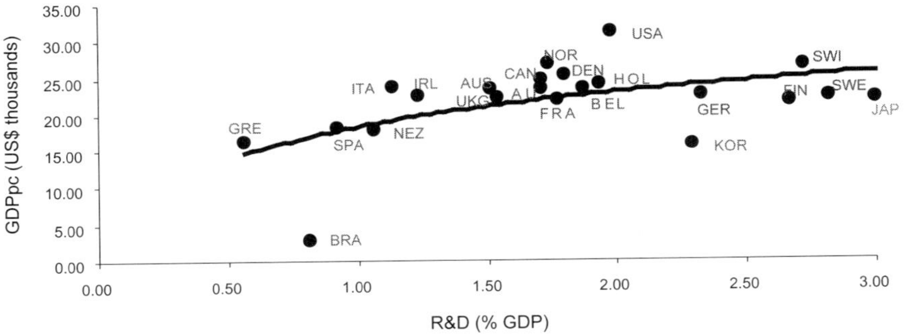

Figure 3: Relationship between R&D and GDPpc from 1991 to 2001 (average values for the decade) (Source: Ruffoni, Zawislak and Lacerda, 2004).

Meanwhile, in terms of technological behavior, different countries have been scrambled, making difficult the identification through the former group division. We could consider a three-fold division. One group, on the right side of the figure, consists of traditional developed countries and some newcomers that are making their début (see, for example, Korea and Finland). The second group is the new pack of middle-field countries. And the third group consists of mostly emerging European economies, such as Ireland, Spain and Greece.

Brazil (BRA) is at an extreme case, since its position remained constant throughout the three consecutive studies: the average R&D investment did not exceed 0.7% of the GDP. Similarly, the GDP per capita remained stalled at below US$ 4,000. Moreover, Figure 3 shows an astonishing draw: Brazilian economy is not only falling behind in terms of innovative efforts vis-à-vis some international competitors, but its development performance is far below the average.

Additional data from the World Bank (Lederman and Saenz, 2003) confirm the pattern. During the same three decades, Brazilian R&D expenditures (as a % of GDP) oscillated between 0.3 (from the 70´s until the end of the 80´s) and 0.85 (in the second half of the 90´s). Moreover, the part of the R&D expenditure that has been financed by the industrial sector has never gone beyond 0.1% of GDP. (i.e., most of the R&D expenditures are from public money).

Clearly, the Brazilian National Innovation System is inefficient in terms of transforming R&D efforts into competitive products. Those inefficiencies have been explained as due to the lack of Science and Technology Policies including university-enterprise collaboration and the dearth of higher educational levels (IBGE, 2005).

Since we have defined the R&D-entrepreneurship-innovation tripod as essential to explain development, could this lack of S&T performance be linked to the existing entrepreneurial profile of Brazilian businessmen?

In the next section, we attempt to shed some light on the Brazilian lack of innovative capabilities.

4. Brazilian Entrepreneurship Profile: Recent Findings

The main characteristic of an *entrepreneur* is that he shall be open to change, to follow contextual signs to guide innovation in a specific social and economic reality. For those countries where revenue is unequally distributed and where policies and politics are unstable with several negative impacts over the economic behavior, to become an *entrepreneur* is much more for self-employment than of a innovative function.

The *Global Entrepreneurship Monitor* highlights a controversial reality. The very problem is that the identified Brazilian entrepreneurs are, as a matter of fact, much more of the so-called "necessity entrepreneurs" than of "opportunity entrepreneurs" (Bosma and Harding, 2006, p 15). The Brazilian entrepreneur is, in reality, in a constant search for mere survival alternatives. But those alternatives are seldom new ones. They seem to be much more of a circumstantial use of opportunities.

This social actor, a kind of "smart guy", will seek social approval rather than economic extraordinary profits (DaMatta, 1997). In this context, it is necessary to understand the entrepreneurial attitude as something that goes far beyond (or falls behind) the realization of economic development. The Brazilian entrepreneurial attitude will point out to collective recognition and self-accomplishment than to value creation.

During 2005, an extensive research[3] has been conducted in order to identify, from the four aforementioned McClelland´s (1972) entrepreneurship dimensions (i.e., self-accomplishment, power, planning and innovation), the profile of the Brazilian entrepreneur.

These four dimensions were divided into 36 indicative questions, exactly as proposed by Lopez Jr. and Souza (2005), in order to verify entrepreneurial attitudes. The questionnaire was directly applied to the respondents and the different data were codified and organized (analyzed) by *Statistical Package for the Social Science* (SPSS 12.0) program. The outcome of the 450 respondents, once object of exploratory analysis, has shown univariate data, which, as Tabachnick and Fidell (2000) suggest, will probably have important impact on the regression solution. By making a factor analysis and by estimating the principal components over the original 36 indicators, the result suggests a new empirical structure with 13 components that can explain 50.66% of the total variation of the data.

Table 1 shows the new factorial structure of the used scale with two components (with their factorial charges) – F1 for a mix of self-accomplishment and power, and F2, for planning —, and the communalities (H2) of the different items, (i.e., the proportion of variance for each variable that can be explained by the new factors). By measuring the internal consistency (Cronbach's Alfa), one can infer from the high values its stability and reliability.

[3] The sample used was composed of 450 businessmen and employees of 33 different small, medium and large service and industrial companies from Brasilia, the Brazilian Federal District. This sample reflects the composition of the Brazilian economic structure.

Table 1: Entrepreneurial Attitude New Components - Source: Souza-Depieri (2005).

ITEMS	F1	F2	H2
05 I define clear and specific long term tasks		.733	.539
08 I am responsible for the on-time conclusion of different tasks	.686		.470
10 I trust on my capability of overcome difficulties	.759		.578
11 I look for new ways of doing things	.632		.451
12 I make clear projections about my business future		.654	.481
13 I get along with my employees to reach the schedule	.697		.503
20 I am constantly renewing my efforts to overcome difficulties	.635		.469
22 I plan my business activities by dividing big tasks into different subtasks		.733	.538
26 I trust on my capability as the very source of business success	.688		.481
27 I use extra efforts to conclude every planned task	.731		.542
28 I use new ideas for problem solving	.683		.487
29 I am constantly defining short term goals		.778	.607
30 I take risks as a way to overpass competition		.661	.439
Eigen Value			
Variance (components)	38.48%	12.18%	
Reliability (Cronbach's Alpha) (components)	0.843	0.763	
Variance (total)	50.66%		
Reliability (Cronbach's Alpha) (total)	0.851		

This factor analysis throws some light on the relationship of three of the entrepreneurial attitude dimensions (i.e., self-accomplishment, power and planning). From this point of view, the analysis can be made in two different levels: the individual (focused on self-accomplishment) and the organizational (linked to different business plan tool kits).

Self-accomplishment is an attitudinal dimension with stronger linkages to the Brazilian cultural traits, since the entrepreneur seems to act as the very responsible for every change, from business routines to the society itself! The group (e.g., employees of a company) will tend to be more committed to its tasks, the more self-accomplished the leader is (in this case, the entrepreneur).

The power dimension of entrepreneurial attitude seems to be quite robust, since the major feature of labor relations in most small and medium Brazilian companies is the hierarchical organization.

The third dimension, planning, explains the human willingness to keep going ahead and make things differently. The general idea is that, to have those things going on inside the company, there is a practical need for business and operational planning.

About the fourth dimension, "innovation", collected data and the factor analysis have shown that is by far the weakest one. By taking in account this lack of innovative ambition, one can better understand the lack of innovative activities (thus, the lack of R&D expenditures and structures).

The detailed interpretation of these considerations suggests that the Brazilian entrepreneur is mostly searching through business ventures, new sources of revenues, financial independence or job alternatives. He is, after all, the necessity entrepreneur. One cannot expect that this entrepreneurial profile will lead to an important increase of R&D expenditures.

5. The Brazilian Missing Link of R&D, Entrepreneurship and Innovation

The low level of both R&D expenditures and innovative outcomes in Brazil seems to be directly related to the absence of an innovative culture and entrepreneurial attitude in what we could call "the Brazilian Businessmen mindset".

The entrepreneur, says Schumpeter (1912), is the economic agent capable of introducing novelty and, thus, to generate national wealth. From a managerial point of view, Schumpeter points out that the role of the entrepreneur should be of introducing new business models, ensuring competitiveness and growth (Souza-Depieri and Souza, 2005). Cultural traits, personal experiences, social, political and economical factors will build an idiosyncratic profile of the national entrepreneurship.

By looking closely to both aforementioned research results, one will find that the Brazilian entrepreneurs seem to be mostly focused on operational management practices as the drivers for running their business. They surely try to align their business strategy with marketing, operations, financial, or human resources strategies. Even more, they work improvising, making gimmicks and preying on a sort of "the Brazilian Way" of doing things (the so-called "*jeitinho*"). What they rarely do is to align business and innovation strategies. What they also rarely do is to transform regular business investments into R&D expenditures.

Whenever they create "new business", they base it on established and successful ventures, where the risks are not high, because they are already known, and the cost of the investment is low. The result is obvious: competition will opt for the low price strategy instead of adding new value through innovation in products or processes. This tends to reduce profitability of the whole sector: Brazilian entrepreneurs work to make "one more business" instead of "one new business".

Without stronger innovative strategic goals, Brazilian companies, lead by the stressed profile of entrepreneurs, seem to operate with very low level of R&D efforts which leads to very low level of development indicators, such as GDP per capita. To sum-up, the Brazilian entrepreneurial experience is less innovative. The missing link seems to be individual cultural traits that put less emphasis on innovation and not in the low level of R&D investments. Perhaps, the State would be able to encourage a change in attitude and supply this missing link. However, we are highlighting psychological traits and personal features that are promoted culturally and that cannot be simply changed by a general policy or enhanced expenditures, at least in the short term. Hopefully, awareness of the problem would encourage Brazilian entrepreneurs to embrace innovation more widely.

References

Archibugi, D. and Coco, A. (2003). A New Indicator of Technological Capabilities for Developed and Developing Countries (ArCo). In: Proc. Conferência Internacional sobre Sistemas de Inovação e Estratégias de Desenvolvimento para o Terceiro Milênio. Globelics. Rio de Janeiro, November.

Bosma , N. and Harding, R. (2006). Global Entrepreneurship Monitor - GEM 2006. Summary Results. London Business School, Babson, London.

Carland J. W., Hoy, F. and Boulton, W. R. (1984). Differentiating entrepreneurs from small business owners: A conceptualization. *Academy of Management Review*, l9(2), pp. 354–359.

Cruz, R. (2005). Valores dos empreendedores e inovatividade em pequenas empresas de base tecnológica. Doctoral dissertation. Escola de Administração, Universidade do Rio Grande do Sul, Porto Alegre.

Damatta, R. (1997). Carnavais, malandros e heróis: para uma sociologia do dilema brasileiro.Rocco, Rio de Janeiro,

De Negri, J. A. and Salerno, M.S. (Eds.) (2005). Inovações, Padrões Tecnológicos e Desempenho das Firmas Industriais Brasileiras, IPEA, Brasília.

Fagerberg, J. and Verspagen, B. (2002). Technology-gaps, innovation-diffusion and transformation: an evolutionary interpretation. *Research Policy*, 31 (8-9), pp. 1291–1304.

Fagerberg, J. (1988) Why growth rates differ. In: Technical Change and Economic Theory (Dosi, G., ed.), Pinter Publishers Limited, London.

Filion, L. J. (1999). Empreendedorismo: empreendedores e proprietários-gerentes de pequenos negócios. *RAUSP - Revista de Administração da Universidade de São Paulo*,34(2), pp. 5–28.

IBGE (2005). Pesquisa Industrial de Inovação Tecnológica – PINTEC 2003. IBGE, Rio de Janeiro.

Lederman, D. and Saenz, L. (2003). Innovation around the World: A Cross-Country Data Base of Innovation Indicators. Escritório do Economista Principal do LCR. World Bank, Washington, D.C.

Lopes Jr., G. and Souza, E. C. L. (2005). Atitude empreendedora em proprietários-gerentes de pequenas empresas. Construção de um instrumento de medida. *REAd.* 11(6). (www.read.adm.ufrgs.br).

Mcclelland, D. C. (1972). A Sociedade Competitiva. Realização e Progresso Social. Expressão e Cultura, Rio de Janeiro.

Nelson, R. R. and Winter, S. G. (1982). An Evolutionary Theory of Economic Change. Belknap, Cambridge, MA.

Ruffoni, J. and Zawislak, P. A. (1998). Knowledge and Economic Development: a comparative study. In: Proc. 7th International Conference on Management of Technology (IAMOT). Orlando: IAMOT.

Ruffoni, J., Zawislak, P. A and Lacerda, J. S. (2004). Uma Análise Comparativa entre Indicadores de Desenvolvimento Tecnológico e de Crescimento Econômico para Grupo de Países In: Proc. XXIII Simpósio de Gestão da Inovação Tecnológica. Curitiba.

Schumpeter, J. A. (1912). The theory of economic development. Harvard University Press, Cambridge, MA.

Schumpeter, J. A. (1942). Capitalismo, socialismo e democracia. Fundo de Cultura, Rio de Janeiro, 1961 (Brazilian translation).

Souza, E. C. L. (2005). Empreendedorismo: da gênese à contemporaneidade. In: Empreendedorismo além do plano de negócio (Souza, E. C. L. and Guimarães, T. A., eds.) , Atlas, Brasília.

Souza-Depieri, C. (2005). Atitude Empreendedora e Cultura: um estudo em organizações brasileiras. Masters Dissertation, Faculdade de Administração, Universidade de Brasília. Brasília.

Souza-Depieri, C. and Souza, E. C. L. (2005). Empreendedorismo e cultura: confusões conceituais. In: Proc. IV EGEPE. Maringá - PR: Universidade Estadual de Maringá.

Tabachnick, B. G. and Fidell, L. S. (2000). Using Multivariate Statistics. Harper-Collins College Publishers, New York.

Section V

Organization Capabilities and Successful Innovation

Chapter 21

Key Elements for Incubating Radical Innovations Successfully

Chintan M. Shah[*], *Marc A. Zegveld*[**], *Leo Roodhart*[***], *and J. Roland Ortt*[**]

[*]*Department of Technology, Policy and Management (TPM), Delft University of Technology (TUD), Jaffalaan 5, 2628 BX Delft, The Netherlands, and Product Development Engineer at Bluewater Energy Services BV*
chintanmshah@gmail.com
[**]*Department of TPM, TUD, the Netherlands*
m.a.zegveld@tudelft.nl, j.r.ortt@tudelft.nl
[***]*Head, Group GameChanger, Shell International, the Netherlands*
leo.roodhard@shell.com

Corporate venturing is a widely accepted mechanism for incubating and developing radical innovations (RI) within large established firms. However, the results remain mixed. We present three elements for an effective corporate venturing (CV) based on a review of the academic literature, a thorough analysis of the successful venture capabilities of Shell, Nokia and IBM, and a study of some venturing initiatives that have been terminated. These three key elements are: carrying out a thorough necessity analysis, achieving clarity of objectives and creating the right ambiance for CV to thrive. We found that the likelihood of a CV initiative to be successful, and to survive, increase if companies are able to apply these three elements. Weakness in, or a lack of, one of these key elements will seriously affect the effectiveness of venturing in the long term.

1. Introduction

Considering the growth potential of radical innovations (RI) (for a definition, please refer to Appendix A), senior leadership at large firms is challenged to tap RI. As a means to incubate and develop RI, establishing a corporate venture unit within large firms is widely propagated by academics and consultants (Block and MacMillan, 1993; Chesbrough, 2003; Leifer et al., 2001; Stringer, 2000). Substantial research has been done on this subject yet several firms have failed in their venturing efforts and booked substantial losses (Burgelman and Valikangas, 2005; Dunn, 1977; Fast, 1978). Many venture units (VU), including those at British Telecom, Lucent, Xerox, Kudu and Vodafone, have closed down; while others are struggling to justify their continued existence. In contrast, venturing at firms like Shell, Nokia and IBM is thriving, and has done well for more than a decade. Since its inception, IBM's Emerging Business Opportunity program has produced 5 multi-billion dollar businesses (including Life Sciences, Pervasive computing and Linux). Over 30% of Shell's Exploratory Research projects have their origin in GameChanger, while Shell Technology Ventures has developed over 20 new start-up companies which bring in strategic returns for Shell. Two of Nokia's four current business groups – 'Multimedia' and 'Enterprise solution' were born from the work done at Nokia's venturing arm – Nokia Venture Organisation.

This gives rise to the following questions: Why do most corporate venturing (CV) initiatives fail, and only some survive? How can a company establish CV that survives successfully? What lessons can executives learn from companies like Shell, Nokia and IBM?

1.1. *Research methodology*

Various researchers have analysed the factors that influence the success and failure of corporate VU(s). On an aggregate level, a few factors emerged that are more influential than others. We clustered them into three key elements – *necessity*, *objectives* and *ambiance*. We used a multiple case research approach (Eisenhardt, 1989; McCutcheon and Meredith, 1993) to validate these factors and refine our understanding. The corporate venturing capabilities (CVC) of Shell, Nokia and IBM were studied and analysed. These three firms were specifically selected because of the success of their CVC as recognised by academics and other professionals. Besides internal venture studies, we took over 20 semi-structured interviews with managers, senior managers and top management at these firms.

1.2. *Corporate venturing capability*

Corporate venturing consists in providing an environment in which ventures can be explored, incubated, turned into projects and provided with the opportunity to demonstrate economic viability (Burgelman, 1985; Leifer et al., 2001; Sykes, 1986). Typically, this is done by creating one or more separate unit(s) within the established firm, separated from the routine structure, processes and evaluation, and semi-autonomous in decision making (Block and MacMillan, 1993). These separate units are called venture units.

The term 'corporate venturing capability' (CVC) is used to highlight the ability of a firm to integrate various isolated pockets of relevant capabilities spread over the organisation, with VU(s) taking a leading role, to develop and commercialise radical innovation projects, and other ventures, on a systematic and continual basis. A firm might have various capabilities within its mainstream operations—for example, expertise in a scientific domain, skills to manage large projects, marketing and distribution network among others. These capabilities can be effectively leveraged to develop new opportunities. A firm may have more than one corporate VU, each contributing in a distinct way. Each VU will have its own objectives and deliverables, yet it forms a part of the overall capability of a firm to manage venture projects effectively.

2. Key Elements to Design a Venturing Capability

Successful venturing capability, diverse objectives and the different venturing instruments deployed by Shell, Nokia and IBM made the case studies of these firms appealing. This section gives a brief overview of the venturing capability at these firms.

2.1. *Venturing capability in practice*

Royal Dutch/ Shell PLC ("Shell") is an integrated oil and gas company. Shell is involved with venturing mainly through two units – GameChanger (GC) and Shell Technology Ventures (STV). GC is a team of managers with seed stage incubation funds to sponsor radical ideas and turn them into a technically working proof-of-concept. STV is a venture capital arm of Shell. STV provides venture capital funds for technologies that are strategic to Shell in the medium and long term. It also serves as a route for commercialising venture that has not found a home within Shell.

Nokia Corporation ("Nokia") is currently the world's largest manufacturer of mobile telephones. Venturing at Nokia was triggered to find new avenues of growth. Several teams are active within Nokia under a broad umbrella of Nokia Venture Organisation (NVO). The overall aim of these venturing teams is to identify and develop new business opportunities that fall outside Nokia's current focus but that fall within the scope of Nokia's strategic agenda. Each team has a unique objective - either of opportunity recognition, incubation or venture capital activity.

IBM is one of the largest IT firm in the world. Venturing at IBM mainly takes place through its Emerging Business Opportunity (EBO) program, while its VC arm —IBM Venture Capital Group (VCG)—plays a unique role. The EBO program is in place to identify, incubate and deploy radical innovations in the form of new growth businesses; the VCG complements EBO management by helping to identify and refine EBO ideas, to identify external partners and to build the ecosystem that is necessary for the success of an EBO.

The basis of the venturing approaches deployed by Shell, Nokia and IBM can be linked to three key elements: necessity analysis, clarity of objectives and right ambiance. In the following section, we elaborate on the meaning of these key elements. In addition, we describe how Shell, Nokia and IBM put these elements in to practice. The firms are put in parallel columns to get an insight into the differences among their approaches. Managers aspiring to create a venturing capability within their firm can learn from the actual practice at these firms.

3. Necessity Analysis

The essence of *necessity analysis* is to evaluate whether a firm needs to develop a distinct venturing capability, whether it is a right time and whether the firm is capable of developing it.

(i) Examining needs. In a small agile firm, for example, where the executives are available for direct communication, establishing a separate VU is often not necessary. In such cases, the project will be supported if the senior executives deem it appropriate. The decisions are made quickly and responsively. This type of environment is not found in most large, established firms (Constantinos and Geroski, 2005). Conversely, a few established firms might not need venturing capability. If a right culture has been established where radical ideas are accepted and are given a chance, for example, in companies like 3M, Virgin, Google and Apple, a separate venturing capability is probably unnecessary. Hence, managers need to decide whether their firm needs a separate VU.

(ii) Analyse if the timing is right. Even if corporate venturing is deemed to be necessary, if the timing is not right then the life of such a capability will be short (Burgelman and Valikangas, 2005). *Timing* implies the presence of a *cash flow* to support RI and the *current growth prospects* of the mainstream business. The most appropriate time to pursue corporate venturing is when mainstream growth is stagnant or negative, and the cash flow is high. For example, considering the status of the automotive industry in 2007, GM is expected to focus on its mainstream business and put its house in order. The time for investing in venturing was perhaps appropriate a few years ago. Toyota, however, has the capability to invest in venturing and it is the right time for the firm to invest in RI.

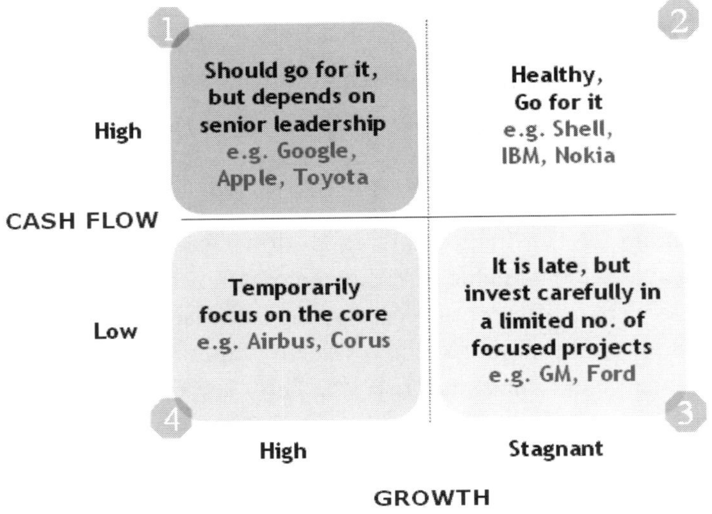

Figure 1: Analysis of timing before establishing a venturing unit.

Figure 1—the cash flow versus growth model—provides a quick overview of whether or not it is a right time for a firm to establish distinct VU. Four scenarios are described with self-explanatory examples.

Table 1: Illustration - 'Necessity analysis' in practice.

Shell	Nokia	IBM
» The closure of fundamental research unit triggered a need for GC program, mainly to foster breakthrough technologies. » GC together with STV targets technologies for new energy in the domains defined by senior management. » Radical ideas for the mainstream targeted at cost cutting, time reduction and to reduce environmental impact are also incubated	» History shows competence of Nokia in making radical changes in its business. To continue its tradition, venturing is considered to be a prime mode for experimenting with new avenues for growth » Venturing is also aimed at providing competitive advantage to the mainstream by fitting breakthrough technologies within the roadmap of business units.	» CEO identified that IBM was consistently missing the emergence of new industries, EBO program is targeted so that IBM does not miss opportunities. » Growth of cash-flow is seen as possible only by creating new high-growth markets

4. Clarity of Objectives

One of the major reasons for the failure or closure of a VU is a lack of clear objectives (Campbell et al., 2003; Chesbrough, 2002; Dunn, 1977). In 10 case studies, Dunn (1977) observed that the venture managers almost unanimously agreed that a lack of clear objectives and key selection criteria forced them to second guess which projects senior management would support. Later, many became discouraged when their projects were summarily rejected by senior management because 'they didn't fit the company strategy'. The actions of the VU(s) can not be focused unless a firm sets out unambiguous *objectives* and *expected deliverables* for the unit(s).

(i) Enterprise first. Objectives are goals that fall within the interest of the company as a single entity, in other words with the 'enterprise first' motive, but that fall beyond the scope of a particular business unit (BU). The 'enterprise first' motive is an essential aspect of the objective. 'Enterprise first' implies seeking what is in the best interest of the firm in all actions internally and while dealing with external parties. In terms of venturing, it means that the venturing and the mainstream units leverage each others capability and that the venturing activities create strategic benefits for the firm.

(ii) The deliverables are tied to the objectives. Deliverables can be financial returns, strategic benefits and/ or developing ventures that fit the roadmap of an existing business unit. For example, if the objective is to incubate radical ideas then the expected deliverables are often aimed at strategic benefits and/or serving as a feeder for R&D or a BU. If the VU is about funding start-ups or generating new businesses, then deliverables are often financial returns and strategic benefits. The deliverables of a VU must be clarified beforehand. Two major pitfalls that led to the closure of Xerox Technology Ventures (XTV) and Lucent's New Venture Group (NVG) were broad objectives and a lack of 'enterprise first' motive (Chesbrough, 2003; Heskett and Kanter, 2000; Hunt and Lerner, 1998).

Table 2: Illustration—'Objectives' for venturing in practice.

Shell	Nokia	IBM
» Overall objective of venturing is to expand the technological and business horizon of current BUs. » Objective of GC is to incubate radical ideas. Any technology that is beyond the scope of a BU (i.e. operating unit), but it is in the interest of Shell to have the technology, GC sponsors it. » STV supports this objective through its VC capability.	» NVO aims to find new avenue of growth through venturing. » Technologies/ platforms that do not have a specific owner are incubated at NVO. Once business potential is identified, the venture is transferred to appropriate BU/ Platforms.	» The main aim of the EBO program is to create new emerging businesses that have the potential to generate revenue of €1 billion within 5 years. Each EBO in itself is a new business.

5. Creating Right Ambiance

Artists describe ambiance as the total environment that is created by the various parts of the composition working together. Likewise, the ambiance for successful venturing is created by the five factors – a visionary executive, a good governing mechanism, committed resources, entrepreneurial managers and an organised process of project transfer. Consider, for example, the ambiance of a good restaurant. The lights, decorations, cutlery, food and even the waiters and their service all combine to form a right ambiance. If one of these factors is missing or misplaced then perhaps it is a sign of just another regular place to eat out. Similarly, if one of the five factors of ambiance is weak or missing altogether, then it will severely affect the capacity of the unit to develop RI projects on a systematic basis.

5.1. *A visionary senior executive involvement*

For the success and survival of a venturing capability, it is imperative that a visionary senior executive, chief executive officer (CEO), chief technology office (CTO), a board member or senior vice-president among others, is responsible (Dunn, 1977; Fast, 1979; Sykes and Block, 1989). This senior executive must not only be able to understand the challenges to be faced with radical innovation projects, but also be committed and genuinely interested in seeing the RI deployed. In the absence of such an individual, when the firm wants to cut costs or if the senior leadership changes, there will be no one to defend the existence of the venturing capability. The presence of such an executive can help when creating a RI culture, acquiring resources, defining objectives, managing interfaces between the mainstream and venture units and managing disappointments when a project, for example, has to be terminated or transferred. In late 2000, Lucent's financial turmoil resulted in a high turnover among its top management. In the absence of a senior executive in charge of New Venture Group, NVG's principals faced a severe challenge in justifying their approach to an entirely new set of senior managers (Chesbrough, 2003).

5.2. *Committed resources*

Commitment of resources (funds, managerial time and organisational home) over a period of time can considerably reduce the unnecessary closure of RI projects or VU(s). 'If a corporation is not willing to commit itself to a five to seven year involvement, then it should not even think of undertaking new ventures' (Roberts, 1980). A cash flow shortage resulted into the closure of British Telecommunications (BT) Brightstar (Batenburg and Julian, 2002; Chesbrough, 2003).

Resource acquisition is one of the major challenges associated with RI projects (Leifer et al., 2000). Convincing business units to fund an RI project at its early stage is often difficult and time consuming. With a modest amount of resources secured, the RI managers can focus their efforts on where they are needed the most. A variety of mechanisms can be deployed to ensure the security of resources.

(i) *Let the venture unit(s) have their own pot of money, which is secured by the centre.*

(ii) *Stimulate collaboration, and acquisition of funds from external sources.* When ensuring fund security, the VU(s) must be stimulated to acquire resources from external agents, e.g. government grants, venture capital firm (VC), corporate VCs, external partnership, and joint ventures. This approach can keep the VU agile and improve its ability to sponsor more projects.

5.3. *Appropriate governing mechanisms*

The foundation of a good governing mechanism is its ability to encourage the VU(s) to achieve their objectives, rather than creating policies and procedures for bureaucratic reasons, e.g. creating various reporting requirements to prepare special analyses for upper management or to take on tasks and attend meetings unrelated to advancing a RI project.

The following mechanisms can be used to give the RI managers enough flexibility and autonomy to achieve the objectives of the VU(s).

(i) *Distinct evaluation criteria for the venture unit(s).*

(ii) *Milestones-based evaluation of RI projects.* RI projects have greater uncertainties and also a larger number of options in terms of directions. New insights often emerge while the project is under development. Identifying milestones over the project's life enables planners to – both, learn from experience about the project's viability and, make adjustments in development plans as necessary. At each milestone, executives must match their assumptions with actual outcomes and determine whether, and how, to proceed to the next milestone (Block and MacMillan, 1985; Sykes and Block, 1989).

(iii) *Quality and not quantity of RI projects.* It is better to have a few high quality projects than to have a high number of low or reasonable quality projects (Dunn, 1977). This issue becomes significantly important as RI projects advance. Development of projects always requires increasing investment of resources, thus competition for scarce resources due to too many projects can lead to failure.

5.4. *Process of project transfer*

RI projects generally process through three phases: incubation, validation and commercialisation, where both mainstream and venture units are involved. The process of project transfer between these phases is often cited as a big challenge. The following approaches can be used to help solve this challenge.

(i) Do not put a wall between the mainstream and the venture unit(s). Maintaining connections with mainstream business units is important, particularly when the motive is to deploy/commercialise the project through a mainstream business unit. Mainstream businesses have valuable knowledge, assets and skills. Keeping too much distance will divorce the RI managers from the most valuable resources.

(ii) Encourage early ownership as far as possible. RI managers must be trained to stimulate early ownership from the business units. Early ownership can be gained by taking into account the opinion of the mainstream managers during the early stages of a venture process and or by involving these managers in 'steering committees' (Burgelman, 1985). GameChanger learned from experience that early involvement of a business unit (BU) without taking money from that BU(s) pocket is very helpful when looking for a home for a project in future.

(iii) Apply open innovation approach (Chesbrough, 2003). All the three case-study firms have embraced the open innovation approach. There is a continuous shift from internally focused to externally embedded idea development. For example, when the Shell GameChanger started, it focused on ideas from internal sources. Idea generation exercises were also limited to internal boundaries. Currently, GC readily accepts ideas from external sources. Continuing its approach, GC has begun to sponsor projects at universities and external enterprises.

5.5. *Importance of human capital*

The important fact that radical innovations are often the result of diligent efforts on behalf of capable and motivated individuals must not be forgotten. It is the people - managers, champions, and team members, who make a radical innovation project work. Firms must place a high emphasis on sourcing and training the right kind of RI managers.

Table 3: Illustration—'Creating right ambiance' in practice.

Shell	Nokia	IBM
a. Senior leadership		
» GC reports to and is evaluated by the VP of the related business unit. Group GC reports to CTO. » STV reports to the board of E&P business.	» VUs report to one of the two heads – NVO Head or VC Head, who in turn reports directly to CSO.	» All the EBO report to and are managed by CSO & CEO. Day to day responsibility for managing EBO belongs to the BU that owns the EBO. » CSO helps to identify and solve the cross-business issues
b. Resources		
» For GC, one quarter of funds come from corporate office, while three quarters come from the BU. » Group GC is totally funded by CTO. » Funds for STV comes from Executive VP of BU.	» Funds for NVO and other VUs come from the senior management. » Funds for the two VC arms partly comes from Nokia and partly from external parties.	» EBO are funded by the benefiting BUs. The BU that has highest amount to gain from an EBO owns it and is responsible for funds. » When such arrangement is not clear, starting funds come from CEO.
c. Governing mechanism		
» Within their limits, GC and STV are autonomous to make decisions. That is, they do not have to ask permission from a BU to fund projects. » Both the units are evaluated differently when compared to a mainstream unit. » Initially GC was open for any sort of ideas including a radical sandwich packing, but now it focuses only on the ideas that fit Shell strategically.	» With a clear objective for each of the VUs, a venture unit is free to make its decision without asking for permissions or looking for funds. » NVO is directly evaluated by the CSO. » A decade ago, when NVO was initiated – it sponsored several projects; but now NVO limits to approximately 10 projects at an early stage and around 10 at the mid-to-late stage	» As the projects are mostly company wide and a large amount of funds, 100's of millions is involved, the decision-making is largely controlled by senior executives. » EBO program started with 7 projects and quickly geared to 18 within a span of 2 years. However, after CEO review the number of EBOs were limited to 7

d. Project transfer process		
» GC and STV works closely with R&D and BU. Both are housed in the same building as the mainstream units. » To encourage early. ownership, GC involves the relevant BU as an advisor at the second stage in the screening process. » GC often collaborates with academia, field experts and competitors to sponsor RI projects.	» The floor used by NVO stands sandwiched between R&D and the head office of BU. This makes communication for NVO managers with that of mainstream easier. » Part of NVO's projects are dedicated to fit roadmap of BU. » NVO often works with academia and start-up firms to develop RI projects.	» BU ownership ensures that a project is directed towards maturity. » Financial mechanisms are put in place to protect an EBO from the vulnerability of expense cuts. » As each EBO is targeted at an emerging industry, collaboration with academia, partners and competitors among others is common.
e. Human capital		
» Senior managers take the charge of managing RI projects. These managers often worked for the mainstream for several years and hence, have a strong network. » Their credibility stimulate team members to work on RI projects. » These managers ensure that team remain motivated, guide them and ensure resources.	» More than projects, Nokia places emphasis on the quality of managers. Managers in charge of NVO often have years of experience in the industry. » Team members participating in RI project are ensured of their career path will not be affected by failure of the project. This in turn stimulates employee participation.	» The CSO places special emphasis on creating a good leadership team for each EBO. » Special innovation broker are designated all around IBM to smoothen the connection among employees.

(i) T-shaped managers. In developing RI projects, managers must be skilled to manage the technical, market and financial feasibility of the projects. On the one hand, they have to deal with scientists; while on the other hand, they have to deal with commercial people (Fast, 1979; Sykes, 1986). We term such managers as 't-shaped' – that is, managers skilled at handling both the technical and the commercial challenges.

6. Conclusions

The three key elements, as described in this article, are a result of lessons learned by observing firms - both successful and not so successful - in their venturing endeavour. It is an essential elixir for executives aspiring to develop RI on a systematic and continual basis. If the three elements are executed suitably, a venturing capability has a higher chance of surviving and being successful; weakness in, or a lack of, one of these key elements seriously affects the effectiveness of venturing in the long term. All the three firms–Shell, Nokia and IBM—have internalised these three elements during the initiation of venturing as well as later in the execution period and produced tangible results. A few observations from the case-study firms are worth noticing:

1. All the three firms have more than one venture unit. They contribute to leverage the firm's overall capability by contributing to systematically incubating, developing and/or commercialising ventures that are beyond the scope of the mainstream business units.
2. Each VU brings with it a unique set of competencies which benefits the mainstream operations. For instance, Nokia Venture Organisation serves as a medium of developing new ventures, collaborating with other firms, working with external experts, pursuing joint development projects and partnering with external stakeholders among others.
3. All the three firms emphasise on the importance of experience of RI managers. Most RI managers at Shell, Nokia and IBM have years of experience in the industry and maintain good contacts with the mainstream. RI managers often cooperate with the mainstream managers and understand well the challenges of the mainstream. Senior leadership involvement and the modest amount of committed resources were clearly visible at these firms.
4. When initiated the VUs started investing in a large number of projects, however they learned from experience, and now focus on only a limited number of quality projects. All the firms were consistent in their efforts. They do not close a VU if it does not produce the desired results in the early years.

5. Each firm essentially practices the elements described in the article; nevertheless they have customised their approach to venturing. A general notion holds that a type of VU, for example a venture capital arm, operates in a similar way at different firms. In contract to this belief, the cases show that the operation of a VU is customised. For instance, the venture capital arm of Shell and Nokia invests in start-ups, while the IBM VCG simply acts as a broker between IBM and the VC community. There also exist a difference in the approach of Shell and Nokia. While Shell invests in start-ups on its own, Nokia partners with other VC firms and pursues joint investments. Arguably, the approach used by each firm's VC arm is aligned to its objectives.

Executives can learn from the approach deployed by these case-study firms; nevertheless a managerial judgement is essential while adopting these elements. BT, Lucent and Xerox could have improved their performance, and perhaps might have survived if they had followed the elements presented in this chapter.

Appendix A

Radical Innovation

Radical innovation, in contrast to incremental innovation, produces something new to the world - a departure from existing technology or methods. While incremental innovations either improve upon something that already exists or reconfigure an existing form or technology to serve some other purpose, radical innovations often displace established technologies and precipitate the decline of companies whose business models are based on them. In many instances, radical innovations create entirely new markets. These markets are initially small, yet have the potential to grow large. The terms breakthrough innovation, discontinuous innovation and disruptive technology are comparable to and are generally a form of radical innovation. From a practical perspective, radical innovations are those with one or more of the following characteristics (Leifer et al., 2000).

(i) Improvements in known performance features of five times or greater. For example, the 'swellable elastomers' technology developed by a Shell GameChanger team dramatically improved the effectiveness of zonal isolation in oil wells, and offers options to replace the traditional use of cement. The microwave oven introduced by Raytheon Company offered a much faster way to heat food products when compared to conventional ovens while maintaining most of the traditional characteristics of oven cooking.

(ii) Offer an entirely new set of performance features. For example, the digital imaging technology used in today's consumer and professional cameras offers a new set of performance features from the chemically coated film technology upon which George Eastman built the Eastman Kodak Corporation over a century ago. The CT scanner introduced by EMI offers 3D medical imaging technology compared to its predecessor's mainly 2D imaging methods. This technology has facilitated huge advances in the field of medicine.

(iii) A 30 percent or greater reduction in costs. For example, Nucor Corporation's steel 'mini-mill' based on a continuous casting process using scrap steel melted in an electric furnace has substantially reduced the cost of steelmaking compared to the traditional process involving huge capital assets and supply chains that stretched back to distant ore mining and coal producing operations. The same can be said for the 'float glass' process introduced by Pilkington Glass which integrated most tasks of glassmaking into a single automated step (Utterback, 1994).

References

van Basten Batenburg, R. and Birkinshaw, J. (2002). BT - Brightstar. London Business School Case.

Block, Z. and MacMillan, I. C. (1993). Corporate Venturing, Creating New Businesses Within the Firm. Harvard Business School Press.

Block, Z. and MacMillan, I. C. (1985). Milestones for successful venture planning. *Harvard Business Review*, Sep–Oct, pp. 184–196.

Burgelman, R. A. (1985). Managing the new venture division: Research findings and implications for strategic management. *Strategic Management Journal*, 6, pp. 39–54.

Burgelman, R. A. and Valikangas, L. (2005). Managing Internal Corporate Venturing Cycles. *MIT Sloan Management Review*, 46, pp. 26–34.

Campbell, A., Birkinshaw, J., Morrison, A. and van Basten Batenburg, R. (2003). The future of corporate venturing. *MIT Sloan Management Review*, 45(1), pp. 30 – 37.

Chesbrough, H. (2002). Making sense of corporate venture capital. *Harvard Business Review*, March, pp. 90–99.

Chesbrough, H. (2003). Open Innovation. Harvard Business School Press.

Constantinos, M. and Geroski, P.A. (2005). Fast Second: How Smart Companies Bypass Radical Innovation to Enter and Dominate New Markets. Jossey-Bass, San Francisco.

Dunn, D. T., Jr. (1977). The rise and fall of ten venture groups. *Business Horizons*, pp. 32–41.

Eisenhardt, K. (1989). Building theory from case study research. *Academy of Management Journal*, 14, pp. 532–550.

Fast, N. D. (1978). The rise and fall of corporate new venture departments. UMI Research Press, Ann Arbor, MI.

Fast, N. D. (1979). The future of industrial new venture departments. *Industrial Marketing Management*, 8, pp. 221–225.

Heskett, M. and Kanter, R. M. (2000). Lucent Technologies New Venture Group. Harvard Business School Case.

Hunt, B. and Lerner, J. (1998). Xerox Technology Ventures, March 1995. Harvard Business School Case.

Leifer, R., McDermott, C. M., O' Connor, G. C., Peters, L. S., Rice, M. P. and Veryzer, R. W. (2000). Radical Innovation: How Mature Companies Can Outsmart Upstarts. Harvard Business School Press.

Leifer, R., O' Connor, G. C. and Rice, M .P. (2001). Implementing radical innovation in mature firms: the role of hubs. *Academy of Management Executive*, 15(3), pp. 102–113.

McCutcheon, D. M. and Meredith, J. R. (1993). Conducting case study research in operations management. *Journal of Operational Management*, 11(3), pp. 239–256.

Roberts, E. B. (1980). New Ventures for Corporate Growth. *Harvard Business Review*, July–August, pp. 134–141.

Stringer, R. (2000). How to manage radical innovation. *California Management Review*, 42(4), pp. 70–88.

Sykes, H. B. and Block, Z. (1989). Corporate venturing obstacles: sources and solutions. *Journal of Business Venturing*, 4(3), pp. 159–167.

Sykes, H. B. (1986). The anatomy of a corporate venturing program: factors influencing success. *Journal of Business Venturing*, 1, pp. 275–293.

Utterback, J. M. (1994). Mastering the Dynamics of Innovation. Harvard Business School Press.

Chapter 22

Rapid Response Capabilities: The Importance of Speed and Flexibility for Successful Innovation

Christoph Grimpe and Wolfgang Sofka***

**Centre for European Economic Research (ZEW), Mannheim, Germany K.U. Leuven, Belgium*

***Centre for European Economic Research (ZEW), Mannheim, Germany; Department for Marketing and Innovation, Institute for Marketing and Media, University of Hamburg, Hamburg, Germany*

Spanish fashion retailer ZARA has been the prototype for a new kind of competitive strategy by beating its rivals through superior flexibility and speed to market new products. We define this organizational ability to learn from the market and respond to it with superior speed as a "rapid response capability". We explore its origins conceptually by drawing arguments from the the capability based view of the firm. Based on a sample of 3,360 German companies our empirical results show that rapid response capabilities are either built around exploiting existing absorptive capacities or exploring options given strong environmental pressures from the technological or demand side, but not a combination of both.

1. Introduction

In the global fight for competitive advantage, many companies, especially in technology-driven industries, seem to have relied heavily on a resource-based strategy (Barney, 1991; Conner, 1991; Peteraf, 1993; Wernerfelt, 1984). This strategy aims at accumulating valuable technological assets combined with an ambitious intellectual property policy. To create a sustained competitive advantage, however, this strategy alone is often not enough. In fact, gaining a head-start over competitors requires timely responsiveness as well as rapid and flexible product innovation. At the same time, competitive pressures from globalization have forced firms to streamline and rationalize their workflow. Many have shifted labor-intensive manufacturing to countries with significantly lower labor costs in order to decrease product prices (Teece *et al.*, 1997). Some however, have created their own approach of coping with this situation. Over the past few years, the Spanish textiles manufacturing company Inditex with its major fashion brand ZARA has successfully turned the predominant industry logic upside down. While traditional fashion companies rely on two collections a year designed and produced in factories all over the world almost nine months before entering stores, ZARA's customers can expect new items every week with an average lead time from design to store delivery of only three weeks. ZARA has been able to transform its dependence on rapidly changing fashion trends and vogues into a competitive advantage and even create own fashion trends. We conceptualize this particular capability as a *rapid response capability* and embed it in the literature on dynamic capabilities (Eisenhardt and Martin, 2000; Hoopes *et al.*, 2003). The development of rapid response capabilities is an important way to overcome competition based on price/cost advantages through speed and flexibility (Berger, 2006). More precisely, we explore its roots and antecedents to discover whether rapid response capabilities qualify as truly dynamic capability.

The goal of this analysis is twofold. On the academic side, we develop a theoretical argument on this particular type of capability and test it empirically. For management practitioners we provide recommendations on how to develop rapid response capabilities. Our study is designed as follows. Section 2 presents our conceptual considerations and the subsequent hypotheses. Section 3 highlights our empirical study to test the latter. The subsequent section 4 provides the results of this quantitative analysis. Based on these results, we discuss these findings on rapid response capabilities in section 5. Section 6 closes with concluding remarks.

2. Theory and Hypotheses

2.1. *Deliberate learning and dynamic capabilities*

Capabilities are organizational processes that bundle strategic resources into unique combinations and constitute superior performance themselves. This follows the basic rationale that competitive advantage does not only arise from the possession of strategic resources but also from the way in which they are used (Penrose, 1959). Several studies argue that capabilities cannot be investigated without considering their relevant context (Atuahene-Gima and Haiyang, 2004; Brush and Artz, 1999): The "when, where and how" resources and capabilities translate into competitive advantage (Priem and Butler, 2001). We follow this stream of literature by discussing the roots of rapid response capabilities and their relevant context.

Management and economics literature has mostly dealt with timing in innovation activities in the context of first mover and follower advantages (see for example Jensen, 2003; Lieberman and Montgomery, 1988; Shankar et al., 1998). This follows a more static perspective: market novelties appear and firms find themselves either on the pioneering or catching-up side. As our concept of rapid response capabilities is dynamic in nature, it combines innovation and imitation. Rapid-response firms like ZARA do not innovate once and reap the benefits from temporary entry barriers for competitors afterwards (first mover). They keep offering new products and services while constantly adjusting to market pressures and opportunities. Their competitive advantage stems neither exclusively from innovation nor imitation but from a combination of both through short feedback and reaction cycles. More precisely, we define rapid response capabilities as organizational routines specifically directed at achieving time compression in a firm's response time to environmental change. It is a unique capability in the sense that its merits originate not primarily from superior performance of individual tasks but instead from sharply reduced response times compared with major competitors. It translates into flexibility, which reduces a firm's exposure to two fundamental risks in innovation: strategic blind spots and technological lock-ins.

The mechanism behind the build-up of rapid response capabilities can be regarded as a continuous and deliberate learning process (Zollo and Winter, 2002). This process describes firms' systematic methods for modifying their operating routines. Such routines constitute stable patterns of organizational behavior and reaction on internal or external stimuli. Routines define predictable as well as interrelated organizational actions e.g. on the order processing for new fashion. However, a second type of routines exists which is typically referred to as search routines (Nelson and Winter, 1982). They deal with changes in the existing set of operating routines.

In a relatively stable environment, operating routines superior to those of competitors can be a source of sustainable competitive advantage. It may even be sufficient to rely on discrete and sporadic changes and improvements in the set of operating routines that may result from a tacit accumulation of experience. However, when the environment turns turbulent and involves rapid changes regarding customer demand, technology or competition, a stable set of routines might no longer be sufficient. Systematic efforts are needed to track the environment and dynamically adjust routines. An accumulation of experience resulting from a repeated execution of routines combined with a trial–and–error proceeding will therefore not be enough for a build-up of rapid response capabilities.

Learning evolves from discursive actions between individuals and groups in the execution of organizational tasks (Levitt and March, 1988; Levinthal and March, 1993). Expressing opinions and individual viewpoints, challenging them and mutually understanding causal linkages — especially in the presence of ambiguities — are a pre-requisite for making implicit or tacit knowledge explicit and hence for enabling collective learning efforts. Knowledge from relevant customers, suppliers, universities etc. has to be made available throughout the company in order to adjust operating routines accordingly and to spread successful action-performance links within the whole organization. Sirmon et al. (2007) have suggested that the effectiveness of this step also depends on environmental munificence, i.e., the degree of availability and accessibility of external resources. The varying munificence of environments might critically affect the potential value of a firm's resources and capabilities. Moreover, munificent environments can support the growth of resources within firms by providing access to complementary, external resources (Baum and Wally, 2003). Those companies that are most efficient in their learning mechanism will reap the benefits in terms of competitive advantage in a given environmental context. In conclusion, a firms' ability to identify promising strategic resources in its environment and integrate them into the existing resource and capability portfolio for superior performance can be considered a capability in itself.

2.2. *Antecedents of rapid response capabilities*

The previous theoretical arguments suggest that rapid response capabilities are truly dynamic in nature. Put simply, they arise from a combination of internal capabilities and the munificence of the environment. We question whether both driving forces of rapid response capabilities necessarily converge. Hence, we develop an evaluation scheme that reflects this aspect by exploring each factor separately as well as their interaction.

2.2.1. *Linking rapid response capabilities with absorptive capacities*

Firms can differentiate themselves through their expertise in synthesizing this information, integrating and combining it with existing knowledge (Henderson and Cockburn, 1994; Kogut and Zander, 1992). An important stream of literature has summarized these capabilities as absorptive capacity (Cohen and Levinthal, 1989, 1990): a firm's ability to identify, assimilate and exploit knowledge from the environment. This differentiation corresponds with the three learning mechanisms in organizational capability development — experience accumulation, knowledge articulation and knowledge codification — but puts a stronger emphasis on exploiting and capitalizing of acquired knowledge. Several studies have linked absorptive capacity to superior firm performance (Landry and Amara, 2002; Love and Roper, 2004; Nadiri, 1993). Absorptive capacities are typically accumulated as a by-product of firms' innovation activities and hence difficult to acquire, imitate or substitute (Amit and Schoemaker, 1993). We extend this view by focusing on the cycle time through all three stages, knowledge identification, assimilation and exploitation, and argue that higher turnover rates can constitute a capability in itself by increasing the efficiency of the whole process, i.e., rapid response capabilities.

Jansen et al. (2005) have recently argued along similar lines by differentiating between potential absorptive capacities (identification, assimilation) and realized absorptive capacity (exploitation). They find that a unique mix of organizational measures is required to balance a broad screening process for valuable ideas with a structured approach towards exploiting them. In conclusion, we derive the following hypothesis:

Hypothesis I: Investments into absorptive capacities enable firms to achieve time compression in their learning engagements and develop rapid response capabilities.

2.2.2. *External pressures and opportunities*

As mentioned before, rapid response capabilities imply change as they inevitably aim at improving operating routines (Collis, 1994; Winter, 2003). This change is necessary to the extent that competitive conditions change. Competitive conditions in turn are largely given by the industry structure with its well known five forces "threat of new entrants", "bargaining power of suppliers", "bargaining power of buyers", "threat of substitute products or services" and, finally, "rivalry among existing competitors" (Porter, 1980). These forces determine the attractiveness of the industry as they exert pressure on the companies but also reveal business opportunities. It is important to note that the industry structure is not completely external to the firm but also a result of a firm's actions and interactions (Porter, 1991). Firms can use their position within an industry to influence the industry structure and take advantage from it.

The more rivals are able to imitate strategies the rougher the climate within an industry gets. Additionally, rivalry is determined by the remaining four forces. When the industry structure is characterized by stable rent appropriation with only minor changes of the competitive environment then the pressure on firms also stays on a rather low level. When the industry structure, however, is continuously altered by the entrance of new competitors, a strong threat by substitute products or a high bargaining power of suppliers and buyers or when existing competitors largely share resources and capabilities then firms typically perceive a high pressure affecting potential value creation (Sirmon et al., 2007). But this pressure also forces firms to learn and develop capabilities to deal with industry turbulence. The higher the rate of change the better capabilities to cope with it have to be developed. Firms unable to do so will ultimately disappear or pull out of the market. Hence, this pressure can also be a learning opportunity to develop rapid response capabilities. This will lead to sustainable competitive advantage to the extent that a firm disposes of rapid response capabilities that cannot be imitated by rivals. Our second hypothesis is thus given as follows:

Hypothesis II: Firms develop rapid response capabilities as they respond to pressure from their competitive environment.

2.2.3. *Conceptualizing rapid response capabilities as dynamic capabilities*

Given the previous discussion, a combination of internal capabilities and external pressures can be envisioned as reinforcing factors for developing rapid response capabilities. In fact, Jansen et al. (2005) find that potential absorptive capacities enhance performance as markets become more dynamic. Rapid response capabilities could therefore be considered as dynamic capabilities (Eisenhardt and Martin, 2000).

However, one may question the truly dynamic nature of the antecedents of rapid response capabilities. Rapid response capabilities may also either evolve based on absorptive capacity (resource driven) or environmental pressure (market driven). We argue that there is a trade-off between them. Capability development is not per se performance enhancing (Sirmon et al., 2007). It is an investment-intensive process with uncertain outcomes (Sapienza et al., 2006). Absorptive capacities are based on experience and hence time. Dynamic shifts in the environment may quickly turn existing competencies obsolete. Hence, a differentiated view is required. Figure 1 summarizes our delineation of rapid response capabilities. Path III follows the dynamic capability logic. However, the environmental context plays a decisive role. Volatile environments increase the likelihood of strategic blind spots or "betting on the wrong horse" as companies invest in specific absorptive capacities. These uncertainties make the cost benefit ratio of such investments less attractive. As a result, rapid response capabilities would be simply a reaction to market pressures (Path II). Then again, stable environments reward investments in absorptive capacities. In this line of reasoning, reducing cycle times for rapid response capabilities stems from efficiency gains based on a reliable and established stock of absorptive capacities (Path I). We propose:

Hypothesis III: There exists an interaction between absorptive capacities and dynamic market environments as the building blocks for rapid response capabilities.

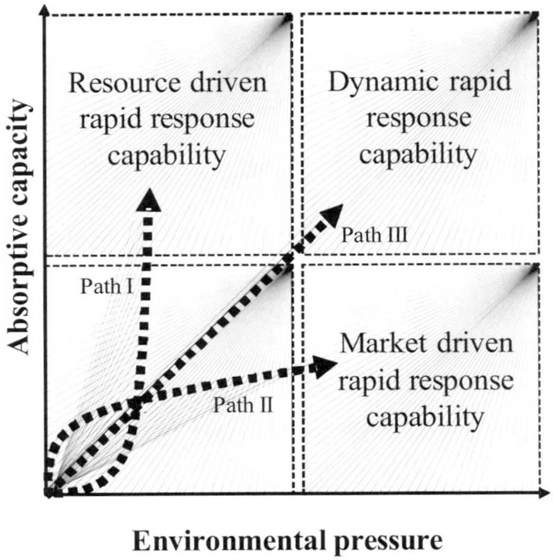

Figure 1: Classification of rapid response capabilities.

3. Methods

3.1. *Data and estimation strategy*

For the empirical part of this analysis we used cross-sectional data from an annual survey on the innovation activities of German enterprises called the "Mannheim Innovation Panel" (MIP) conducted by the Centre for European Economic Research (ZEW). The methodology and questionnaire used by the survey, which is targeted at enterprises with at least five employees, are the same as those used in the European Union's Community Innovation Survey (CIS). For our analysis we used the 2005 survey which covers the three-year period 2002–2004. About 5,200 firms in manufacturing and services responded to the survey and provided information on their innovation activities. The sample was drawn using the stratified random sample technique. A comprehensive non-response analysis showed no systematic distortions between responding and non-responding firms with respect to their innovation activities. For a more detailed description of the dataset and the survey see Rammer et al. (2005). We utilize these data to measure the concepts presented above.

Our dataset without missing values contains data on 3,360 firms located in Germany. Very few companies collect data on the cycle time of their innovation activities. We therefore rely on the self assessment of heads of R&D departments and innovation management on whether they established rapid response capabilities. More precisely, the survey contains the question: 'Did your organizational innovation activities lead predominantly to a reduction in response time to customer or supplier requirements?' From the total sample, 779 firms did and we interpret this approach as the establishment of rapid response capabilities. This indicator is the dependent variable in all subsequent steps of the analysis; the remaining 2,581 serve as the comparison group.

We will subsequently estimate two probit models since our dependent variable is binary in nature (Baum, 2006). This allows us to identify factors which significantly increase a company's probability to pursue rapid response capabilities while controlling for other firm characteristics (e.g. industry effects). We will rely on interaction terms to separate additive effects from interactive ones. Interaction terms follow a straightforward rationale (Aiken and West, 1993). A regression equation of the form Y=b1X+b2Z+b0 allows testing for linear, additive effects of X on Y and Z on Y respectively. An interaction term producing Y=b1X+b2Z+b3XZ+b0 allows for additional insights. Firstly, if b3 is significant then Y depends jointly upon X and Z. Secondly, if b1 and/or b2 are significant there is a separate effect of X on Y (or Z on Y) apart from the mitigating factor XZ.

3.2. *Exogenous Variables*

3.2.1. *Measuring absorptive capacity*

Absorptive capacities are developed through the conduct of R&D activities. We capture their effect in line with the literature (Cohen and Levinthal, 1990; Rothwell and Dodgson, 1991) through variables on the two major inputs for innovation activities: R&D expenditures (as a share of sales) and the expertise of employees (share of employees with college education divided by industry average). Given our analytical framework, we are especially interested in accumulation process of absorptive capacities. We add therefore an additional dummy variable for indicating whether R&D activities are performed on a continuous basis. Hypothesis I would be supported if the coefficients of the absorptive capacity variables are positive and significant.

3.2.2. *Measuring environmental pressure*

Environmental challenges and opportunities have been most prominently elaborated by Porter (1985). We rely on Porter's model with the following principles:

(1) Competitor behavior is difficult to predict.
(2) Threat from market entry of new competitor is high.
(3) Rapid changes in technology occur frequently.
(4) Market life cycles of products and services are short.
(5) Close substituting products exist.

Respondents were asked to rate the prevalence of each of these factors for their business on a four point Likert scale. In the next step, we generated a scale of environmental pressure through principal factor analysis and varimax rotations on these items. We retain one factor with an eigenvalue larger than one. The factor analysis produces a satisfactory Kaiser–Meyer–Olkin measure of sampling adequacy of 0.62.

3.2.3. *Control variables*

We control for several other factors: Regional differences between East and West Germany, company size (number of employees in logs and in squared terms to control for the effect of especially large firms), industry effects (grouped NACE2, see table A1 in the appendix for details) and technological stability (through the share of sales with unchanged products).

Descriptive details of the data are provided in table A3 in the appendix. Rapid responding firms are on average twice as large as the control group and operate more frequently in medium-high tech manufacturing (e.g. automotives) and less frequently in distributive services (e.g. transportation). Interestingly, they are more likely to perform R&D continuously but invest lower shares of their turnover on it. Finally, they are exposed to higher levels of environmental pressures especially from technology changes and product obsolescence.

4. Results

The analysis is split up into two separate models shown in Table 1. While model 1, our baseline case, only estimates the main effects of absorptive capacity and environmental pressure on the development of rapid response capabilities, model 2 includes the interaction term that serves as a basis for describing rapid response capabilities as a dynamic capability.

Generally speaking, our results show a high stability across the different models. Starting with the main effects in model 1 we observe no significant impact of two of the variables that make up absorptive capacity: formal education of employees and R&D intensity. In contrast to that, continuous R&D engagement as third indicator of absorptive capacity is positive and significant. Obviously, there is a strong emphasis on the experience effect with its long-term accumulation of knowledge. This seems to shape absorptive capacities in a way that is relevant for building rapid response capabilities. We can hence confirm our first hypothesis. Regarding the impact of environmental pressure we can observe a positive and significant effect, too. Hypothesis II can therefore be supported as well.

Model 2 includes the effect of the interaction term that is made up of the significant variable continuous R&D engagement as our measure of absorptive capacities and environmental pressure. However, we do not find a significant effect. Evidently, there are additive effects of absorptive capacities and environmental pressure but no interaction of both. This also implies that rapid response capabilities do not necessarily stem from a combination of both which would have qualified them as a truly dynamic capability. Hypothesis III has thus to be rejected.

Furthermore, we included control variables in our analysis. Their effects vary across the four models only to a very limited extent. The results show that particularly large firms measured in terms of the number of employees are more likely to develop rapid response capabilities. An explanation might be that as firms grow larger they have to be more goal-oriented in improving their speed and flexibility while smaller firms are — at least to some degree — flexible anyway. Moreover, there is a negative significant effect of sales of existing products which serves as a measure for technological dynamics. Evidently, the lower this share of sales and consequently the higher the technological dynamics the more rapid response capabilities are propelled which is in line with our previous argumentation. Finally, we included industry effects into the analysis that are hardly significant with the exception of high-tech manufacturing companies that exhibit a negative effect on the build-up of rapid response capabilities. Results are shown in Table A1 of the Appendix.

5. Discussion

Our empirical results do not support the idea that rapid response capabilities are a dynamic capability. Rapid response capabilities are developed through persistent R&D engagements or highly dynamic environmental pressures, but not a combination of both. We discuss each capability separately and return to the reasons behind this branching in the final synthesis.

5.1. *Resource-driven rapid response capability*

The positive effect of absorptive capacity on rapid response capabilities stresses the importance of prolonged R&D commitments. Current investments in R&D projects and personnel have no significant impact. This supports the general accumulation aspect of absorptive capacity (Cohen and Levinthal, 1990). What is more, we find that firms that engage consistently in innovation activities develop routines and capabilities that cannot be readily acquired on factor markets (Amit and Schoemaker, 1993). This supports our view of capability building as a continuous and deliberate learning mechanism. Firms with established competencies and routines find it easier to reduce cycle times for individual innovation projects. This is achieved by streamlining the knowledge accumulation, articulation and codification steps within the learning process. In other words, capabilities need to be "tightened" to ensure their efficiency (Sirmon et al., 2007). Hence, this facet of rapid response capabilities is born out of efficiency gains from experience effects.

5.2. *Market-driven rapid response capability*

With regards to environmental dynamics, we find that they propel the development of rapid response capabilities. Firms deal with these uncertainties in their environment by developing rapid response capabilities that allow flexible solutions and prevent strategic blind spots as well as technological lock-ins. When implemented effectively, this can produce a series of competitive advantages (Sirmon et al., 2007). Hence, rapid response capabilities of this type have primarily a kind of insurance function. While resource-driven strands of rapid response capabilities exploit existing internal assets, market-driven ones have more of an exploratory purpose of external assets (March, 1991), which is necessitated and potentially rewarded by dynamics in the environment depending on its munificence (Baum and Wally, 2003). However, it might also be possible that under extreme environmental uncertainty it might not be enough to rapidly respond but also to direct capabilities at the development of a new technology that might itself create environmental pressure for competitors (Sirmon et al., 2007).

5.3. *Interaction of resource- and market-driven rapid response capabilities*

While both resource- and market-driven rapid response capabilities make intuitively sense the most striking result of our analysis lies in the fact that they do not interact or converge. Then again, equating speed with flexibility may be questionable in the first place.

Helfat and Peteraf (2003) describe the process of capability branching when external factors are sufficiently strong to alter existing development trajectories. We argue that the market-driven type of rapid response capabilities is in effect a branch of capabilities based on efficiency, i.e., long term R&D commitments and the resulting absorptive capacities. Building absorptive capacities requires continued resource commitments. It necessitates significant investments that have to be balanced with expected outcomes (Sapienza et al., 2006). If technological and demand uncertainties are high, lock-ins are dangerous because knowledge stocks may depreciate quickly. The overall cost/benefit ratio turns less favorable. As a result, firms return to an exploratory market-driven strategy that hedges their options until the fog clears. Part of this strategy is staying flexible and keeping investments in specialized absorptive capacities at a minimum level. If the technology and demand landscape becomes more predictable, though, investments in targeted absorptive capacities produce promising competitive assets which can be exploited in the future. Part of this exploitation is clearly superior speed in adapting products and processes (Baum and Wally, 2003). Put simply, overly exploration in stable environments is a waste of resources; exploitation in dynamic environments is risky. Based on this assessment, it is not surprising that we find two ways to rapid response capabilities: Speed in a sense of accelerating the exploitation of existing knowledge stocks (resource-driven) and speed in terms of the flexibility for securing future trajectories (market-driven). A combination of both is difficult to envision.

5.4. *Practical implications*

Environmental dynamics are not a factor under management discretion although firms can chose which market to enter. Hence, management recommendations have to center around the investments into building absorptive capacities. Our results indicate that rapid response capabilities are born out of long term engagements. Once technological routines have been established they can be tightened for more efficient execution. Then again, these investments have to be balanced with technological volatility and demand uncertainty. If the latter are high, lock-ins have to be avoided in favor of rapid response initiatives for flexible exploration. We suggest a generic three step process for dynamic environments. Companies should enter such markets with a focus on flexibility with basic investments in absorptive capacities. As certain products or sub-markets mature decisively, long term commitments are advisable. Turning these engagements into efficient rapid response capabilities is only the final part of this process.

Table 1: Results of the probit models.

Variables	Model1 Coeff.	Model2 Coeff.
Employees with graduate education divided by industry average (ratio)	0.01 (0.02)	0.01 (0.02)
R&D expenditures as a share of sales (%)	0.00 (0.00)	0.00 (0.00)
Continuous R&D engagement (dummy)	0.21*** (0.07)	0.22*** (0.07)
Environmental pressure (scale)	0.17*** (0.03)	0.19*** (0.04)
Interaction term Continuous R&D and Environmental pressure scale		-0.09 (0.07)
Location East Germany (dummy)	-0.05 (0.05)	-0.04 (0.05)
Employees (no. in logs)	0.12** (0.05)	0.12** (0.05)
Employees (no. in logs, squared)	0.00 (0.01)	0.00 (0.01)
Share of sales with existing products (%)	-0.01*** (0.00)	-0.01*** (0.00)
Industry dummies	yes	yes
Constant	-0.63*** (0.17)	-0.64*** (0.17)
Observations	3,360	3,360
R^2	0.09	0.09
Wald chi2(14)	162.61	162.78
P>0	0.00	0.00

* significant at 10%; ** significant at 5%; *** significant at 1%.
Robust standard errors in parentheses. Industry dummy results reported in table A1.

6. Concluding Remarks

The goal of this study was to determine the antecedents of rapid response capabilities, embed them into the literature and test them empirically We acknowledge important limitations in this study, which may offer promising routes for future research projects. First, we can only report empirical results for Germany. Comparisons with other established (e.g.

USA, Japan) as well as emerging economies could provide important additional insights. What is more, we work with a comprehensive dataset which is nevertheless only available as a cross section. As time has shown to be an important factor in this context, longitudinal studies could shed more light on the build-up process of capabilities over time. Regarding our conceptualization it has to be noted that we do not address performance effects of rapid response capabilities. Although we raised the importance of gaining competitive advantage from such capabilities we did not analyze the impact on firm performance. However, while creating value for customers, a firm must also generate profits to be distributed to the owners. Future research should hence focus on the performance effect of rapid response capabilities.

Appendix

A1: Probit results: Industry dummies

Variables	Model I	Model II
	Coef.	Coef.
High tech manufacturing (dummy)	-0.20*	-0.19*
	(0.10)	(0.10)
Medium-high tech manufacturing (dummy)	-0.02	-0.03
	(0.08)	(0.08)
Distributive services (dummy)	-0.10	-0.09
	(0.07)	(0.07)
Knowledge intensive services (dummy)	-0.03	-0.03
	(0.08)	(0.08)
Technological services (dummy)	-0.11	-0.11
	(0.09)	(0.09)

* significant at 10%; ** significant at 5%; *** significant at 1%. Robust standard errors in parentheses. For full regression results see Table 1.

A2: Industry breakdown

Industry	NACE Code	Industry group
Mining and quarrying	10 – 14	Other manufacturing
Food and tobacco	15 – 16	Other manufacturing
Textiles and leather	17 – 19	Other manufacturing
Wood / paper / publishing	20 – 22	Other manufacturing
Chemicals / petroleum	23 – 24	Medium high-tech manufacturing
Plastics / rubber	25	Other manufacturing
Glass / ceramics	26	Other manufacturing
Metal	27 – 28	Other manufacturing
Manufacture of machinery and equipment	29	Medium high-tech manufacturing
Manufacture of electrical equipment and electronics	30 – 32	High-tech manufacturing
Medical, precision and optical instruments	33	High-tech manufacturing
Manufacture of motor vehicles	34 – 35	Medium high-tech manufacturing
Manufacture of furniture, jewellery, sports equipment and toys	36 – 37	Other manufacturing
Electricity, gas and water supply	40 – 41	Other manufacturing
Construction	45	Other manufacturing
Retail and motor trade	50, 52	Distributive services
Wholesale trade	51	Distributive services
Transportation and communication	60 – 63, 64.1	Distributive services
Financial intermediation	65 – 67	Knowledge-intensive services
Real estate activities and renting	70 – 71	Distributive services
ICT services	72, 64.2	Technological services
Technical services	73, 74.2, 74.3	Technological services
Consulting	74.1, 74.4	Knowledge-intensive services
Motion picture/broadcasting	92.1 – 92.2	Knowledge-intensive services
Other business-oriented services	74.5 – 74.8, 90	Distributive services

A3: Descriptive statistics

Variables	Full sample		Rapid responders		Control group	
	Mean	S.D.	Mean	S.D.	Mean	S.D.
Employees (no.)	442.44	5,082.74	725.64	5,328.91	356.96	5,003.98
Share of sales with existing products (%)	86.77	22.71	80.67	25.84	88.61	21.35
Employees with graduate education (%)	19.80	24.29	20.01	23.08	19.74	24.64
Employees with graduate education divided by industry average (ratio)	0.97	1.10	1.01	1.02	0.96	1.12
R&D expenditures as a share of sales (%)	7.49	149.92	5.11	37.71	8.21	169.80
Continuous R&D engagement (dummy)	0.26	0.44	0.37	0.48	0.22	0.42
Environmental pressure (scale)	-0.02	0.81	0.13	0.79	-0.07	0.81
Competitor moves are hardly predictable (dummy)	0.17	0.37	0.17	0.37	0.17	0.37
New competitors threaten market position (dummy)	0.15	0.35	0.16	0.36	0.14	0.35
Product technology changes rapidly (dummy)	0.09	0.29	0.14	0.35	0.07	0.26
Products become rapidly obsolete (dummy)	0.07	0.25	0.10	0.30	0.06	0.23
Easy substitution with competing products (dummy)	0.25	0.43	0.27	0.44	0.24	0.43
Demand forecasting is difficult (dummy)	0.21	0.41	0.23	0.42	0.20	0.40
Location East Germany (dummy)	0.34	0.47	0.31	0.46	0.34	0.48
Medium-high tech manufacturing (dummy)	0.13	0.34	0.17	0.37	0.12	0.32
High tech manufacturing (dummy)	0.07	0.26	0.08	0.27	0.07	0.25
Distributive services (dummy)	0.18	0.39	0.14	0.35	0.19	0.40
Knowledge intensive services (dummy)	0.12	0.32	0.11	0.31	0.12	0.32
Technological services (dummy)	0.13	0.33	0.12	0.33	0.13	0.33
Observations	3,360		779		2,581	

References

Aiken, L.S. and West, S.G. (1993). Multiple regression: Testing and interpreting interactions, Newbury Park.

Amit, R. and Schoemaker, P.J.H. (1993). Strategic assets and organizational rent, *Strategic Management Journal*, 14 (1), pp. 33–46.

Atuahene-Gima, K. and Haiyang, L. (2004). Strategic decision comprehensiveness and new product development outcomes in new technology ventures, *Academy of Management Journal*, 47 (4), pp. 583–597.

Barney, J.B. (1991). Firm resources and sustained competitive advantage, *Journal of Management*, 17 (1), pp. 99–120.

Baum, J.R. and Wally, S. (2003). Strategic decision speed and firm performance, *Strategic Management Journal*, 24, pp. 1107–1129.

Baum, C.F. (2006). An introduction to modern econometrics using stata, College Station (TX).

Berger, S. (2006). How we compete: What companies around the world are doing to make it in today's gobal economy, New York.

Brush, T.H. and Artz, K.W. (1999). Toward a contingent resource-based theory: the impact of information asymmetry on the value of capabilities in veterinary medicine, *Strategic Management Journal*, 20 (3), pp. 223–250.

Cohen, W.M. and Levinthal, D.A. (1989). Innovation and learning: The two faces of R&D, *The Economic Journal*, 99 (397), pp. 569–596.

Cohen, W.M. and Levinthal, D.A. (1990). Absorptive capacity: A new perspective on learning and innovation, *Administrative Science Quarterly*, 35 (1), pp. 128–152.

Collis, D.J. (1994). Research note: How valuable are organizational capabilities? *Strategic Management Journal*, 15 (8), pp. 143–152.

Conner, K.R. (1991). A historical comparison of resource-based theory and five schools of thought within industrial organization economics. Do we have a new theory of the firm? *Journal of Management*, 17 (1), pp. 121–154.

Eisenhardt, K.M. and Martin, J.A. (2000). Dynamic capabilities: What are they? *Strategic Management Journal*, 21 (10/11), pp. 1105–1121.

Helfat, C.E. and Peteraf, M.A. (2003). The dynamic resource-based view: Capabilities life cycles, *Strategic Management Journal*, 24, pp. 997–1010.

Henderson, R. and Cockburn, I. (1994). Measuring competence? Exploring firm effects in pharmaceutical research, *Strategic Management Journal*, 15 (Special Issue Winter), pp. 63–84.

Hoopes, D.G., Madsen, T.L. and Walker, G. (2003). Guest editors' introduction to the special issue: Why is there a resource-based view? Toward a theory of competitive heterogeneity, *Strategic Management Journal*, 24, pp. 889–902.

Jansen, J.J.P., Van den Bosch, F.A.J. and Volberda, H.W. (2005). Managing potential and realized absorptive capacity: How do organizational antecedents matter? *Academy of Management Journal*, 48 (6), pp. 999–1015.

Jensen, R. (2003). Innovative leadership: First-Mover advantages in new product adoption, *Economic Theory*, 21 (1), pp. 97–116.

Kogut, B. and Zander, U. (1992). Knowledge of the firm, combinative capabilities, and the replication of technology, *Organization Science*, 3 (3), pp. 383–397.

Landry, R. and Amara, N. (2002). Effects of sources of information on the novelty of innovation in Canadian manufacturing Firms: Evidence from the 1999 Statistics Canada Innovation Survey.

Leonard-Barton, D. (1992). Core capabilities and core rigidities: A paradox in managing new product development, *Strategic Management Journal*, 13 (5), pp. 111–125.

Levinthal, D.A. and March, J.G. (1993). The myopia of learning, *Strategic Management Journal*, 14, pp. 95–112.

Levitt, B. and March, J.G. (1988). Organizational learning, *Annual Review of Sociology*, 14, pp. 319–340.

Lieberman, M.B. and Montgomery, D.B. (1988). First-Mover advantages, *Strategic Management Journal*, 9, pp. 41–58.

Love, J.H. and Roper, S. (2004). Knowledge sourcing, innovation and performance: A preliminary analysis of Irish innovation panel data, *Aston Business School Working Paper*, Birmingham.

March, J.G. (1991). Exploration and exploitation in organizational learning, *Organization Science*, 2 (1), pp. 71–87.

Nadiri, I.M. (1993). Innovations and technological spillovers, National Bureau of Economic Research, NBER Working Paper, No. 4423, Cambridge, MA.

Nelson, R.R. and Winter, S.G. (1982). An evolutionary theory of economic change, Cambridge, MA.

Penrose, E.T. (1959). The theory of the growth of the firm, New York.

Peteraf, M.A. (1993). The cornerstones of competitive advantage: A resource-based view, *Strategic Management Journal*, 14 (3), pp. 179–191.

Porter, M.E. (1980). Competitive strategy. Techniques for analyzing industries and competitors, New York, London.

Porter, M.E. (1985). Competitive advantage: Creating and sustaining superior performance, New York.

Porter, M.E. (1991). Towards a dynamic theory of strategy, *Strategic Management Journal*, 12 (8), pp. 95–117.

Priem, R.L. and Butler, J.E. (2001). Is the resource-based 'view'a useful perspective for strategic management research? *Academy of Management Review*, 26 (1), pp. 22–40.

Rammer, C., Aschhoff, B., Doherr, T., Peters, B. and Schmidt, T. (2005). Innovationsverhalten der deutschen Wirtschaft - Indikatorenbericht zur Innovationserhebung 2004, ZEW / Infas, Mannheim.

Rothwell, R. and Dodgson, M. (1991). External linkages and innovation in small and medium-sized enterprises, *R&D Management*, 21, pp. 125–137.

Sapienza, H.J., Autio, E., George, G. and Zahra, S.A. (2006). A capabilities perspective on the effects of early internationalization on firm survival and growth, *Academy of Management Review*, 31 (4), pp. 914–933.

Shankar, V., Carpenter, G.S. and Krishnamurthi, L. (1998). Later mover advantage: How innovative late entrants outsell pioneers, *Journal of Marketing Research*, 35, pp. 54–70.

Sirmon, D.G., Hitt, M.A. and Ireland, R.D. (2007). Managing firm resources in dynamic environments to create value: Looking inside the black box, *Academy of Management Review*, 32 (1), pp. 273–292.

Teece, D.J., Pisano, G. and Shuen, A. (1997). Dynamic capabilities and strategic management, *Strategic Management Journal*, 18 (7), pp. 509–533.

Wernerfelt, B. (1984). A resource-based view of the firm, *Strategic Management Journal*, 5 (2), pp. 171–180.

Winter, S.G. (2003). Understanding dynamic capabilities, *Strategic Management Journal*, 24, pp. 991–995.

Zander, U. and Kogut, B. (1995). Knowledge and the speed of the transfer and imitation of organizational capabilities: An empirical test, *Organization Science*, 6 (1), pp. 76–92.

Zollo, M. and Winter, S.G. (2002). Deliberate learning and the evolution of dynamic capabilities, *Organization Science*, 13 (3), pp. 339–351.

Chapter 23

Innovation Process Evaluation: From Self-Assessment to Detailed Technology Audit

Laure Morel and Vincent Boly
ERPI (Equipe de Recherche des processus innovatifs, EA 3767) ENSGSI-INPL
8, rue Bastien Lepage, 54010 Nancy Cedex, France
laure.morel@ensgsi-inpl-nancy.fr
vincent.boly@ensgsi-inpl-nancy.fr

We define two types of technological audit: a self-assessment of the innovation process and an in-depth audit. The self-assessment leads to a basic but pertinent evaluation of a company scores regarding three categories: strategy, piloting, igniting. Through our questionnaire it is possible finally to check which are innovating practices existing in company compared to its innovative system, and the usual actions which they make to support innovation. The in-deep evaluation audit is based on a sophisticated multi-objective system of optimization of the innovative capacity of an industrial system. The evaluation was tested in an SME in the chemical sector.

1. Introduction

A common theme in Industry today is that innovation appears as "the solution" to ensure companies' survival and prosperity in a competitive environment.

But what are we really talking about? Even if we can reach a consensus on one aspect of the innovation concept (it is both a result and a process of evolution), the theme of their evaluations is more complex and fuzzier due to the diversity and nature of the information needed to create metrics (Prager, 2005; Betbeze, 2006). Indeed, innovation is, at microeconomic level, a search to adapt a new activity to maintain sales growth, to ensure company stability and expansion on the market (Dert, 1997; Afuah, 1999; Turriago, 2002). At the macroeconomic level, countries implemented actions in order to activate, to encourage and to develop innovation. This implies to make the necessary adjustments needed to enhance it by implementing specific innovation policies and support programs to activate the whole Nation system. (Amable et al., 1997 ; Levy et Woessner 2006 ; Betbeze, 2006). Consequently, it becomes important to know if the actions carried out to increase the capacity of innovation at the firm level are efficient and effective; it's to say, if subsequently, the resources brought for the deployment of the innovation in a company prove to be justified. As a result, many researches were done in order to develop tools allowing innovation performance evaluation (Canibano et al., 2000; Yam et al., 2004; Corona, 2005; Guan et al., 2006; Walters, 2007).

After a positioning of the innovation concept definitions, we will propose a metric of innovation process. Then, based on a 5 years research work, we will show how, using a systemic modeling process of innovation in three poles: Strategy, Piloting, Igniting (Morel, 1998; Grandhaye et al., 2001; Boly, 2004), a company can have a rapid self assessment tool to evaluate its overall innovation performance. Then, based on the previous results, we can ask the researcher for a multicriteria technological innovation process audit more complete and more targeted: product, process or organization (Colette and Siarry, 2002).

1.1. *Innovation: Concept definition and modeling*

Some authors underline the strategic and economic aspect of innovation (OCDE, 2005). For them, innovation is linked to the successful launching of a new product on its market. Other authors describe innovation as an informational or decisional process, by privileging the individual study of projects (Ayerbe, 2006). These visions of innovation can be linked to schemes in which innovation is a succession of unitary operations for "processing" an idea to transform it into a new product

(Guellec, 2003). On the other hand, some research on cognition tackles innovation by questioning traditional modes of reasoning and the development of new representations of objects. Out of the box thinking and resulting ruptures of paradigms would thus constitute the major elements of innovation (Buckler and Zien, 1996; Booz et al., 2006). One can also cite works considering innovation as the process of adjustment between firms seen as complex adaptive systems and an evolutionary environment (Diedrichs et al., 2006).

Facing this multitude of approaches, our aim is thus to elaborate a representation enabling the description of innovating processes and which take into account the different visions we summarized above. As a result, our fundamental hypothesis will consist in considering innovation as a process (Morel, 1998; Boly, 2004; Aune, 2007). Indeed, innovation can be considered as a non-linear chain linked model (Kline and Rosenberg, 1986). Its outcome is a specific, tangible and describable object. As a result, innovation is a process and so has all the proprieties associated with this concept:

(i) A temporal dimension including boundaries, finality and information flows. In our case, the boundaries relate to the beginning and the end of a given project. Moreover, an innovation process puts into relation several activities that require time and data exchanges. Finally, it induces a change in short-term strategy as well as in long term strategy oriented toward seizing new opportunities (Adams et al., 2006)

(ii) A relational dimension including the organisational routines. An innovation process mainly corresponds to a knowledge creation process. A paradox can be highlighted between optimization (reinforcing technical capacities and so generating routines) and newness (changing the referential at a global level in the firm in order to break with routine and as a result favor creativity) (Albert et al., 2005)

(iii) A productive dimension: that leads to the transformation of resources into products. Innovation is an added value process that consumes material and immaterial resources in order to transform an idea into a new product or service (Tomkovick and Miller, 2000; Millier, 2005)

(iv) A cooperative dimension contains in the sharing of knowledge process. An innovation process necessitates both collective and organizational learning, because innovation is located in each person involved and has to be capitalized and shared for another use (Ozman, 2006; Adams et al., 2006).

In particular, the innovation process may be defined with the following characteristics:

(i) It is an organized process dependant on the surrounding context. In fact, the nature and quality of both the process and its results are highly dependent on the external environment of the company and on the culture of employees (internal environment).
(ii) It is an uncertain process. As a result, a rational approach cannot be used. As technological innovation emerges from *in situ* observations, we propose to choose a constructivist way of managing the innovative process (that is to say, step by step and through an adaptive mode) (Boly et al., 2003).
(iii) It is a complex process where numerous informational and decisional flows are interconnected (Cooper, 2001).
(iv) It is a federative process, i.e., all company departments are concerned. More generally, Innovation involves all relations between the company and its partners.

So the innovation process is fundamentally multi-dimensional and global. As a result, we will propose to use a systemic approach for modeling it. Based on the work of (Le Moigne, 1977), we have proposed a systemic process modeling of innovation (combine the two sentences) (Figure 1).

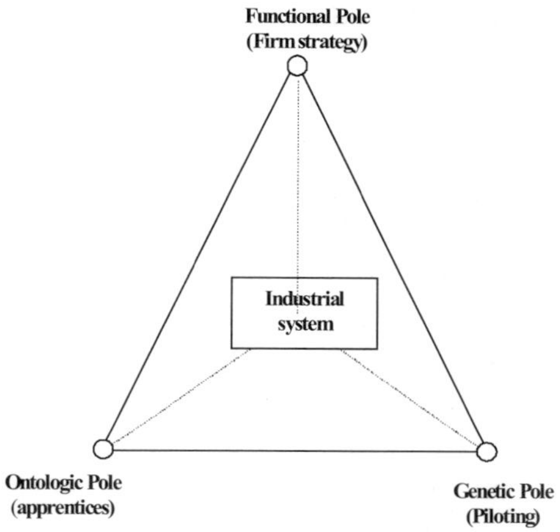

Figure 1: Systemic process modeling of innovation (Morel, 1998).

This leads to a 3-pole analysis of the process of innovation:

(i) A functional pole: this represents the firm strategy, i.e., the technological strategy it wants to develop (organizational, product or process),
(ii) A genetic pole: it is the **piloting** process, i.e., the mechanism of evolution and the associated indicators,
(iii) An ontological pole: it is the human part with the persons and their capacity of reasoning and learning.

Later, (Grandhaye et al., 2001) showed that we can also consider that:

(i) The functional pole is in fact driven by the **strategy** a company wants to follow,
(ii) The ontological pole is the place where the potential of development can be **ignited.**

The combination of the two previous proposals leads to the SPI model: Strategy, Piloting, and Igniting we will use for constructing our audit model.

2. Measuring Innovative Capabilities

Many authors have addressed the challenge of evaluating a firm's technical innovation capability and performance. Most of the scientific literature we have analyzed about innovation metrics (Yam et al., 2004; Prager, 2005; Guan et al., 2006; Betbeze, 2006) measure the level of innovation capabilities of a firm using a set of outputs (number of patents, number of new product introduction, number of new markets...). Our approach is more about measuring the potential capability as an intrinsic property of the whole system to produce those innovation outputs. As a result, during the last years, many efforts was done to better define the role of innovation and to quantify its various forms and impacts. The systemic viewpoint on innovation was made to reconsider implication of technology in economy. To invest in R&D is mostly for the long term rather than the development of new products or short-term innovations. Furthermore, many factors influence the innovating activities at the companies. To better understand how and where innovation occurs, we need to develop new statistics and indicators for the next years. Indeed, the globalization of economic activities requires standardized measurement of innovation (Canibano et al., 2000; Guellec, 2003; Walters, 2007).

While many contributions to a "measurement" of innovation were based principally on economic, financial and managerial viewpoints, our proposal is to evaluate, concretely and objectively, quantitative and qualitative parameters to describe the general situation of an innovating system. We use the Potential Innovation Index (PII), which was developed during a PhD work (Corona, 2005). The PII is based on a set of 13 fundamental practices that Boly has proposed (2004) as necessary to drive and implement innovative projects. Note that each of the thirteen practices is subdivided in fifteen sub-practices on average.

(i) **Practice 1**: Innovation actors work to develop projects and technology evolution with **design tasks.**
(ii) **Practice 2**: A follow-up of each innovative project is essential.

(iii) **Practice 3**: A global supervision of new innovative projects (budget, deadline...) must be led with an **integration of the strategic dimension** dictated by the management team.
(iv) **Practice 4**: Within **the project's portfolio**, the direction ensures **coherent management** between different initiatives.
(v) **Practice 5**: The management team and project managers have to **control** and **receive feedback** on innovation processes in order to develop the practices of the actors.
(vi) **Practice 6**: Suitable context and working conditions have to be created to stimulate innovation.
(vii) **Practice 7**: An optimum allocation of human resources is needed to favor the innovation process.
(viii) **Practice 8**: **Moral support** by the management team and the project managers to the innovation process participants.
(ix) **Practice 9**: **An environment of collective learning** has to exist for the actor, as project progresses.
(x) **Practice 10**: An effort of **capitalisation on the know-how and knowledge** acquired during the former projects must be done, know-how which will be used for forthcoming projects.
(xi) **Practice 11**: **Survey Tasks** (technological, competitive, economic, managerial, intelligence) must be organised in order to open-up the company to the environment.
(xii) **Practice 12**: The management team has to **manage the networks** in which the firm is integrated.
(xiii) **Practice 13**: **New ideas** from research, marketing or those proposed by the employees must be continuously collected using **creativity**, in order that future projects emerge.

Innovative companies develop these thirteen practices, totally or partially, whether formally or informally.

The PII has been computed by using a multi-criteria analysis decision support approach. In particular, it uses a compensatory technique from the Multi-attribute utility theory. It is defined as:

$$PII = \sum_{i=1}^{n} w_i G_i(p_i) \quad \text{and,} \quad \sum_{i=1}^{n} w_i = 1 \tag{1}$$

where:

PII is the score of the Potential Innovation Index of a company
p_i is the development degree of the practice i,
w_i are the importance value (weight) of practice p_i
n is the innovation practice number
i is the number of the practice

In addition, G is the set of utility functions associated to each practice:

$$G_i(p_i) = \frac{p_i}{P_i} \tag{2}$$

An important aspect of the PII is that, not only its absolute or relative score, but also the opportunity to take into accounts different kinds of interactions between innovation practices.

Application of the above formalism provides a classification of a group of enterprises with respect to their innovation capability based on an adaptation of the classification of "attitudes towards the future" due to Michel Godet (1997). This last one classifies companies according to 4 strategic attitudes: Proactive, Preactive, Reactive, and Passive:

(i) **Proactive** companies are the most dynamic and most offensive. They create technological changes and have a long-term vision. As a consequence, they are the ones that control the competitive environment.

(ii) **Preactive** companies do not initiate changes, but can anticipate them by the use of a very active system of technology watch. They are dynamic and offensive companies but their strategic vision is for the medium term.

(iii) **Reactive** are companies that react to the dynamics of their environment. This means that their only drivers for technological change come from concrete demands from the market. Their vision of economic planning is short-term.

(iv) **Passive** are companies with a defensive attitude in front of disturbances of the environment, i.e., that they think only of surviving.

On the basis of the above (i.e., the innovation process modeling in three poles, our innovation measurement system IIP and the Godet classification), we present the framework of our audit model.

2.1.1. *Construction of the audit model*

Before constructing the model, the scope of innovation audit must be defined. As defined by (Cheese et al., 1996), an audit can have two dimensions, a Process audit and a Performance audit. The process audit is centered on the individual processes that are necessary for innovation practices to be developed. The performance audit focuses on the effects of each individual process on the overall technological process and its impact on competitiveness.

Our framework contributes the process audit along two dimensions: a self-assessment method of the innovation capability of a company and, if desired, an in-deep evaluating innovation process audit.

The self-assessment method results from the aggregation of the 13 practices according to the SPI model.

STRATEGY	0%	100%
Practice 3: integration of the strategic dimension		
Practice 11: to open-up the company to the environment		
Practice 12: to manage the networks		
Practice 10: capitalisation of know-how and knowledge acquired		
PILOTING	0%	100%
Practice 1: projects and technology evolution with design tasks		
Practice 2: follow-up of each innovative project		
Practice 4: coherent management between different initiatives		
Practice 5: to develop the practices of the actors		
IGNITING	0%	100%
Practice 6: to stimulate innovation		
Practice 7: competence allocation		
Practice 8: Moral support		
Practice 9: An environment of collective learning		
Practice 13: New ideas emergence		

Figure 2: Innovation self-assessment method.

Then, each of the categories (Strategy, Piloting and Igniting) is evaluated through a questionnaire comprising 30 sets of statements (Boly et al., 2000) (see the Appendix). Each category is evaluated using the answers to 10 answers in average. The mode of answer is based on a colored table (Table 1).

I totally agree	**I agree ....**
X	X
I have a moderate opinion	**I totally disagree ......**
X	X

Table 1: The colored table of the questionnaire.

This leads to a very quick and simple evaluation based on an average count of the number of questions per category. Depending on the result, a company can choose to continue through an in-deep innovation process audit through a questionnaire with 130 questions. Table 2 synthesizes the number of closed questions by practice as well as the relative weight for each practice of the technological innovation given by experts in MOT.

The use of the IIP index leads to an evaluation of the innovative system of a company through the Godet classification previously defined and so defines multicriteria technological scenarios. This point was presented in (Corona et al., 2005; Morel and Camargo, 2006).

Table 2: Questions and weight by practice.

Technological Innovation Practices		Questions	Weight
P1	Conception	20	38
P2	Project Management	11	3
P3	Integrating Strategic	12	20
P4	Project Portfolio Management	8	1
P5	Organization of Innovation Task	7	10
P6	Feedback on Innovation Process	5	2
P7	Allowance of Necessaries Competences	5	1
P8	Incentive to Innovating	6	2
P9	Memorizing Know-how	3	4
P10	Technological Watch	15	2
P11	Networks	18	2
P12	Collective Training	6	5
P13	Capitalization of Ideas and Concepts	14	10
	Σ	130	100

2.2. *Experimentation and discussion*

We tested our proposal with the help of a panel of 20 French SMEs. We will present the results with one SME specialized in manufacturing chemical products.

We have previously shown that (Morel et al., 2007):

(i) A passive company is centered in —terms of know-how and resources allocation— around the practice P1 in order to maintain the current products.

(ii) A reactive company must be mainly centered in term of know-how and associated means around the practices P1, P2, P4, P5, P7, P8. This is in order to induce incremental product innovation and so to favor the optimization of its products portfolio.

(iii) A preactive company must be mainly centered in term of know-how and associated means around the practices P1, P2, P3, P4, P5, P6, P7, P8, P9, P11 for incremental product and process innovations.

(iv) A proactive company is using the thirteen practices to develop incremental innovations as well as radical innovations.

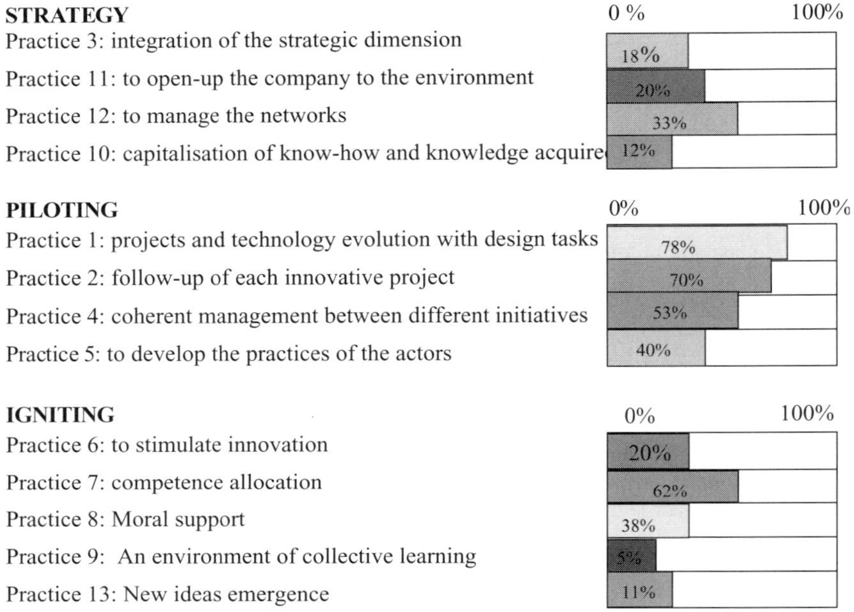

Figure 3: Result of our sample.

As a result, we can conclude that the SME in our experimentation case is of the reactive type. Accordingly, it has to improve first practices 1, 2, 4, 5, 7 and 8 in order to get the better optimization process on its class. The aim is to favor incremental innovations by a better use of the present means. No investments are needed; only a better use of the current resources is possible. We can also suggest to this company to give less importance to P12, which is a practice not important for that kind of firm. It is a waste of resources (both human and financial). Thus, it is possible to re-allocate that kind of investment on practices which are very strategic for the reactive type (P4, for example which is not well developed for the moment).

Then, if the company so desires, we can follow with an in-deep innovation process audit based on a multi-objective system of optimization of the innovative capacity of an industrial system (Morel et al., 2007). This leads to an evaluation at two levels, a global index of the innovative capacity on the basis of the PII and a unitary cost of operation. This allows a decision-maker to have a real benchmark evaluation of his company in its innovation reference class. As a result, he will be able to affect the necessary means (financial or human resources) on the practices to be developed in order to reach the best compromise between product-process innovation and costs.

3. Conclusions

We have defined two technology audits, a self assessment and an in-depth evaluation. The self-assessment is a basic but pertinent evaluation of a company scores on three dimensions: strategy, piloting, igniting. Through our questionnaire, it is possible determine what innovating practices are existing in company compared to its innovative capabilities classification and recommend some typical actions to improve their support for innovation. The in-depth audit is based on a sophisticated multi-objective system of optimization of the innovative capacity of an industrial system. Depending on resources and time a company is able to give to an audit process, we can propose an adapted solution.

Appendix

Note that due to confidential agreements, we will just present one third of our questionnaire.

We have devised a very simple self-assessment test comprising 30 statements to which we simply ask you to agree, agree strongly or disagree, disagree strongly.

Read carefully each of the sets of statements given, then select and mark the one that most closely reflects your own view of your firm's current normal practice. Use the colored table for answering.

1. ***To innovate on a long term is to ensure tasks of technical, organizational conception ...***

totally agree	agree	moderate opinion	totally disagree

2. ***To innovate on a long term is to have a system of project management (delay, costs, project review, planning...)***

totally agree | agree | moderate opinion | totally disagree

3. ***To innovate on a long term is to make evolve or to enrich the conception methods of new products (products, processes or organization)***

totally agree | agree | moderate opinion | totally disagree

4. ***To innovate on a long term is to define the development sense of the firm and to take into account the strategic choices in the decisions made on each project.***

totally agree | agree | moderate opinion | totally disagree

5. ***To innovate on a long term is to manage the coherence within the portfolio of projects (for the firms who lead several projects)***

totally agree | agree | moderate opinion | totally disagree

6. ***To innovate on a long term is to rethink and to readjust regularly the organization of the firm in order to study and then adopt the launched innovation (interservice team ...)***

totally agree | agree | moderate opinion | totally disagree

7. ***To innovate on a long term is to implement actions in order to develop key-capacities of innovation (to be creative, to question routine, to integrate technical, financial, juridical, marketing, data...) by training, recruitment, internal management...***

totally agree | agree | moderate opinion | totally disagree

8. ***To innovate on a long term is to implement a collective training (by analyzing how each action could bring new knowledge).***

totally agree | agree | moderate opinion | totally disagree

9. ***To innovate on a long term is to create or take advantage of crisis situations in order to urge innovation.***

totally agree agree moderate opinion totally disagree

10. ***To innovate on a long term is to capitalize knowledge and know-how of old projects in order to valorize better the investment in study for future projects.***

totally agree agree moderate opinion totally disagree

References

Adams, R., Bessant, J. and Phelps, R. (2006). Innovation management measurement: A review, *International Journal of Management Reviews*, 8(1), pp 21–47.

Afuah, A. (1999). La dinámica de la innovación organizacional. Oxford University Press México, Mexico.

Albert, P., Martin, M. and Tanguy, C. (2005). Capacité d'innovation des entreprises et insertion dans les réseaux : une classification des entreprises agroalimentaires bourguignonnes. Symposium International « Territoire et enjeux du développement régional », Lyon, 9–11 March.

Amable, B., Barré, R. and Boyer, R. (1997). Les systèmes d'innovation à l'ère de la globalisation, Economica, Paris.

Aune, J.-L. (2007). L'innovation, un processus à instaurer, *Industries et Technologies*, N°889.

Ayerbe, C. (2006). Innovations technologiques et organisationnelle au sein de PME innovantes : complémentarité des processus, analyse comparative des mécanismes de diffusion, université de Nice Sofia Antipolis, *Revue Internationale PME*, 19(1), pp.1-28

Betbeze J.-P. and Saint-Etienne C. (2006). Une stratégie PME pour la France. La Documentation Française, Paris.

Boly, V., Morel, L. and Renaud J. (2003). Towards a constructivist approach to technological innovation management: An overview of the phenomena in French SMEs. In: International Handbook on Innovation (Shavinina, L. V., ed.), pp. 790–803, Elsevier, Oxford.

Boly, V. (2004). Ingénierie de l'innovation : organisation et méthodologies des entreprises innovantes. Hermès Science Publications – Lavoisier, Paris.

Booz Allen Hamilton (2006). R&D : l'argent ne fait pas toujours le bonheur !, une étude du cabinet Booz Allen Hamilton auprès des 1000 sociétés ayant dépensé les budgets R&D les plus élevés en 2005.

Buckler, S. and Zien, K. (1996). From Experience, the Spirituality of Innovation : learning from stories. *Journal of Product Innovation Management*, 13(5), pp. 391–405.

Cañibano, L., García-Ayuso, M. and Sánchez, M. P (2000). Shortcoming in the measurement of innovation: Implications for accounting standard setting. *Journal of Management and Governance*, 4, pp. 319–342.

Chiesa, V., Couhhlan, P. and Voss, C .A. (1996). Development of a Technical Innovation Audit. *Journal of Production Innovation Management*, 13, pp. 105–136.

Collette, Y. and Siarry, P. (2002). Optimisation Multiobjectif. Eyrolles, Paris.

Cooper R. G. (2001). Winning at New Products, (3rd Edition). Perseus Books, Cambridge, MA.

Corona, A J. R. (2005). Innovation et métrologie : une approche en terme d'Indice d'Innovation Potentielle. Thèse de Doctorat, Institut Nationale Polytechnique de Lorraine.

Corona, A. J. R, Morel-Guimaraes, L. and Boly, V. (2005). A Methodology to Measure the Innovation Processes Capacity in Enterprises. Proceedings of IAMOT 2005, Vienna, 8p.

Dert, F. (1997). L'art d'innover ou la conquête de l'incertain. Maxima, Paris.

Diedrichs, E., Engel, K. and Wagner K. (2006). European innovation management landscape, assessment of current practices in Innovation Management Consulting Approaches and Self-Assessment Tools in Europe to define the requirements for future "best practices", IMP^{3}rove, European Commission Directorate General Enterprise and Industry, Europe Innova paper n°2, Novembre.

Godet, M. (1997). Manuel de prospective stratégique, Tome 2, L'art et la méthode. Dunod, Paris.

Grandhaye, J.-P., Tani, M. and Guidat, C. (2001). Le management par la valeur et l'impulsion par les Hommes, vers une méthodologie robuste, *Revue Française de Gestion Industrielle*, 20(2), pp. 75–86.

Guan, J. C., Yam, R., Mok, C. K. and Ma, N. (2006). A study of the relationship between competitiveness and technological innovation capability based on DEA models, *European Journal of Operational Research*, 170, pp. 971–986.

Guellec, D. (2003). Mesurer l'innovation, quelques leçons de l'expérience de l'OCDE, *INSEE méthodes*, n°105.

Kline, S. and Rosenberg, N. (1986). An overwiew of innovation. In: *The Positive Sum Strategy* (Landau R. and Rosenberg, N., eds.), National Academy Press, Washington.

Le Moigne, J. L. (1977). La théorie du Système Général—Théorie de la modélisation. P.U.F, Paris.

Levy, R. and Woessner, R. (2006). Le territoire français en tant que Système Régional d'Innovation, document de travail No. 2006-24, Université Louis Pasteur, Bureau d'économie théorique et appliquée.

Millier, P. (2005). Modèle synthétique des conditions de succès d'un projet d'innovation, Cahiers de recherche/Working Papers, The European Institute for Lifelong Learning, http://www.jinnove.com/upload/documentaire/emlyon_succes_innov.pdf.

Morel, L. (1998). Proposition d'une Ingénierie Intégrée de l'Innovation vue comme un processus permanent de création de valeur. Thèse de Doctorat, Institut Nationale Polytechnique de Lorraine.

Morel, L. and Camargo, M. (2006). Comparison of multicriteria analysis techniques to improve the innovation process measurement, IAMOT 2006, Beijing, China, 8 p.

Morel, L., Camargo, M. and Fonteix, C. (2007). Integrating Product Innovation Degree and Technological Strategy. Using Constrained multi-criteria optimisation for a Polymerisation Process. Congrès de la Société Française de Génie de Procédés. Saint Etienne, October 11–13.

OCDE, EUROSTAT, (2005), Manuel d'Oslo, Principes directeurs pour le recueil et l'interprétation des données sur l'innovation, 3e édition, La mesure des activités scientifiques et technologiques, Éditions OCDE.

Ozman, M. (2006) Networks and Innovation : A Survey of Empirical Literature, document de travail No°2006-07, Bureau d'économie théorique et appliquée, February.

Pascale, R. T. (1999). Surfing the edge of chaos, *Sloan Management Review*, 40(3), pp. 83–94.

Prager, J-C. (2005). Le management stratégique des régions en Europe, Tome I : les enjeux et les stratégies. Rapport disponible auprès de l'Agence pour la diffusion de l'information technologique, Paris, France.

Tomkovick, C. and Miller, C. (2000). Perspective-riding the wind: managing new product development in an age of change, *Journal of Product Innovation Management*, 17, pp. 413–423.

Turriago, H.A. (2002). Gerencia de la innovación tecnológica. Collection Guías Empresariales. Alfaomega Colombiana, Bogotá, Colombia.

Walters, H. (2007). An Official Measure of Innovation, *Business Week Online*, 4 April.

Yam, C. M., Guan J. C., Pun J. C. and Tang P.Y. (2004). An audit of technological innovation capabilities in Chinese firms: some empirical findings in Beijing, *China Research Policy*, 33, pp. 1123–1140.

Section VI

Technology Foresight and Forecasting

Chapter 24

An Integrative Approach to Disruptive Technology Forecasting in Companies

Marion A. Weissenberger-Eibl and Stephan Speith
University of Kassel
Department of Innovation and Technology Management
Nora-Platiel-Str. 4, 34109 Kassel, Germany

This chapter presents an integrative approach to disruptive technology forecasting based on technology roadmapping and indicator-based forecasting. We discuss our experiences with the approach in the case of two disruptive technology projects in two firms.

Our basic proposition is that roadmapping disruptive technological progress is a process with three principal functions: information analysis, strategic anticipation and decision-making. We propose a six-step process to accomplish these functions. Database analysis is combined with expert judgement to provide the status quo of the technology field under study. The contexts of possible future applications are developed jointly by managers and technology experts. Finally, alternative strategies toward the future application contexts are constructed.

The approach helps to structure the disruptive innovation process. Information from different sources can be combined and strategies documented for planning further actions. The joint imagining of possible future application contexts leads to a shared understanding among technology experts and managers. Strategic anticipation and thinking in alternatives was stimulated, especially when critical events were included in the discussion. Finally, additional valuable information on the scientific bases of the technologies was generated.

1. Introduction

With the advent of the NBIC-technologies (nano, bio and information technologies, cognitive sciences), academics and executives realized the limitations of traditional forecasting methods in the context of disruptive technologies (Kostoff et al., 2004). Conventional approaches either do not account for the dynamics and openness of emerging technologies with regard to possible applications (e.g., the traditional roadmapping exercises) or fail to provide short- and medium-term guidance for decision-making (e.g., scenario planning).

Not surprisingly, research on new methods to forecast disruptive technologies has experienced an upsurge of interest. These new methods especially try to capture the dynamics and the high complexity of disruptive innovation. Examples can be found in journal special issues on disruptive technology roadmapping (*Technology Forecasting and Social Change* 2004, Volume 71) and technology indicators for emerging technologies (*Technology Forecasting and Social Change* 2006, Volume 73; *Long Range Planning* 2004, Volume 37) in publicly funded research projects like ATBEST (Rip and Propp, 2005) and SocRobust (Kets et al., 2003).

Nevertheless, these new approaches have some shortcomings which are addressed in this paper. First, most methods cover only fragments of the forecasting process. We assume that technology forecasting is an integrative process that fulfils the functions of information analysis, strategic anticipation and decision-making (Reger 2001). The majority of indicator-based approaches focus on the information analysis function, whereby strategic anticipation and decision-making are overlooked. On the other hand, traditional technology roadmapping stresses the decision-making and strategic anticipation functions without linking both to new developments in information analysis. These isolated views of technology forecasting harm decision-making in the context of disruptive technology in two ways: either the utilization of the generated information in decision-making processes is sub-optimal or strategic anticipation and decision-making lack an adequate information base. The ATBEST project (Propp and Rip 2005), the "Umbrella Approach" by Noori et al. (1999) and the approach taken by Kostoff et al. (2004) are among the few exceptions. Second, these exceptions focus on disruptive technology forecasting on an industry level, thereby neglecting the level of individual companies and networks (Rip and Propp, 2005; Kostoff et al., 2004). We transfer the positive experience made with disruptive technology forecasting from a sectoral level to a corporate context. We draw heavily on recent advances in disruptive technology roadmapping and indicator-based forecasting of emerging technologies to create a more holistic approach. The approach presented here addresses technology managers who deal with disruptive technology projects.

The systematic approach of our research was to apply the forecasting framework to two disruptive technology projects in two companies and to modify the framework based on the experiences made. The first project related to the development and commercialization of a new technology. The second project dealt with the transfer of an existing technology to a completely new application context. The objective was to provide companies with an approach to forecast and plan for disruptive technologies.

The chapter proceeds as follows. In the next section, core concepts of central importance to our disruptive technology forecasting approach are introduced. Next, a six-step process to forecast disruptive technologies in companies is described. We present our experiences with the approach in section 4. The final section contains a discussion of the results and some promising areas for further research.

2. Definitions and Core Concepts

This section introduces definitions and core concepts which are of central importance to the disruptive technology forecasting approach.

2.1. *Technology forecasting and strategy*

Technology forecasting is a process activity (Technology Futures Analysis Methods Working Group, 2004; Lichtenthaler, 2004; Reger, 2001; Rip and Propp, 2001; Noori et al., 1999; Martin, 1995). Technology forecasting has three functions: information analysis, strategic anticipation and decision-making. This exercise should not be restricted to technological information only and should also take social influences into account. Thus, technology forecasting and strategic management of technology are closely entwined in the context of disruptive technologies (Reger, 2001).

2.2. *Disruptive technology*

Disruptive technology is a technological change that modifies the bases of competition and requires a reconfiguration of competencies and resources (Danneels, 2004; Bower and Christensen, 1995; Abernathy and Clarke, 1985). Disruptive technologies can emerge through one of three scenarios. New scientific knowledge is transferred to the economic sphere as a technological innovation. Second, an existing technology is applied to new application domains (Adner and Levinthal, 2002). The fusion of formerly separated technologies into a new technology represents a third scenario (Kodama, 1992). Disruptive technologies provide new technological opportunities and ultimately lead to new technological systems with new sets of actors and social rules. These systems are very dynamic and highly complex. All in all, environments of disruptive technological progress differ fundamentally from periods where the bases of technological competition remain the same. Depending on the scenario of disruptive technological innovation, companies either need to find new applications for a given technology (scenario two) or need to develop and commercialize new technologies (scenarios one and three).

2.3. *Co-evolution*

Disruptive technological evolution is characterized by co-evolutionary phenomena. New actors and networks emerge, new social institutional rule-sets have to be created, new knowledge has to be developed and diffused, new organizational forms become established and new social patterns arise (Weissenberger-Eibl, 2006 and 2005; Malerba, 2005; Geels, 2004; Nelson, 1998; Christensen and Rosenbloom, 1995). The problems for technology management can best be summarized in Geels' words: "There is not just one kind of dynamic [...], but multiple dynamics which interact with each other" (Geels, 2004, p. 909). Traditional planning approaches do not account for the complex dynamics of disruptive technological innovation.

2.4. *Path dependencies*

The direction in which a new technology develops is relatively open until a lock-in occurs and the technology develops incrementally along a specific path (David, 1982). Phases of disruption are followed by phases of incremental changes along defined trajectories. Once a disruptive technology is locked into such a path, its further development is greatly restricted. Thus, technology planning in disruptive technology environments (i.e., in the "fluid" phases) underlies a dilemma between exploitation and exploration of technological paths. If a disruptive technology is transferred into a specific path too early, its full potential may not be obtainable (Rip and Propp, 2005; van Merkerk and van Lente, 2005). Disruptive technology planning needs to take the problem of lock-ins into sub-optimal technological paths into account.

2.5. *Disruptive technology roadmapping*

We extend Walsh's definition and view roadmapping as a forecasting and strategic management process (Walsh, 2004:166). Its aim is to bring together and visualize technology-related information from experts and databases, to provide a shared understanding and to produce strategic decisions (Weissenberger-Eibl and Speith, 2006; Kostoff et al., 2004; Walsh, 2004; Phaal et al., 2004; Rinne, 2004; Kostoff and Schaller, 2001). While traditional roadmapping exercises draw paths and milestones towards predefined target applications, disruptive technology roadmapping emphasizes the uncertainty and openness of innovation. Furthermore, the disruptive technology roadmapping approach presented in this paper accounts for the factors causing the dynamics of disruptive innovation, e.g. co-evolving networks, social rule-sets (Kets et al., 2003), knowledge bases and technological infrastructure.

To sum up, the disruptive technology roadmapping approach presented in this chapter addresses technology managers who face highly dynamic and complex environments. More specifically, the framework provides companies with a way to structure information analysis, anticipation and decision-making in an uncertain, highly dynamic environment. The forecasting and strategic planning process incorporates the openness of disruptive innovation processes and accounts for phenomena of disruptive technological progress (e.g. co-evolution and path dependencies). The following sections document our approach and the experiences we made with its application in two companies.

3. An Approach to Forecast Disruptive Technologies in Companies

What follows is a detailed description of the six process phases including goals, procedure, people to be involved and methods employed. The process is made up of a positional audit, an identification of possible application contexts, a gap analysis, an identification of alternative technological paths, a decision-making phase and an evaluation phase.

The forecasting process starts with a positional audit to figure out the state-of-the-art in technology development. The guiding question for this phase is "Where do we start from?" In order to quickly obtain an overview of the status quo of technology development, we employ database analysis. The use of large datasets has sparked growing interest in the past years (Kostoff, 2006; Daim et al., 2006; Porter and Cunningham 2006; Spinardi and Williams 2005; Technology Futures Analysis Methods Working Group, 2004; Menon and Tomkins, 2004). This is mainly due to the ever rising amount of data accessible, improved algorithms and high performance soft- and hardware. In order to gather information on new technologies, databases like ISI Web of Knowledge or EI Compendex (for scientific and engineering literature), STN or EI Patents (for patent data) and Lexis Nexis (for popular press articles) can be used. Keywords for a detailed search are distilled from interviews and the results are discussed and refined together with technology experts (Porter and Cunningham, 2005). After the first phase, alternative technologies, stakeholders, ideas about research needs and possible application contexts will be known.

In the second process phase, possible future application contexts are described and analysed in detail. "Which future application contexts are possible?" is the lead question in this phase. Additionally, a detailed description of these contexts is elaborated. The basic thought behind this step is that a technology which is embedded into an application context is characterized by a specific actor and network structure, specific social institutional rules, a certain knowledge base and a technological infrastructure (e.g., plants, machinery, complementary products). The project SocRobust follows a similar approach, with a restriction to the institutional environment (Kets et al., 2003). Use is made of disciplined imagining and strategic visioning to determine a set of applications including their contexts (Schwair, 2001; Noori et al., 1999). Information about applications and their contexts stems from expert interviews and workshops. Technologies and application contexts are visualized in a three-level roadmap comprising applications, products and technologies.

Having defined the upper roadmap level ("application context") and the lower level ("technology"), the third step consists of a gap analysis. The central question is "What are the gaps between the status quo and the future applications?" Based on the detailed description of future applications, gaps can be identified concerning the actor and network structure, institutional rule-set, knowledge base and technological infrastructure. One gap could be, for example, an expected technology-specific regulation in an application domain. We also expect some possible barriers to be revealed during this process, e.g. the lack of political commitment to regulate a new domain. The barriers are preliminarily visualized for each combination of technology and application context. Furthermore, alternative technologies are evaluated and mapped against the technology under analysis. Expert workshops in particular are held to complete the third phase.

The fourth step of the process deals with imagining alternative strategies to bridge the gap between application and technology. We view this exercise as a combination of prospective forecasting and backcasting from the application contexts (Dortmans, 2005; Weissenberger-Eibl, 2004; Schwair, 2001). These ways are described in terms of potential products and measures to be implemented for commercialization. Having defined ways to reach the application contexts, another set of barriers is identified for each combination of application and technology. Furthermore, alternatives to each application context are imagined in expert workshops to account for the uncertainty in technology development. Thinking in alternatives confronts technology experts and decision makers with the uncertainty in disruptive technologies. To support this process, critical events are defined which could harm or boost the closure of the gap between the status quo and the future context. In the fourth phase, the effect of critical events on alternative technological solutions will also be analyzed.

Based on the alternative strategies imagined, a decision for one or more strategies is achieved in stage five. Before a decision is made, workshop participants are prompted to highlight the flexibility inherent in each strategy. In other words, the degree to which a company is dependent on a specific technological path when following a strategic option will be made explicit. The more dependencies there are, the higher the risk of a lock-in to a path and the higher the costs of switching paths when other paths become more attractive (e.g. in terms of costs or performance). From a defined strategy, further steps and responsibilities of each participant can be defined. During the decision-making phase, it is important to stress the fact that the defined strategy will most likely not be realized in its predefined way.

In light of new information concerning the state of technology or new developments with an influence on the project, the decision will be evaluated. Since the available information in disruptive technology environments potentially changes very fast, the process should be repeated after a predefined time span.

4. Application of Disruptive Technology Forecasting in Two Companies

The systematic approach of the project was to apply the forecasting framework introduced above to two disruptive technology projects in two companies and to modify the framework based on the experiences gathered. Before the actual project started in both companies, disruptive technology projects had been identified, based on the criteria specified in section two. Critical to this definition was the commitment of core stakeholders (e.g. shareholders, CEOs) to the technology. Next, project team members representing different functional areas were selected from within the companies (e.g. R&D, planning, sales, controlling).

4.1. *Company 1*

The first project was the development and commercialization of a new technological innovation. Interviews led to the identification of alternative technologies and keywords to conduct database analysis subsequently. Database research showed that the technology was still in a state of scientific research, with patenting activity still very low. A set of about 600 publications was distilled from ISI Web of Knowledge and analyzed using a proximity map. We employed the software RefViz™ from Adept Scientific to cluster publications and to visualize thematic priorities.

No alternative technologies could be identified which were unknown to the company, but two papers pointed to a potential application not envisaged by the company beforehand. The project proceeded to describe the two basic application contexts in detail. It became obvious that the "new" application context ("Application Context 1") was too unknown to the company to provide a description. The project team decided to continuously monitor progress with regard to the application context 1 and to concentrate on context 2. The description of the application context led to the definition of gaps to be closed. Workshop participants rated the gap between the present and future knowledge base to be most significant. The actor-network structure was identified as another gap to be bridged.

Afterwards, the project team began with the identification of possible strategies to bridge the gap between future and present states (Step 4). During this exercise, conflicts about the usefulness of possible strategies arose between participants. After having identified a large set of potential strategies, we formed clusters of strategies that were perceived to be relatively similar. The process of reaching a consensus between workshop participants was the most time-consuming part of the roadmapping process. However, the time invested was experienced as extremely positive across the whole group because a shared understanding of the present situation and possible goals had been reached. With the clusters of strategies at hand, we started looking at barriers as identified before. We defined actions to be taken to bypass the barriers under the dominant strategy; e.g. to avoid an unstable production process for product A2, increased manpower will be concentrated on R&D activities in this area. No strategies could realistically be identified for application context 1. Therefore, an "owner" was defined for strategies that were directed towards application context 1. Finally, we tried to identify dependencies that would arise from following the dominant strategy. Admittedly, these dependencies were hard to imagine for technology experts. Only one dependency could be identified concerning the use of a specific form of machinery, but participants could not imagine a plausible alternative to the equipment.

4.2. *Company 2*

The second project dealt with the transfer of technology to a completely new application context. After the kick-off workshop, interviews were conducted to specify the possible application contexts and the potential technological alternatives. Database research confirmed that the technology itself was ready to market, with patenting activity on a constantly high level. We employed the software RefViz™ from Adept Scientific to cluster technical documents from engineering journals we obtained from searches in EI Compendex. No alternative technologies could be identified which were unknown to the company. The project proceeded to describe four basic application contexts in detail and three alternative technologies

All applications could be realized within five years in the future. The first application of technology B in the second context was envisaged taking place within two years in the future. From the twelve technology-application combinations possible, workshop participants identified seven as being attractive for further development. The remaining five contexts were disregarded. The actor-network structure was pointed out as the major gap to be bridged. Participants also found large gaps between future and present institutional environments and technological infrastructures.

We began with the identification of possible strategies to bridge the gap between future and present states (Step 4). In the second company, the process of reaching a consensus between workshop participants was the most time-consuming part of the roadmapping process, too. We defined actions to be taken to bypass the barriers in each strategy; e.g. to avoid the production of new components, increased manpower will be concentrated on development activities to use existing material. Finally, we tried to identify dependencies that would arise from following the dominant strategy. The dependencies were very obvious for the different technologies and therefore easy to identify by technology experts. Interestingly, thinking about dependencies led to the identification of alternative strategies. More precisely, the applicability of one technology in an application context was seen as realistic after being rejected before.

5. Discussion

All in all, the forecasting exercise was rated useful by the companies. First, the approach provided a guideline and supported navigating in a "fuzzy" environment. Second, the use of automated information processing, together with expert judgements and the visualization of search results in technology roadmaps was deemed useful because it provided an objective picture of the state-of-the-art. Participants found the approach useful since strategies were jointly developed, visualized and discussed. Third, the construction of alternative strategies helped decision-makers to adapt to the uncertainty inherent in disruptive technology projects. Strategic anticipation and thinking in alternatives was stimulated, especially when critical events were included in the discussion. Fourth, additional valuable information was generated. In the first company, we identified an additional application to be explored. In the second company, new information led to the rejection of potential application contexts previously preferred by the firm. In this way, new information resulted in an adjustment of the companies' development activities.

The disruptive technology forecasting approach shows two major areas for improvement. First, the process is still very time-consuming. This included much iteration between each process phase since information was often revealed "bit by bit". Potentially, the approach can be implemented faster when the participants have more routine. Further research should concentrate on ways to speed up the process. Second, it was hard for the participants in the exercise to imagine dependencies. Therefore, it should be investigated how technology managers can better analyze and manage the risks of lock-ins and path dependencies.

Acknowledgements

The authors would like to acknowledge research grants provided by the HA Hessen Agentur GmbH under the European Social Funds (HA-Project-No. 120/06-01). The authors highly appreciated Hashem Sherif and Jeff Butler's invaluable comments which helped to improve this article. We are grateful for the comments of an anonymous referee as well.

References

Adner, R. and Levinthal, D.A. (2002). The Emergence of Emerging Technologies, *California Management Review*. 45(1), pp. 50–66.

Abernathy, W.J. and Clark, K.B. (1985). Innovation: Mapping the Winds of Creative Destruction, *Research Policy*. 14, pp. 3–22.

Bengisu, M. and Nekhili, R. (2006). Forecasting Emerging Technologies with the Aid of Science and Technology Databases, *Technological Forecasting and Social Change*. 73, pp. 835–844.

Bower, J.L. and Christensen C.M. (1995). Disruptive Technologies: Catching the Wave, *Harvard Business Review*. January-February, pp. 43–53.

Christensen, C.M. and Rosenbloom, R.S. (1995). Explaining the Attacker's Advantage: Technological Paradigms, Organisational Dynamics, and the Value Network, *Research Policy*. 24, pp. 233–257.

Daim, T.U., Rueda, G., Martin, H. and Gerdsri, P. (2006). Forecasting Emerging Technologies: Use of Bibliometrics and Patent Analysis, *Technological Forecasting and Social Change*. 73, pp. 981–1012.

Danneels, E. (2004). Disruptive Technology Reconsidered: A Critique and Research Agenda, *Journal of Product Innovation Management.* 21(4), pp. 246–258.

Dortmans, P.J. (2005). Forecasting, Backcasting, Migration Landscapes and Strategic Planning Maps. *Futures*, 37, pp. 273–285.

Dosi, G.A. (1982). Technological Paradigms and Technological Trajectories, *Research Policy.* 11, pp.147–162.

Geels, F.W. (2004). From Sectoral Systems of Innovation to Socio-technical Systems – Insights about Dynamics and Change from Sociology and Institutional Theory, *Research Policy.* 33, pp. 897–920.

Kets, A., Burger, H. and de Zoeten-Dartenset, C. (2003). Experiences with SocRobust at ECN. Petten: Energy Research Centre of the Netherlands.

Kodama, F. (1995). Emerging Patterns of Innovation – Sources of Japan's Technological Edge. Harvard Business School Press, Boston.

Kostoff, R.N. and Schaller, R. R. (2001). Science and Technology Roadmaps, *IEEE Transactions on Engineering Management.* 48 (2), pp.1 32–143.

Kostoff, R.N., Boylan, R. and Simons, G.R. (2004). Disruptive Technology Roadmaps, *Technological Forecasting and Social Change.* 71, pp. 141–159.

Kostoff, R.N. (2006). Systematic Acceleration of Radical Discovery and Innovation in Science and Technology, *Technological Forecasting and Social Change.* 73, pp. 923–936.

Lichtenthaler, E. (2004). Technology Intelligence Processes in Leading European and North American Multinationals, *R&D Management.* 34 (2), pp. 121–135.

Malerba, F. (2005). Sectoral Systems of Innovation: A Framework for Linking Innovation to the Knowledge Base, Structure and Dynamics of Sectors, *Economics of Innovation and New Technologies.* 15(1-2), pp. 63–82.

Martin, B.R. (1995). Foresight in Science and Technology, *Technology Analysis and Strategic Management.* 17(2), pp. 139–168.

Menon, A. and Tomkins, A. (2004). Learning about the Market's Periphery: IBM's WebFountain, *Long Range Planning.* 37, pp. 153–162.

Nelson, R.R. (1998). The Co-evolution of Technology, Industrial Structure, and Supporting Institutions. In: Dosi, G., Teece, D.J. and Chytry, J. (Eds.): Technology, Organization, and Competitiveness - Perspectives on Industrial and Corporate Change. Oxford University Press, Oxford, pp. 319–335.

Noori, H., Munro, H., Descza, G. and McWilliams, B. (1999). Developing the 'Right' Breakthrough Product/Service: An Application of the Umbrella Methodology to Electric Vehicles. Part B, *International Journal of Technology Management.* 17(5), pp. 563–579.

Phaal, R., Farrukh, C. and Probert, D. (2004). Customizing Roadmapping, *Research Technology Management.* March-April, pp. 26–37.

Porter, A.L. and Cunningham, C.W. (2005). Tech Mining - Exploiting New Technologies for Competitive Advantage. Wiley, New York.

Propp, T. and Rip, A. (2005). Assessment Tools for Breakthrough and Emerging Science and Technology - Literature Review. In Project ATBEST - Final Activity Report, Deliverable 1, pp. 21–77.

Reger, G. (2001). Technology Foresight in Companies: From an Indicator to a Network and Process Perspective, *Technology Analysis and Strategic Management*, 13(4), pp. 533–553.

Rinne, M. (2004). Technology Roadmaps: Infrastructure for Innovation, *Technological Forecasting and Social Change.* 71, pp. 67–80.

Rip, A. and Propp, T. (2005). Can Open-ended Roadmapping Address the Tension between Organisational Effectiveness and Strategic Flexibility. Presentation at the INIR Workshop, Enschede, June 2005.

Schwair, T.M. (2001). Inventing the Future, Not Only Predicting the Future – Futures Research at Siemens AG, Corporate Technology, *Futures Research Quarterly.* 17(3), pp. 35–41.

Spinardi, G. and Williams, R. (2005). New and Emerging Science and Technology and Their Assessment. In: Project ATBEST - Final Activity Report, Deliverable 2, pp. 79–115, Enschede.

Technology Futures Analysis Methods Working Group (2004). Technology Futures Analysis: Towards Integration of the Field and New Methods, *Technological Foresight and Social Change.* 71, pp. 287–303.

Utterback, J.M. (1994). Mastering the Dynamics of Innovation. Harvard Business School Press, Boston.

van Merkerk, R.O. and van Lente, H. (2005). Tracing Emerging Irreversibilities in Emerging Technologies: The Case of Nanotubes, *Technology Forecasting and Social Change.* 72, pp. 1094–1111.

Walsh, S.T. (2004). Roadmapping a Disruptive Technology: A Case Study. The Emerging Microsystems and Top-down Nanosystems Industry, *Technology Forecasting and Social Change.* 71, pp.161–185.

Weissenberger-Eibl, M. (2004). Unternehmensentwicklung und Nachhaltigkeit. 2nd Edition. Cactus Group, Rosenheim.

Weissenberger-Eibl, M. (ed.) (2005). Gestaltung von Innovationssystemen. Cactus Group, Kassel.

Weissenberger-Eibl, M. (2006). Wissensmanagement in Unternehmensnetzwerken. 2nd Edition. Cactus Group, Kassel.

Weissenberger-Eibl, M. and Speith, S. (2006). Flexibles Roadmapping - Eine Methode für die Vorausschau und Technologieplanung im Umfeld technologischer Durchbrüche. in: Gausemeier, J. (Ed.), Vorausschau und Technologieplanung. HNI-Verlagsschriftenreihe, Heinz Nixdorf Institut, Paderborn, pp. 396–424.

Chapter 25

Quadratic-Interval Innovation Diffusion Models for New Product Sales Forecasting

Fang-Mei Tseng

Department of International Business, Yuan Ze University 135, Yuan Tung Rd., Chung-li, Taiwan 320, fmtseng@saturn.yzu.edu.tw

An appropriate sales forecasting method is vital to the success of a business firm. The logistic model and the Gompertz model and a series innovation diffusion models which is based on the Bass model (Bass, 1969) are usually adopted to forecast the growth trends and the potential market volume of innovative products. All of these models rely on statistics to explain the relationships between dependent and independent variables, and use crisp parameters. However, fuzzy relationships are more appropriate for describing the relationships between dependent and independent variables; these relationships require less data than traditional models to generate reasonable estimates of parameters. Therefore, we have combined fuzzy regression with the logistic and Gompertz models to develop a quadratic-interval Gompertz model and a quadratic-interval logistic model, and we applied the models to three cases. Our practical application of the two models shows that they are appropriate tools that can reveal the best and worst possible sales volume outcomes.

1. Introduction

It is crucial for corporations to develop innovative products if they are to maintain or improve their competitive advantage. According to the Product Development and Management Association (PDMA), 42.4% of company profits come from new products launched in last 3 years (Wiebe and Eng, 2006). A correct evaluation of sales volume is also important when businesses introduce products from foreign countries into local markets, because it allows the firms to appropriately allocate resources. Therefore, it is extremely important to find an appropriate new products sales forecasting model that provides estimates of the product's diffusion speed and total market potential. The speed of diffusion represents the sales increase in any period, while the market potential is the product's maximum total sales after its introduction. From these two variables, the amount of time required for a return of the initial investment can be determined.

There is a long history of research on diffusion theory, although early work focused on issues of sociology, epidemic contagion, biology, and ecology. The logistic model was one of the first introduced to study innovation diffusion. Verhulst developed the logistic growth model for population growth forecasting purposes in 1843 (Frank, 2004). The logistic and Gompertz models are also popular in technology forecasting. While, Bass (1969) integrates the modified exponential model (Fourt and Woodlock, 1960) and logistic model (Mansfield, 1961) to propose the new product growth model. This model is one of the more well-known and widely used model of developing the curve of product life cycle, and provides to forecast the sales for the timing of initial purchase of new products. Diffusion models typically require at least 6–10 observations to generate reasonable parameter estimates (Heeler and Hustad, 1980). This is a problem for new products, because sales data are insufficient. Moreover, the logistic model and the Gompertz model explain the deviations between estimation and observations through measurement error, which is problematic because the data are precise values that do not include measurement errors.

In addition, if a phenomenon under consideration does not have stochastic variability but is also uncertain in some sense, it is more natural to seek a fuzzy functional relationship for the given data, which may be either fuzzy or crisp. That is to say, a fuzzy phenomenon should be modeled by a fuzzy functional relationship. Fuzzy regression analysis was first proposed by Tanaka et al. (1982), who used a fuzzy linear system as a regression model to solve a fuzzy environment problem and to avoid modeling error. Tanaka and Lee (1998) proposed an interval regression analysis based on a quadratic programming approach. This quadratic programming approach produces more diverse spread coefficients than a linear programming approach, and integrates the property of central tendency in least squares analysis and the possibility property in fuzzy regression. Kim et al. (1996) found that fuzzy regression can be a viable alternative to statistical linear regression in estimating regression parameters when the data set is insufficient.

Therefore, this chapter proposes two quadratic-interval growth-diffusion models that combine quadratic-interval regression with the Gompertz model and the logistic model to solve a fuzzy relationship between explanatory and response variables and to provide forecasts of sales to decision makers. These models require fewer observations than traditional innovation diffusion models. We applied the models to three cases to demonstrate their performance, and found that they make good forecasts and appear to be appropriate tools.

This chapter is organized as follows: in Section 2, we review growth models and the quadratic interval regression model; in Section 3, the quadratic interval growth models are formulated and proposed; in Section 4, the quadratic-interval growth-diffusion models are applied to three cases; and conclusions are discussed in Section 5.

2. Innovation Diffusion Models and Fuzzy Regression Model

To explain the proposed models, innovation diffusion models and the fuzzy regression model are described in the following sections.

2.1. *Innovation diffusion models*

Innovation diffusion models have been used in marketing to capture the life-cycle dynamics of a new product. This chapter adopts three popular diffusion models in which the diffusion of new technology follows an S-curve function: the logistic model and the Gompertz model. These two models represent epidemic contagion models; the S-curve of Gompertz model is asymmetric and the S-curve of logistic model is symmetric.

Gompertz model is described by the function (Islam and Meade, 1997)

$$Y_t = m\exp(-\exp(-a + bt)) + e_t, \tag{1}$$

where Y_t is the number of agents that have adopted the new technology at time t, a is the timing of the initial adoption, b is the speed of adoption, and m is the total number of potential adopters.

The logistic model is described by the function (Gruber, 2001)

$$Y_t = m/(1 + \exp(-(b_0 + b_1 t))) + e_t, \tag{2}$$

where, Y_t is the number of agents that have adopted the new technology at time t, b_0 is the timing of the initial adoption, b_1 is the speed of adoption, and m is the total number of potential adopters.

The Gompertz and the logistic models are similar, but represent different evolutionary patterns. If the dynamics of the diffusion process is such that growth is quite rapid in the early phase and relatively slow when approaching the saturation level, the Gompertz model is best because it attains its maximum rate of growth at an earlier phase than does the logistic model (Botelho and Pinto, 2004).

Other researchers have focused on the performance of these forecasting methods. Meade and Islam (1995) used data from 25 time series to compare the forecasting performances of 17 growth curve models, including the logistic, Gompertz, and the Bass models and found that the logistic and the Gompertz models significantly outperformed complex models. Rai et al. (1998) analyzed the fitting and forecasting ability of the exponential, the logistic, and the Gompertz model by applying them to the global diffusion of the internet, and found that the exponential model was the most accurate. Teng et al. (2002) examined the diffusion patterns of 20 information technology (IT) innovations using the exponential, the logistic, the Gompertz, and the Bass models. They concluded that the Bass model best described the diffusion pattern of IT innovations. In summary, different models perform better under different circumstances.

2.2. *Quadratic-interval regression model*

The basic idea of fuzzy regression theory is that the residuals between estimators and observations are produced by uncertainties in the model parameters rather than by measurement errors, and a possibility distribution is used to deal with practical observations.

A generalized model of fuzzy linear regression is as follows (Tanaka et al., 1982):

$$y_j = A_0 + A_1 x_{1j} + \cdots A_n x_{nj} = \sum_{i=1}^{n} A_i x_{ij} = \boldsymbol{x}_j' \boldsymbol{A}., \tag{3}$$

where, $x_j = (1, x_{1j}, \cdots, x_{nj})'$ is a real input vector of independent variables, n is the number of variables, and $A = (A_0, \cdots, A_n)'$ represents a vector of the fuzzy parameters in the model. Instead of using a crisp value, the ith fuzzy parameter A_i after the L-type fuzzy numbers of Dubois and Prade (1982), $(\alpha_i, c_i)_L$, the possibility distribution is

$$\mu_{Ai}(A_i) = L\{(\alpha_i - A_i)/c_i\}, \tag{4}$$

where, L is a membership function type. Fuzzy parameters in the form of triangular fuzzy numbers are used,

$$\mu_{\tilde{B}_i}(A_i) = \begin{Bmatrix} 1 - \dfrac{|\alpha_i - A_i|}{c_i}, \alpha_i - c_i \le A_i \le \alpha_i + c_i \\ 0, \quad otherwise \end{Bmatrix}, \tag{5}$$

where, $\mu_{\tilde{B}_i}(A_i)$ is the membership function of the fuzzy set which is represented by parameter A_i, α_i is the center of the fuzzy number, and c_i is the width or spread around the center of the fuzzy number.

According to Zadeh's extension principle (1965), the membership function of the fuzzy number $y_j = \boldsymbol{x}_j' \boldsymbol{A}$ can be defined by a membership function using pyramidal fuzzy parameter $\boldsymbol{A}$, as follows:

$$\mu_{\tilde{Y}}(y_j) = \begin{cases} 1 - |y_j - x_j' \boldsymbol{\alpha}| / \boldsymbol{c}' \, |\boldsymbol{x}_j| & for\ \boldsymbol{x}_j \neq 0, \\ 1 & for\ \boldsymbol{x}_j = 0, y_j = 0, \\ 0 & for\ \boldsymbol{x}_j = 0, y_j \neq 0. \end{cases} \tag{6}$$

where $\boldsymbol{\alpha}$ and $\boldsymbol{c}$ denote vectors of model parameter values and spreads, respectively, for all of the model parameters, and j denotes the j-th observation, $j = 1,2,\cdots,m$.

Finally, this method uses the criterion of minimizing the total vagueness (where is it defined?) and the sum of squared distances between the estimated output centers and the observed output, $\boldsymbol{S}$, which reflects both properties of least squares and possibilistic approaches (Tanaka and Lee, 1998).

$$\text{minimize} \qquad S = k_1 \sum_{j=1}^{m} (y_j - \boldsymbol{\alpha}\boldsymbol{x}_j')^2 + k_2 \sum_{j=1}^{m} \boldsymbol{c}'|\boldsymbol{x}_j||\boldsymbol{x}_j|'\boldsymbol{c}, \quad (7)$$

where, $\sum_{j=1}^{m} |\boldsymbol{x}_j||\boldsymbol{x}_j|'$ is a $(n+1)\times(n+1)$ symmetric positive definite matrix and k_1 and k_2 are weight coefficients.

A matrix is a positive definite if and only if all of the eigenvalues of the matrix are positive. The weight coefficients k_1 and k_2 in Eq. (7) have an important role in formulating fuzzy regression models. For example, if we use a large value of k_1 compared to k_2, a more central tendency is expected, i.e., the obtained central regression line would tend to be the regression line obtained by least squares regression. However, if we use a large value of k_2 compared to k_1, we reduce the fuzziness of the model.

At the same time, this approach also considers that the membership degree (Eq. (6)) of each observation y_j is greater than an imposed threshold possibility, as $h, h \in [0,1]$. This criterion simply expresses the fact that the fuzzy output of the model should 'cover' all of the data points $y_1, y_2, \cdots, y_m$ to a certain level, h. The value of the h level that is chosen will influence the widths, $\boldsymbol{c}$, of the fuzzy parameters:

$$\mu_Y(y_j) \geq h \; \forall j = 1,2,\cdots,m \;, \quad (8)$$

where. the index j denotes the j-th observation. Finding the interval regression parameters is formulated by Tanaka and Lee as a quadratic programming problem (1998):

$$\text{minimize} \qquad S = k_1 \sum_{j=1}^{m} (y_j - \boldsymbol{\alpha}'\boldsymbol{x}_j)^2 + k_2 \sum_{j=1}^{m} \boldsymbol{c}'|\boldsymbol{x}_j||\boldsymbol{x}_j|'\boldsymbol{c}$$

subject to

$$\boldsymbol{x}_j\boldsymbol{\alpha} + (1\text{-}h)\boldsymbol{c}'|\boldsymbol{x}_j| \ge y_j,\ j = 1,2,\cdots,m,$$
$$\boldsymbol{x}_j\boldsymbol{\alpha} - (1\text{-}h)\boldsymbol{c}'|\boldsymbol{x}_j| \le y_j,\ j = 1,2,\cdots,m,$$
$$\boldsymbol{c} \ge 0 \qquad (9)$$

where $\boldsymbol{\alpha}' = (\alpha_0, \alpha_1, \cdots, \alpha_n)$ and $\boldsymbol{c}' = (c_0, c_1, \cdots, c_n)$ are vectors of unknown variables.

Kim et al. (1996) found that fuzzy linear regression is a viable alternative to statistical linear regression in estimating regression parameters when the data set is insufficient to support statistical regression analysis and /or when the regression model is inappropriate. However, fuzzy regression should not used when the data is of poor quality (i.e., when there are outliers in the data or when the data are highly variable).

3. Quadratic-interval Diffusion Models

I have developed two quadratic-interval diffusion models, a quadratic-interval Gompertz model and a quadratic-interval logistic model based on the fuzzy regression model. The required data size is similar to that of fuzzy regression analysis, and therefore is smaller than in traditional innovation diffusion models. This is illustrated in Sections 3.1–3.2.

3.1. *Quadratic-interval Gompertz model*

The quadratic-interval Gompertz model was constructed using the Gompertz model and Tanaka's interval regression (Tanaka and Lee, 1998).

In order for the simple Gompertz model to meet conventional simple regression, Eq. (1) was transformed as follows:

$$\begin{aligned} Y_t &= m\exp(-a\exp(-bt)) + e_t \\ z_t &= \log(\log Y_t - \log Y_{t-1}) = -bt + (\log(a\exp b - a)) \\ &= c + (-b)t, \end{aligned} \tag{10}$$

where $c = \log(a\exp b - a)$,
$a = \exp\alpha / (\exp b - 1)$, and
$m = \exp(\log Y_t + a\exp(-bt))$.

A quadratic-interval Gompertz model is described with a fuzzy parameter:

$$z_t = A_0 + A_1 t = \langle \alpha_0, c_0 \rangle + \langle \alpha_1, c_1 \rangle t = \boldsymbol{t}' \boldsymbol{A}. \tag{11}$$

According to Eq. (11) and using the extension principle, the membership function of the fuzzy number $z_t = A_0 + A_1 t$ can be defined by a membership function using the pyramidal fuzzy parameter $\boldsymbol{A}$, as follows:

$$\mu_{\tilde{z}}(z_t) = \begin{cases} 1 - |z_t - \boldsymbol{t}'\boldsymbol{\alpha}| / \boldsymbol{c}' |\boldsymbol{t}| & for\ \boldsymbol{t} \neq 0, \\ 1 & for\ \boldsymbol{t} = 0, z_t = 0, \\ 0 & for\ \boldsymbol{t} = 0, z_t \neq 0. \end{cases} \tag{12}$$

where $\boldsymbol{\alpha}$ and $\boldsymbol{c}$ denote the vectors of the model parameter values and the spreads, respectively, for all of the model parameters.

Finally, this method requires minimization of the total vagueness and the sum of squared distances between the estimated output centers and the observed output, S, defined as

$$\text{minimize} \qquad S = k_1 \sum_{t=1}^{m} (z_t - \boldsymbol{\alpha}'\boldsymbol{t})^2 + k_2 \sum_{t=1}^{m} \boldsymbol{c}' |\boldsymbol{t}| |\boldsymbol{t}|' \boldsymbol{c}. \tag{13}$$

Simultaneously, this approach takes into account the condition that the membership degree of each observation z_t is greater than an imposed threshold possibility as $h, h \in [0,1]$. This criterion expresses the fact that the fuzzy output of the model should 'cover' all of the data points $z_1, z_2, \cdots, z_m$ to a certain level of h. The selection of the h level value influences the widths, $\boldsymbol{c}$, of the fuzzy parameters:

$$\mu_{\tilde{z}}(z_t) \geq h \; \forall t = 1,2,\cdots,m. \tag{14}$$

The index t refers to the number of nonfuzzy data used in constructing the model. Finding the fuzzy regression parameters was formulated by Tanaka and Watada as a linear programming problem:

$$\begin{aligned} &\text{minimize} \quad S = k_1 \sum_{t=1}^{m} (z_t - \boldsymbol{\alpha}'\boldsymbol{t})^2 + k_2 \sum_{t=1}^{m} \boldsymbol{c}'|\boldsymbol{t}||\boldsymbol{t}|'\boldsymbol{c} \\ &\text{subject to} \\ &\boldsymbol{t\alpha} + (1\text{-}h)\boldsymbol{c}'|\boldsymbol{t}| \geq z_t, \quad t = 1,2,\cdots,m, \\ &\boldsymbol{t\alpha} - (1\text{-}h)\boldsymbol{c}'|\boldsymbol{t}| \leq z_t, \quad t = 1,2,\cdots,m, \\ &\boldsymbol{c} \geq 0 \end{aligned} \tag{15}$$

where $\boldsymbol{\alpha}' = (\alpha_0, \alpha_1)$ *and* $\boldsymbol{c}' = (c_0, c_1)$ are vectors of unknown variables.

The procedure of the quadratic interval Gompertz model is as follows:

Step 1: Fit the Gompertz model using the available sets of observations, i.e., the input data is considered nonfuzzy. According to the concept derived by Savic and Pedrycz (1991), the result of this step is that the optimum solutions of the parameters $\boldsymbol{\alpha}^* = (\alpha_0^*, \alpha_1^*)$ *and* z_t^* are used as input data in Step 2, where $\alpha_0^* = \hat{c}, \alpha_1^* = -\hat{b}$.

Step 2: Determine the minimal fuzziness using the criteria Eq. (15) and $\alpha^* = (\alpha_0^*, \alpha_1^*)$. The number of constraint functions is the same as the number of observations, a concept derived by Savic and Pedrycz (1991).

3.2. *Quadratic-interval logistic model*

The quadratic-interval logistic model was constructed along with the logistic diffusion model and Tanaka's interval regression (Tanaka and Lee, 1998).

In order for the logistic model to meet conventional simple regression, Eq. (2) was transformed as follows:

$$Y_t = m/(1+\exp(-(b_0+b_1t))+e_t$$
$$Y_t/m = \pi_t = 1/(1+\exp(-(b_0+b_1t))+e_t$$
$$z_t = \log\left(\frac{\pi_t}{1-\pi_t}\right) = \boldsymbol{tb}\,. \tag{16}$$

A quadratic-interval logistic model is described with a fuzzy parameter:

$$z_t = A_0 + A_1 t = \langle \alpha_0, c_0 \rangle + \langle \alpha_1, c_1 \rangle t\,. \tag{17}$$

Finding the fuzzy parameter of the quadratic-interval logistic model uses the same procedure as in the quadratic-interval Gompertz model, as follows:

Step 1: Fit the logistic model using the available sets of observations, i.e., the input data are considered nonfuzzy and used the estimated parameters, m, b_0, and b_1. According to the concept derived by Savic and Pedrycz (1991), this step produces an optimum solution of the parameter m, $\boldsymbol{\alpha}^* = (\alpha_0^*, \alpha_1^*) = (\hat{b}_0, \hat{b}_1)$ and π_t^*, which are used as input data sets in Step 2.

Step 2: Calculate the logistic mean function z_t. Substituting the result of the parameters from Step 1 into Eq. (16), we obtain the estimated logistic mean function, z_j, $\forall j = 1,2,\cdots,m.$.

Step 3: Determine the minimal fuzziness using the Eq. (15) and.

$$\boldsymbol{\alpha}^* = (\alpha_0^*, \alpha_1^*)$$

4. Empirical Results

The performance of the two quadratic innovation diffusion models were compared with two popular new product sales forecasting models, the Gompertz and the logistic using three data sets: the inventory of cars in the Netherlands, cellular phones in Portugal, and worldwide personal computer (PC) demand. The details are in Sections 4.1–4.3.

4.1. *The inventory of cars in the Netherlands*

I used the data set of Franses (1994), who proposed a Gompertz curving fitting method and used the inventory of cars in the Netherlands from 1965 to 1989 to examine the performance of the model. The smoothed series is depicted in Figure 1. I used his data set and research results and to compare the logistic model and our proposed models. The procedures of the proposed models are described in Sections 4.1.1–4.1.2.

4.1.1. *Building the quadratic-interval Gompertz model*

The two steps of quadratic interval Gompertz model are as follows.

Step 1: Fit the Gompertz model: The estimation results of Franses' (1994) Gompertz curving fitting model in Eq. (1) are $\hat{a}=1.5$ $\hat{b}=0.104$, and $\hat{m}=5962$, and the forecasts of the 95% confidence intervals are shown in Figure 1.

Step 2: Determine the minimal fuzziness.

According to Eq. (15), and by setting $(\alpha_0, \alpha_1) = (0.7841,\text{-}0.104)$ h=0 , the following quadratic interval Gompertz model is obtained using the LINGO package software (1999) and the estimated equation is shown in Eq. (18) and the forecasts are shown in Figure. 1.

$$z_t = \langle 0.7841, 0.18828\rangle + \langle -0.104, 0.09948\rangle_t \tag{18}$$

4.1.2. *Building the quadratic interval logistic model*

The three steps of quadratic interval logistic model are as follows.

Step 1. Fit the logistic model: we applied the logistic model in Eq. (2) to estimate parameters. The estimation results are $\hat{b}_0 = -1.274$, $\hat{b}_1 = 0.166$, $\hat{m} = 5520.62$ and the forecasts of 95% confidence intervals are shown in Figure 2.

Step 2. We apply Eq. (16) to calculate z_t.

Step 3. Determine the minimal fuzziness.

According to Eq. (15), and by setting $(\alpha_0, \alpha_1) = (-1.2704, 0.166)$ h=0, the following quadratic-interval logistic model is obtained using the LINGO package software (1999); the estimated equation is shown in Eq. (19) and the possible upper and lower bounds are shown in Figure 2.

$$z_t = (-1.274, 0.0645) + (0.166, 0)t. \tag{19}$$

4.1.3. *Comparisons*

The parameter estimation from the different models is shown in Tables 1 and 2 and the confidence interval and possibility interval curve are shown in Figures 1 and 2. The interval of the quadratic-interval Gompertz model is the narrowest and all of the actual data are located in the interval of the quadratic-interval Gompertz model; however, all of the actual data are not located in the 95% confidence interval of the Gompertz model. This means that the quadratic-interval Gompertz model has greater prediction capability. However, the interval of the quadratic-interval logistic model is larger than the 95% confidence interval of the logistic model. These forecasting intervals show that the quadratic-interval Gompertz model is the most appropriate for predicting the best and worst possible sales volumes.

4.2. *Cellular phone subscribers in Portugal*

To demonstrate the performance of quadratic-interval innovation models, we applied the four models to time series depicting cellular phone subscribers in Portugal, used by Botelho and Pinto (2004). They used the exponential growth model, the Gompertz model, and the logistic model to examine the diffusion pattern of cellular phones in Portugal, and found the logistic model to be the best. Their time-series data of the cumulative number of subscribers runs from Quarter 4 of 1995 to Quarter 2 of 2000, and is depicted in Figure 3.

We have omitted parameter estimation procedures; the results of the parameter estimation are shown in Tables 1 and 2, and the confidence interval and possibility interval curve are shown in Figure 3. Because of the diffusion process of this time series, growth is initially slow and then relatively rapid during the maturing phases; this results in poor forecasting performance by the Gompertz and the quadratic-interval Gompertz model. We have omitted the forecasting interval curves in Figure 4. According to Figure 3, the interval of the quadratic- interval logistic model is narrower than the 95% confidence interval of the logistic model and all of the actual data locate in the intervals, which means that the quadratic-interval Bass model has greater predictive ability.

4.3. *Worldwide PC demand*

To demonstrate the performance of the quadratic-interval innovation models, we applied the six models to the third time series, worldwide PC demand from 1981 to 1999 (Figure 4). The data were used in an examination of PC demand using the Bass model (Bass, 1999).

We have omitted the procedures of parameter estimation; the parameter estimations are shown in Tables 1 and 2, and the confidence interval and the possibility interval curve are shown in Figure 6. According to Table 1, the parameter estimation of the Gompertz model is not significant; we therefore don't discuss the Gompertz and quadratic-interval Gompertz model in this data set.

According to Figure 4, the interval of the quadratic-interval logistic model is narrower than the 95% confidence interval of the logistic model; all of the actual data locate in the intervals.

4.4. *Comparisons and discussion*

The forecasting performance comparisons of the four models in the three data sets reveal no convincing evidence that one model outperforms than the others; performance depends on the specific time series pattern. In the first set of time series data, the inventory of cars in the Netherlands (Figure 1), the diffusion process produces rapid growth during the early phase, and relatively slow growth when approaching the maturing phases. The quadratic-interval Gompertz model has the best forecasting performance. However, in the other two time series, shown in Figures 3 and 4, the diffusion processes produce initially slow growth, then relatively rapid growth during the maturing phases. In this situation, the quadratic-interval logistic model outperforms the other models; the Gompertz model is unsuitable for these two data sets. Because these three data sets have no outlying data and the data are not highly variable, a quadratic-interval innovation models outperformed the others.

Table 1: The Parameter estimates of Gompertz and logistic models

		Stock of cars		Cellular phones		Worldwide PC	
	Parameter	Estimate	R^2	Estimate	R^2	Estimate	R^2
Gompertz	m	5962.00**	0.857	15552.02	0.987	Not convergence	
	a	1.500**		8.4348**			
	b	0.104**		0.2374			
Logistic	m	5520.62**	0.9975	6736.448**	0.9975	290.898**	0.989
	b_0	-1.274**		-3.6025**		-4.308**	
	b_1	0.166**		0.2563**		0.1965**	

Table 2: The Parameter estimates of Quadratic interval Gompertz and Quadratic interval logistic models

	Stock of cars		Cellular phones	Worldwide PC
	Parameter	Estimate	Estimate	Estimate
Quadratic Gompertz	$<\alpha_0, c_0>$	<0.7841,0.1882>	<-0.3369,1.5014>	Can't calculate
	$<\alpha_1, c_1>$	<-0.104,0.0995>	<-0.2374,0>	
Quadratic logistic	$<\alpha_0, c_0>$	<-1.274,0.0645>	<-3.6025,0.4137>	<-4.308,0.4137>
	$<\alpha_1, c_1>$	<0.166,0>	<0.2563,0.001>	<0.1965,0.001>

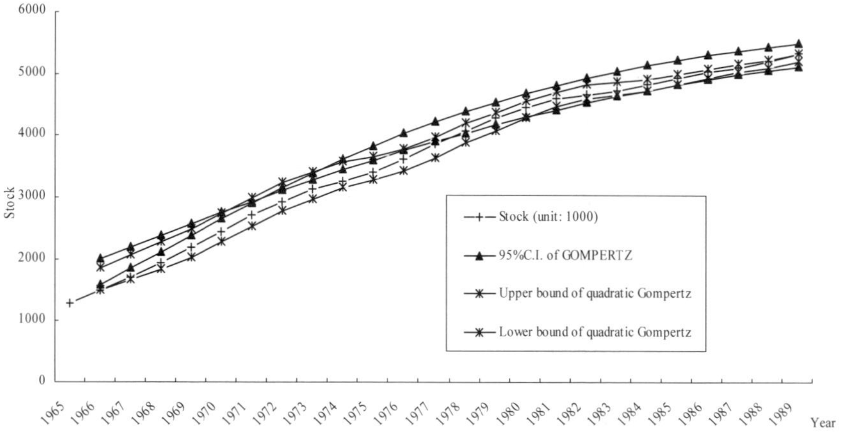

Figure 1: The forecasts of the Gompertz and Quadratic Interval Gompertz model for stock of cars.

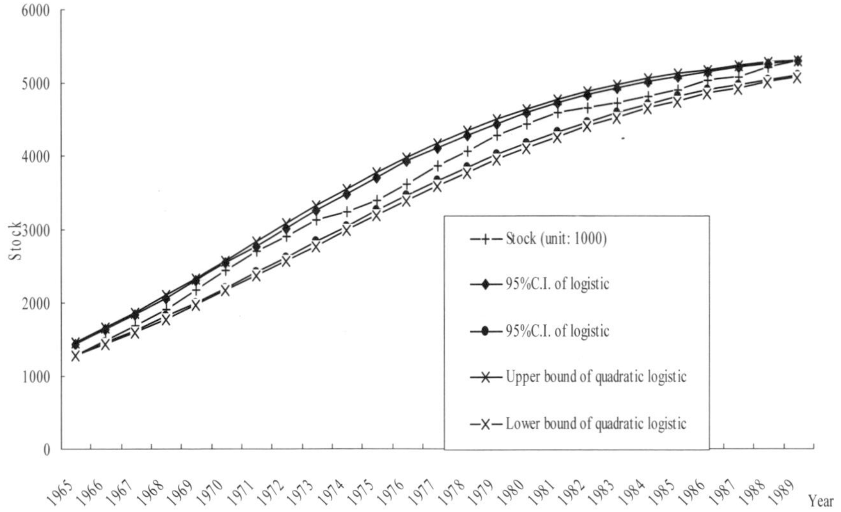

Figure 2: The forecasts of the logistic and quadratic interval logistic model for stock of cars.

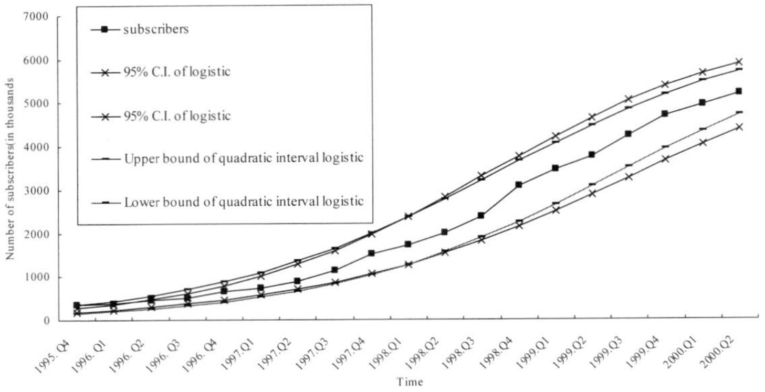

Figure 3: The forecasts of the cumulative number of cellular phone subscribers in Portugal.

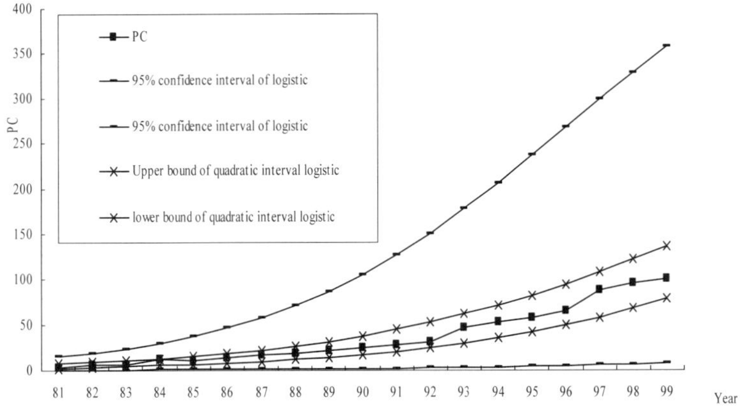

Figure 4: The forecasts of world-wide PC demand.

5. Conclusions

We have combined Tanaka's quadratic-interval regression model with the basic concepts of the Gompertz and logistic models, to create two new innovation diffusion models, the quadratic-interval Gompertz model and the quadratic-interval logistic model. The possibilistic regression is formulated to obtain the smallest interval system, including all of the selected data, so that it can provide the possible interval. Therefore, if the data are not sufficient, quadratic-interval diffusion models are potentially useful tools. However, when there are outlier data or when there is high variability in the data, the quadratic-interval diffusion models should not be used.

We then used them to forecast sales performance in three sample data sets. The empirical analyses show that quadratic-interval diffusion models can be applied to new product sales forecasting using the sales histories of similar products, and can reveal the best- and worst-case sales volume outcomes. Moreover, when the diffusion process of the time series produces rapid early growth and relatively slow growth later in the series, the quadratic-interval Gompertz model appears to be the most suitable. In contrast, when the diffusion process produces slow early growth and relatively rapid later growth, the quadratic-interval logistic model performs better. In practice, we suggest that decision makers draw scatter diagrams to determine the diffusion patterns, and then choose the appropriate diffusion model from both conventional and quadratic diffusion models.

In the future, combining and how to combine the forecasts of different models are more accurate are good issues.

Acknowledgment

This work was partially supported by funding from the Nation Science Council of the Republic of China (NSC 94-2416-H-155 -008)

References

Bass, F. M. (1969). A new product growth model for consumer durables. *Management Science*, 15, pp. 215–227.

Bass, F. M. (1999). Diffusion theory in marketing: a historical perspective, www.utdallas.edu/~mzjb/bass.ppt.

Bayus, B. L. (1993). High-definition television: assessing demand forecasts for a next generation consumer durable. *Management Science*, 39(11), pp. 1319–1333.

Botelho, A. and Pinto, L. C. (2004). The diffusion of cellular phones in Portugal. *Telecommunications Policy*, 28, pp. 427-437.

Dubois, D. and Prade, H. (1980). Theory and Applications, Fuzzy Sets and Systems. Academic Press, New York.

Frank, L. D. (2004). An analysis of the effect of the economic situation on modeling and forecasting the diffusion of wireless communications in Finland, *Technological Forecasting and Social Change,* 71, pp. 391–403.

Franses, P. H. (1994). Fitting a Gompertz curve. *Journal of Operational Research Society,* 45, pp. 109–113.

Fourt, L. A. and Woodlock, J. W. (1960). Early prediction of market success for grocery products. *Journal of Marketing,* 25, pp. 31–38.

Griliches, Z. (1957). Hybrid corn: An exploration in the economics of technological change, *Econometrica*, 25, pp.501–522.

Gruber, H. (2001). Competition and innovation. the diffusion of mobile telecommunications in central and eastern Europe. *Information Economics and Policy*, 13, pp. 19–34.

Heeler, R. and Hustad, T. (1980). Problems in predicting new product growth for consumer durables. *Management Science,* 26(10), pp. 1007–1020.

Islam, T. and Meade, N. (1997). The diffusion of successive generations of a technology: a more general model. *Technological Forecasting and Social Change,* 56, pp. 49–60.

Kim, K. J, Moskowitz, H. and Koksalan, M. (1996). Fuzzy versus statistical linear regression. *European Journal of Operational Research,* 92, pp. 417–434.

LINGO, (1999). Optimization Modeling With LINGO, LINDO systems Inc, Chicago.

Mansfield, E. (1961). Technical change and the rate of imitation. *Econometrica*, 29(4), pp. 741–766.

Rai, A., Ravichandran, T. and Samaddar, S. (1998). How to anticipate the internet's global diffusion. *Communications of the ACM*, 41, pp. 97-106.

Savic, D. A. and Pedrycz, W. (1991). Evaluation of fuzzy linear regression models. *Fuzzy Sets and Systems*, 39, pp. 51-63.

Schmittlein, D. C. and Mahajan, V. (1982). Maximum likehood estimation for an innovation diffusion model of new product acceptance. *Marketing Science,* 1(1), pp. 57-78.

Tanaka, H., Uejima, S. and Asai, K. (1982). Linear regression analysis with fuzzy model, *IEEE Trans, Systems Man Cybernet.* , 12, pp. 903–907.

Tanaka, H. and Lee, H. (1998). Interval regression analysis by quadratic programming approach, *IEEE Transactions on Fuzzy Systems*, 6, pp. 473–481.

Teng, J. T. C., Grover, V. and Gutter, V. (2002). Information technology innovation: general diffusion patterns and its relationships to innovation characteristics. *IEEE Transactions on Engineering Management*, 49, pp.13–27.

Wiebe and Eng. (2006). http://www.productinnovators.com/greatlakes/events/documents/ Wiebe. Pdf accessed 1 January 2007.

Zadeh, L. A. (1965). Fuzzy sets;*Information and Control*, 8, pp. 338–353.